Environmental Science and Engineering

The ultimate goal of this series is to contribute to the protection of our environment, which calls for both profound research and the ongoing development of solutions and measurements by experts in the field. Accordingly, the series promotes not only a deeper understanding of environmental processes and the evaluation of management strategies, but also design and technology aimed at improving environmental quality. Books focusing on the former are published in the subseries Environmental Science, those focusing on the latter in the subseries Environmental Engineering.

Ipsita Dipamitra Behera · Alok Prasad Das
Editors

Impact of Petroleum Waste on Environmental Pollution and its Sustainable Management Through Circular Economy

Springer

Editors
Ipsita Dipamitra Behera
Department of Chemical Engineering
Indira Gandhi Institute of Technology
Odhisha, India

Alok Prasad Das
Department of Life Sciences
Rama Devi Women's University
Bhubaneswar, Odisha, India

ISSN 1863-5520 ISSN 1863-5539 (electronic)
Environmental Science and Engineering
ISBN 978-3-031-48219-9 ISBN 978-3-031-48220-5 (eBook)
https://doi.org/10.1007/978-3-031-48220-5

This Springer imprint is published by the registered company Springer Nature Switzerland AG
The registered company address is: Gewerbestrasse 11, 6330 Cham, Switzerland

Paper in this product is recyclable.

Contents

Chapter 1
In-Depth Coverage of Petroleum Waste Sources, Characteristics, Environmental Impact, and Sustainable Remediation Process

Deeptimayee Pal and Sujit Sen

Abstract The massive and intricate petroleum refining industry converts crude oil into various products, such as gasoline, diesel, jet fuel, and heating oil. The refining sector contributes significantly to the world economy by producing multiple goods for home use, industry, and transportation. The refining process, however, also has a significant quantity of trash and contaminants, which may harm the environment. The refining industry is subject to various rules and standards designed to reduce emissions and safeguard the environment to minimise these effects. The refining sector has also been pressured to reduce carbon emissions and lessen climate change's consequences. Some businesses have responded by investing in greener technology like renewable energy and carbon capture and storage. Spent oil and gasoline waste can be generated during the exploration, production, transportation, refining, processing, distribution, and consumption of oil and gas. Some common characteristics of petroleum waste include toxicity, ignitability, corrosivity, reactivity, and volatility. The contamination of soil and water, disruption of the aquatic food chain, harmful effects on marine species directly, habitat destruction, and impact on human health are just a few of the most severe repercussions. A multifaceted strategy is required to solve this issue, including environmentally friendly remediation techniques such as bioremediation, chemical treatments, air sparging, soil vapour extraction, thermal methods, filtration, novel approaches, and prevention. Therefore, it is essential to conduct ongoing research and develop innovative, sustainable remediation methods to address the problem of petroleum waste and safeguard the environment.

Keywords Petroleum wastes · Oily water · Surface and groundwater pollution · In-situ remediation · Bioremediation

D. Pal · S. Sen (✉)
Department of Chemical Engineering, National Institute of Technology Rourkela, Rourkela 769008, Odisha, India
e-mail: sensujit@nitrkl.ac.in

I. D. Behera and A. P. Das (eds.), *Impact of Petroleum Waste on Environmental Pollution and its Sustainable Management Through Circular Economy*, Environmental Science and Engineering, https://doi.org/10.1007/978-3-031-48220-5_1

1.1 Introduction

The last ten years have seen unprecedented growth in unconventional oil and gas production, resulting in widespread changes in the structure of the petroleum industry's supply and demand, its approach to recovery, and its technological advancements (Jia et al. 2012). The petroleum industry plays a significant role in the global economy, providing raw materials and energy to other sectors. Petrochemicals, lubricants, and fuels derived from petroleum are vital for contemporary living. The sector includes the exploration, production, processing and refining, transportation, and marketing of oil and gas. The petroleum sector is complicated, globally distributed, and forever evolving. The industry is divided into upstream, midstream, and downstream operations, each crucial to the energy supply chain. Upstream activities include exploration and production, whereas midstream operations include refining, transportation, and storage. Lastly, marketing and distribution are downstream activities (Aalsalem et al. 2018). The sector is extensively regulated, with laws and standards covering every facet of the industry. Global trends and events, such as the rise in energy consumption and the trend toward renewable energy sources, also impact this industry. It is a fascinating and dynamic sector because of its wide variety of operations, intricate rules, and constantly shifting worldwide trends.

The prime objective of the petroleum industry is to provide the world population with energy and fuels, accomplished by producing crude oil and natural gas, which have several global energy sources. In addition to electricity and heat, these energy sources consist of gasoline, diesel, jet fuel, and kerosene. In addition to lubricants, petrochemicals, and plastics, it is accountable for various other products. Several sectors use petroleum products, including transportation, industry, agriculture, and construction. According to the International Energy Agency (IEA), World Energy Balances 2020, the total percentage share of petroleum consumption by various major end-use sectors in 2020 includes transportation, non-energy uses, aviation, industries, navigation, others (agriculture, commercial and public services), residential, and railways as illustrated in Fig. 1.1. The petroleum sector also creates a vast array of petrochemical products used to manufacture plastics, paints, detergents, and fertilisers, among other things (Jia et al. 2012). The petroleum sector is crucial for global energy production. Exploration, drilling, refining, and distribution are just a few of the many activities and operations involved in the sector. Hence, a substantial amount of waste is produced by these activities.

1.2 Petroleum Wastes

"Petroleum waste" refers to various oil and gas extraction byproducts. These wastes include spent oil, oily sludge, oily water, drilling muds, drill cuttings, drilling fluids, and other byproducts of oil and gas production. Additional petroleum waste products include lubricating, hydraulic, and other petroleum-based products. Petroleum

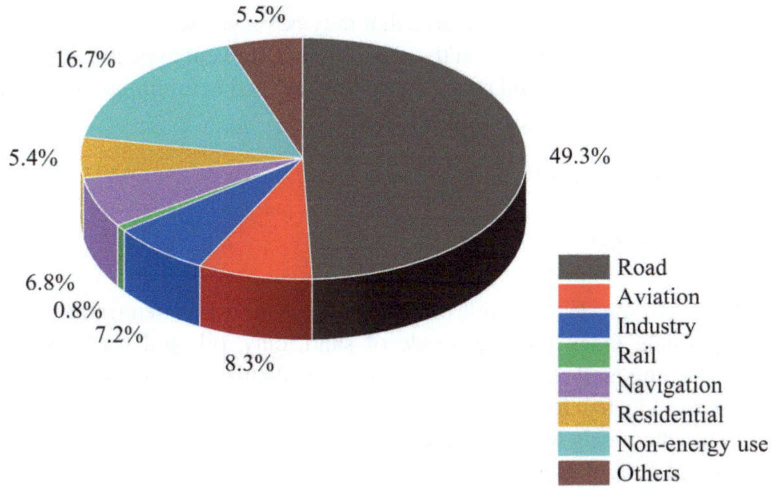

Fig. 1.1 Petroleum consumption by end-use sectors

waste is a form of hazardous waste that, owing to its potential to contaminate the environment and damage human health, demands special handling and disposal.

1.2.1 Used Oil

Any synthetic or petroleum-based oil used as a lubricant for various applications is called "used oil" in this context. Used engine oil and used hydraulic oil are the two main kinds of used oil. Used engine oil is defined as oil utilised in an internal combustion engine but polluted with soot, metal particles, dirt, and other combustion byproducts (Jia et al. 2012). Hydraulic fluid additives, such as rust inhibitors and anti-wear compounds, may contaminate used hydraulic oil, which has been utilised in hydraulic systems. Several industries generate used oil, including the automotive, industrial, and construction sectors. Used oil is generated from various sources, such as engine oil changes, brake fluid changes, and lubricant changes. Oil spills, leaks, and maintenance activities also produce it.

After used oil has been generated, it may either be recycled and utilised again or appropriately disposed of. An essential component of the petroleum industry is the recycling of spent oil. Usually, used oil is recycled to create lubricants, fuel oils, and other goods. Used oil that has been recycled may be used as a lubricant in metallurgical processes and to make asphalt. The Environmental Protection Agency regulates the removal of spent oil (EPA) (Nichols 2001). Used oil is a hazardous waste governed by the Resource Conservation and Recovery Act (RCRA). Used oil must be disposed of at authorized hazardous waste facilities and kept in containers designated as hazardous waste. The recycling of used oil is crucial to the petroleum industry. Used oil is a valuable resource that may be recycled and used again to save energy, lessen its adverse effects on the environment, and cut expenses. Used oil is a useful product; hence, it should be appropriately handled to minimise environmental harm and ensure its continuing industry usage.

1.2.2 Oily Sludge

Oily sludge is a by-product of crude oil processing resulting from refining and extraction. It is a thick, black material made of sand, clay, oil, and water. Across the processing, storing, and refining operations for crude oil, oily sludge is produced in the petroleum industry. It is produced due to chemical processes that break down crude oil into chemicals when it comes into contact with air, water, and sediments. Oily sludge has a high viscosity and is contaminated heavily with metals, hydrocarbons, and other organic substances. The presence of oily sludge during petroleum processing may cause several operational and environmental issues (Hu et al. 2013). It may build up in containers, pipelines, and other machinery, reducing operational effectiveness and raising maintenance and repair expenses.

Moreover, oily sludge may pollute the environment, harming the water, land, and air. Aquatic species may even perish due to their detrimental effects on marine life. The petroleum industry must take various measures to prevent the accumulation of oily sludge. Initially, it's essential to identify and, if feasible, eliminate the sources of oily sludge, including replacing outdated pipelines and vessels, enhancing equipment maintenance, monitoring, and minimising the usage of tainted water sources. If feasible, implement water treatment systems and improve oil and water separation. Finally, to lessen the quantity of oily sludge in the system, oil–water separators, coalescers, and other filtering devices should be considered (Guerin 2002). Lastly, the oily sludge must be disposed of appropriately. The most prevalent disposal techniques are recycling, incineration, and landfilling. The most widely used form of disposal is landfilling, which risks damaging the ecosystem. Oily sludge may be effectively burned out, although doing so might be expensive and require specialised equipment. Recycling is preferable since the oily sludge may be recovered or employed as a fuel source. These steps may save operating expenses while contributing to environmental protection.

1.2.3 Oily Water

The petroleum industry refers to wastewater containing oil, grease, and other suspended materials as "oily water"; the wastewater results from regular maintenance tasks and byproducts of industrial activities. Oil, grease, and other pollutants are often found in this water but must be purified before discharge. Oil and grease are two principal contaminants in oily water due to the substantial use of petroleum-based products in manufacturing (Singh and Shikha 2019). During manufacturing and maintenance, petroleum products, including gasoline, diesel, lubricants, and other industrial chemicals, are discharged into the environment. These contaminants can reach rivers and streams and cause significant harm to aquatic life. Many issues arise when oil and grease are present in oily water. One of the most critical issues is that they may form a film on the water's surface, which lowers how much sunlight can reach the water. As a result, the quantity of accessible oxygen in the water may decrease, impacting the well-being of aquatic life.

Moreover, it may reduce photosynthesis, which is necessary for plant development. In addition, grease and oil may impede the water's ability to absorb nutrients from its surroundings. They can hinder the passage of oxygen to the water, resulting in a decline in the water's quality and output. Several additional issues might also be brought on by oil and grease. They may make the water more acidic, encouraging the development of bacteria and other microorganisms that harm both people and animals.

Moreover, they can cause metal pipes and other things to erode, which might lessen their lifespan. It is crucial to use various techniques to treat oily water appropriately. Using a skimmer or another physical separation procedure is one of the most popular approaches. This procedure aims to remove the oil and grease from the water. After being separated, the oil is filtered out of the water and treated to eliminate impurities. Using a chemical treatment technique is another way to remediate greasy water. During the operation, a chemical must be introduced to the water to disperse the oil and grease and reduce their concentration. The residual impurities are then removed from the water during treatment. Using a biological treatment technique is the last way to treat greasy water. Introducing microorganisms into the water will degrade the oil and grease, reducing the concentration. The bacteria will subsequently change the residual contaminants into safe molecules to release into the environment (Yu et al. 2017). These techniques make it feasible to lessen the toxins in oily water and guarantee that it is secure enough to be released into the environment. Before disposal, ensuring the water is clean and free of any harmful contaminants is crucial.

1.2.4 Drilling Mud

Drilling mud is a kind of fluid used during the drilling process in the petroleum industry. It is crucial to the drilling process because it lubricates and cools the drill bit, prevents the drill string from becoming stuck in the wellbore, and transports cuttings from the wellbore's bottom to the surface. The components of drilling mud include clay, water, and additional additives. Although the clay lubricates the drill bit, the additives boost the fluid's density and viscosity. The specific drilling process determines the kind and quantity of additives used in drilling mud. Barite and bentonite are the most frequent additives used in drilling mud. Barite is often added to drilling mud to improve density, preventing the drill string from becoming trapped in the wellbore. Bentonite is a clay form that minimises friction and lubricates the drill bit (Yap et al. 2011). Typically, drilling mud is pumped through the drill string and wellbore to lubricate, cool the drill bit, and remove cuttings from the well bottom. The mud is cycled around the drill bit by being pumped down the drill string and into the wellbore. Chips and other debris are produced when the drill bit cuts through the rock. These fragments are known as cuttings, and they are suspended in the drilling mud as it is pumped back to the surface from the wellbore. The drilling mud is cleaned and filtered at the surface to remove any solid cuttings or other contaminants. The mud is then recirculated back down the wellbore to lubricate the drill bit and prevent the drill string from becoming stuck. Without drilling mud, keeping the drill bit cold and removing cuttings from the well's bottom would be far more complex (Yap et al. 2011).

1.2.5 Drilling Cutting

Drilling cuttings are solid pieces of rock and other substances produced during oil and gas drilling (Charles and Sayle 2010). As the drill bit grinds through the rock formations, these particles accumulate and are often deposited in the mud that is cycled around the drill bit. This mud is returned to the surface and divided into cuttings and muck. The trimmings are afterwards discarded or recycled. The sort of rock formations found at different depths is revealed by the drilling cuttings, making them a crucial part of the drilling process. The amount of oil and gas extracted from the well may be more accurately estimated. In addition, cuttings are used to monitor the drill bit's development and identify possible issues. Typically, cuttings from drilling are categorised into soft and hard. Soft cuttings often consist of clays and shales, while hard cuttings consist of sandstone, limestone, and other more complicated substances. Often, smooth trimmings are simpler to handle since they are lightweight and may be sliced into smaller pieces.

Nevertheless, hard cuttings are more challenging to handle and need specialist equipment for separation. Managing drilling cuttings is crucial to oil and gas extraction since these particles may include harmful substances such as heavy metals, hydrocarbons, and other pollutants. These toxins may pollute groundwater sources and harm the ecosystem if not handled appropriately; hence, drilling cuttings must be adequately managed. A cuttings disposal system is the most prevalent method for controlling drilling cuttings (Hu et al. 2021). Typically, this system consists of a series of tanks, pumps, and filters that separate the cuttings from the mud and then store them in a secure area. The trimmings are afterwards discarded following municipal requirements. In addition to their management, there are other ways to recycle drilling cuttings. For instance, specific trimmings may be reused as fillers for road building or as garden soil cover. In addition, these particles may be used to manufacture fertiliser or burned to provide electricity. The management and recycling of drilling cuttings is an essential aspect of the oil and gas industry, as it contributes to protecting the environment and the execution of the drilling process safely and effectively. By controlling and reusing these particles effectively, firms can limit the possibility of environmental contamination and guarantee that the drilling process is completed safely.

1.2.6 Drilling Fluid

An essential component of the petroleum industry is drilling fluid, used to cushion the drill bit to lessen vibration and wear, lubricate and cool the drill bit, make cuttings removal easier, keep the drill bit upright, enhance the rate of penetration (ROP) by decreasing friction, and retain the drill bit in an upright posture. Water, clay, and chemicals are often found in drilling fluids. Although the chemicals used in drilling fluids rely on the intended fluid qualities, the kinds of clay used in drilling fluids depend on the type of formation being drilled. The most popular style of drilling fluid is water-based drilling fluid. The types of clay utilised vary depending on the drilling formation. Water-based drilling fluids include compounds in addition to clay employed to modify the physical and chemical characteristics of the mud (Deville et al. 2011). Oil, clay, and chemicals are all components of oil-based drilling fluids. When water is not accessible or when drilling through rocks that are not amenable to water, oil-based drilling fluids are employed. Water-based drilling fluids are economical, whereas oil-based drilling fluids provide more excellent lubrication, enabling quicker drilling speeds. Oil-based fluids also decrease fluid loss and are better at suspending drill debris. Drilling fluids with a synthetic basis are made from synthetic fluids, clay, and chemicals. While synthetic-based fluids are more costly than water- and oil-based ones, they provide better lubrication and drill-cutting suspension.

There is an inverted emulsion and foamed drilling fluids and drilling fluids based on water, oil, and synthetic fibres. Although foamed drilling fluids comprise an amalgam of air, water, and chemicals that produce foam, inverted emulsion drilling fluids comprise an oil-in-water emulsion. Both liquids are employed when conventional water-based, oil-based, and synthetic-based fluids are ineffective. The fluid must be monitored and adjusted during the drilling process regardless of the kind being utilized (Nas et al. 2009). It is necessary to keep an eye on the drilling fluid's characteristics to ensure it operates as it should and to detect any changes that could occur as the drilling operation progresses. Also, it is necessary to modify the drilling fluid's characteristics to ensure it is appropriate for the formation being drilled. Chemical additions, clay concentration changes, drilling fluid viscosity or density changes, and clay concentration changes are all examples of adjustments.

1.3 Sources of Petroleum Wastes

Petroleum waste is a byproduct of producing, refining, and consuming products with a petroleum base. It comprises hazardous substances that might endanger the environment if not handled properly. Petroleum waste may be derived through several processes, including extraction, processing, transportation, storage, and oil and gas consumption.

- Oil and gas exploration, drilling, and production generate petroleum waste. This waste may consist of drilling mud, generated water, drilling cuttings, and chemical additives used throughout the drilling operation.
- Petroleum waste is produced throughout the refining process due to the processing, storage, and manufacture of petroleum products. Sludge, tainted filter cakes, and dangerous compounds used in refining are examples of this waste.
- The transportation of crude oil and petroleum products generates waste, including discharges during loading, unloading, and transporting petroleum products.
- The storage of crude oil and petroleum products generates petroleum waste, which comprises spills, leaks, and other emissions during petroleum product storage.
- Petroleum waste is also produced when petroleum products are consumed, including spills, leaks, and other emissions that occur when petroleum-based goods are used.

Any petroleum waste forms may severely influence the environment if not correctly handled. It is crucial to ensure that all petroleum waste is dealt with, stored, and disposed of in compliance with local, state, and federal rules to decrease the potential for injury. The environment can be safeguarded by taking the necessary measures to manage petroleum waste effectively, assuring everyone a healthy future.

1.3.1 Production

The production of petroleum products entails a vast array of operations that might generate several types of waste. These activities include exploration, drilling, refining, transporting, and selling. Each of these processes can develop a variety of debris, ranging from solids and liquids to toxic compounds. The production sector of the petroleum industry is responsible for obtaining crude oil from the ground and changing it into usable goods for customers. This industry is relatively energy-intensive, using various processes requiring substantial energy and materials.

As a consequence of these activities, a large variety of waste is generated, which may seriously affect the environment and public health. The production sector of the petroleum industry generates hazardous waste (Leemann 1988) at the highest frequency, including compounds like benzene and toluene, which are employed as drilling solvents, as well as oil, grease, heavy metals, and other pollutants. These hazardous elements may be discharged into the environment through air, water, and land sources, causing air and water pollution and soil and groundwater contamination. In addition, the production sector of the petroleum industry releases non-hazardous waste, including refinery sludge, drilling mud, and other process wastes (Asim et al. 2021). These wastes are often constituted of organic and inorganic components and, if not properly managed, may also contribute to environmental damage. With the use of machinery and equipment, the production sector of the petroleum industry also creates waste; this involves using diesel engines, hydraulic pumps, and other equipment essential for the drilling and refining operations. These engines and pumps produce exhaust fumes and other pollutants that may be emitted into the atmosphere, in addition to oil and grease that can pollute water resources. In addition, the production sector of the petroleum industry produces waste in the form of drilling muds, which are used to lubricate and excellent drilling equipment. These muds include a range of contaminants, including heavy metals and other pollutants, which, if mishandled, might be released into the environment. By transporting crude oil and other goods, the production sector of the petroleum industry also generates trash; this involves the usage of tanker ships, tanker trucks, and other vehicles capable of emitting pollutants into the air and water. In addition, transporting these items may produce trash in the form of packing materials, such as plastic bags and containers, which can also cause pollution if not handled effectively. In addition to having garbage via the disposal of its products, the petroleum production sector also creates waste; this includes waste oil disposal, which may consist of heavy metals and other contaminants, as well as refinery sludge and the different process by-products. These wastes may be discharged into the environment if not effectively handled, resulting in air and water pollution and soil and groundwater contamination. Thus, the petroleum production sector needs strict waste management regulations and procedures to reduce environmental implications (Ite et al. 2013).

Moreover, the extraction of petroleum products may lead to the emission of harmful chemicals. During extraction, dangerous compounds might be released into the environment, including lead, mercury, and polycyclic aromatic hydrocarbons (PAHs). These contaminants may build in the ecosystem and have long-lasting effects on human health and the environment. Moreover, manufacturing plastics from petroleum resources may lead to hazardous chemicals such as phthalates and Bisphenol A (BPA) emissions. These compounds may seep into the environment and have many adverse health impacts (Nowak et al. 2019).

1.3.2 Refining

Petroleum waste is a byproduct of the refining process in the petroleum industry. It is generated through numerous refining steps, including separating crude oil's distinct components, distilling the various fractions, and processing them into fuel-grade products. During the separation of crude oil, waste products such as water, sediment, and sludge are produced (Kokal 2005). Several physical and chemical techniques separate crude oil's constituents throughout the separation process. The use of centrifuges and settling tanks, for example, assists in separating the heavier components of oil from the lighter ones. The separated elements are discarded as waste. The distillation process further separates the crude oil's various fractions. The crude oil is heated to a particular temperature throughout this operation. Once crude oil is heated, the various components of the oil evaporate and are then collected and sorted into fractions. These fractions are refined into gasoline, diesel, and kerosene. During this procedure, ash, emulsions, and oils are formed as byproducts. As a waste, these items are discarded. Lastly, the production of fuel-grade goods from the fractions also creates waste. The traces are broken down and recombined to produce products with certain qualities. In addition to solvents, metals, and sludge, this process also creates waste products. Often, these materials are discarded as waste (Shah et al. 2011). The refining process in the petroleum industry creates various waste products. These substances include water, silt, sludge, ash, emulsions, oils, solvents, and metals. While some of these items may be recycled, reused, or processed for reuse, they are generally discarded as waste.

Moreover, toxic compounds are released into the environment while refining petroleum products. Refineries generate vast quantities of waste, which includes sulfur-containing compounds, metals, and other substances. These contaminants may damage the air, soil, and water and cause respiratory and other health issues. Thus, petroleum businesses must establish efficient waste management systems to reduce the quantity of waste created during refining.

1.3.3 Transportation

Oil and gasoline spills are the most prevalent waste generated while transporting petroleum products. Many reasons, including mechanical breakdowns, human errors, and natural calamities, may trigger spills. The environmental impact may vary from minimal to catastrophic, depending on the amount of the spill. Various mechanical faults, such as broken valves, seals, and pipes, may result in spills. Human error is also a typical cause, with operators failing to lock valves properly or execute required maintenance. Natural calamities like hurricanes and tropical storms may also produce spills. Oil spills may harm the ecosystem (Kardena and Helmy 2015). Oil consists of various organic chemicals, which, if released, may pollute land, water, and air.

Since oil decreases oxygen levels and damages the hypersensitive gills of fish and other aquatic organisms, oil spills may harm marine life. In certain instances, the oil may even harm the livestock. Since the oil evaporates into the air and is inhaled by people and other animals, oil spills may also contribute to air pollution, resulting in various respiratory issues, including coughing, wheezing, etc. In addition, oil spills may pollute the soil, making it harder to cultivate crops. In addition to oil spills, petroleum products may generate other types of waste. Fuel may be spilled on the ground during transportation, endangering the land and water. In addition, gasoline evaporation may contribute to air pollution. Lastly, petroleum products may discharge harmful substances into the atmosphere. These substances may be ingested by plants, animals, and people, causing various health issues. Transporting petroleum products may produce many wastes. Spills may cause damage to the environment, while fuel can taint soil and water and release toxic compounds. Another important cause of environmental harm from petroleum products is oil spills. Accidental oil spills may also occur during oil extraction and transportation. Oil spills may have disastrous consequences on marine ecosystems because they can damage aquatic life, disrupt marine species' habitats, and contaminate the water supply (Zhang et al. 2019).

Moreover, oil spills may cause coastal erosion and damage adjacent places. It is essential to efficiently maintain vehicles and follow all safety regulations to reduce the amount of waste created. In addition, operators must be aware of their surroundings and avoid spills and other types of waste.

1.3.4 Storage and Handling

Petroleum products are essential to the functioning of contemporary civilization, with applications ranging from transportation to manufacturing and processing. Yet, their storage and management might produce various waste products. The generation of these waste products may be broadly categorized into two groups: those created during the petroleum products' storage and those generated by accompanying operations, such as emissions from tankers and storage tanks. While storing petroleum products, it is vital to safeguard the natural world from any contamination. Storage tanks must be periodically examined, cleaned, and maintained to generate various waste products, including oil, solvents, and sludge. These wastes may include toxic chemicals such as heavy metals, volatile organic compounds, and polycyclic aromatic hydrocarbons, which, if not disposed of properly, may be damaging to human health and the environment. Tankers, which transport petroleum products to storage tanks, also contribute to the formation of waste during petroleum product storage (Rana 2008). Tankers are often outfitted with valves and other equipment that might cause oil or other chemicals to spill during transit. These leaks can damage the environment and cause the emission of volatile organic compounds and other air pollutants. In addition to the waste generated by tankers, storage tanks may generate various waste products. For instance, the tanks may be lined with steel, plastic, rubber, and other synthetic materials, which may degrade over time and release their components into the environment.

Similarly, the tanks may contain varied degrees of contaminants, such as polycyclic aromatic hydrocarbons, that may be discharged into the environment. In addition, storage tanks may produce waste during the unloading of petroleum products; this might involve the discharge of oil, solvents, and other liquids into the environment and the emission of volatile organic compounds and other air pollutants. All storage tanks and tankers must be regularly maintained and inspected to reduce the risk of contamination.

1.3.5 Consumption

In the petroleum industry, the consumption industry is accountable for the refining, distributing, and consuming petroleum products. This sector is responsible for manufacturing extensive goods, including gasoline and diesel fuels, lubricants, asphalt, and petrochemicals. The utilization of these products generates a substantial amount of waste, which has a significant influence on the environment. The consumer sector of the petroleum industry generates the most significant quantity of hazardous waste; this includes polluted oils, fuels, and other petroleum-based goods, such as lead, mercury, and arsenic. Most of these hazardous pollutants are generated as crude oil is refined into gasoline, diesel, and other petroleum-based products (Demirbas et al. 2016). These hazardous compounds may leach into the environment, pollute soil and water, and severely harm people and the ecosystem. In addition to hazardous waste, the consumption sector of the petroleum industry generates wastewater and solid waste (Ramírez-García et al. 2019). This solid waste may also significantly impact the ecosystem since it can occupy important landfill space and possibly leak harmful substances into the environment. Air pollution from the consumer sector of the petroleum industry may have significant adverse effects on human health and the environment. Particulate matter, nitrogen dioxide, sulfur dioxide, and volatile organic compounds are examples of air pollutants produced by the petroleum industry's consuming sector. These pollutants may contribute to the formation of smog, which can impair visibility, irritate the lungs, and trigger other respiratory issues. In addition to releasing greenhouse gas emissions, the consumer sector of the petroleum industry may also contribute to climate change. Carbon dioxide, generated by burning petroleum-based products, is one of the primary contributors to climate change.

Moreover, the manufacture of petroleum products may contribute to the emission of methane and nitrous oxide, among other greenhouse gases. In conclusion, the consumer sector of the petroleum industry may create a substantial amount of hazardous and non-hazardous waste, which can considerably influence the environment. This waste may comprise hazardous chemicals, wastewater, solid waste, air pollution, and greenhouse gas emissions, all of which can contribute to environmental and human health deterioration. The petroleum industry must minimize the waste it produces and ensure that it is properly managed and disposed of to have as minimal impact as possible on the environment.

1.4 Characteristics of Petroleum Wastes

Petroleum waste characteristics may vary considerably depending on the source of the waste as well as the type of petroleum product involved. Toxicology, biodegradability, flammability, ignitability, corrosivity, volatility, and mobility are some prevalent characteristics of petroleum waste (Fu et al. 2021). Petroleum wastes are categorized into two categories: hazardous and broadly non-hazardous. A comparative

Table 1.1 Petroleum wastes characteristics

Characteristics	Hazardous petroleum wastes	Non-hazardous petroleum wastes
Toxicity	High levels of toxicity	Typically, less toxic
Biodegradability	Not easily biodegradable	Bit easily biodegradable
Volatility	Volatile	Less volatile
Ignitability	Can be flammable	Not much flammable
Corrosivity	Highly corrosive	Lesser corrosive
Reactivity	Can be reactive	Generally, not much reactive

study of the characteristics of hazardous and non-hazardous petroleum wastes is listed in Table 1.1.

1.4.1 Hazardous Petroleum Wastes

Hazardous petroleum wastes include petroleum wastes with components that may pose a health risk to humans and the environment. Heavy metals, PAHs, polychlorinated biphenyls (PCBs), and volatile organic compounds (VOCs) may be detected in hazardous petroleum wastes. When hazardous petroleum wastes are discharged into the environment, they may harm air, water, and land quality. Oil spills, storage tank breaches, and pipeline ruptures are the most typical sources of hazardous petroleum waste discharged into the environment (Santella et al. 2010). These sorts of emissions have the potential to contaminate water sources and destroy ecosystems. If breathed or swallowed, the compounds in hazardous petroleum waste may be hazardous to humans. Once discharged into the environment, toxic petroleum wastes may be challenging to remove and have long-lasting repercussions. It is crucial to prevent the emission of hazardous petroleum wastes to avert these consequences; using proper storage and disposal methods and adopting rules and standards to ensure the safe handling and storage of hazardous petroleum wastes can accomplish it. The Environmental Protection Agency (EPA) has established various laws and guidelines to prevent the discharge of hazardous petroleum wastes into the environment. The Oil Pollution Act, the Resource Conservation and Recovery Act, and the Clean Air Act are examples of these policies and standards. These laws and guidelines require facilities that create hazardous petroleum waste to handle and dispose of the waste safely. In addition, there are several ways to limit the discharge of hazardous petroleum wastes into the environment. They include lowering the quantity of gasoline and diesel fuel used in cars, recycling old oil and other petroleum-based goods, and disposing of hazardous petroleum wastes in an environmentally responsible manner. These measures may reduce the quantity of toxic petroleum waste discharged into the environment, protecting human health and the environment.

1.4.2 Non-hazardous Petroleum Wastes

Non-hazardous petroleum wastes do not immediately endanger human health or the environment. Typically, these types of wastes are produced during the manufacture, refining, shipping, and storage of petroleum products. These wastes may include byproducts of these activities, such as oil-filled tanks and containers, drilling mud, sludges, and oil spills. Typically, non-hazardous petroleum wastes are classed as "solid" or "liquid" wastes. Solid wastes may include oil-filled tanks and containers, gasoline tanks and drums, filters, rags, wipes, and absorbents. Oily water-generated water and wastewater polluted by petroleum-based products are examples of liquid wastes. Proper management of non-hazardous petroleum wastes is necessary to limit their environmental effect. Identifying, collecting, and disposing of these wastes safely and responsibly constitute adequate management. In general, non-hazardous petroleum wastes should be segregated from ordinary waste streams, kept in suitable containers, and disposed of according to municipal, state, and federal requirements (Li et al. 2021). In rare instances, non-hazardous petroleum wastes may also be reused or repurposed; for instance, recycling the used oil and oil filters into new lubricants or fuel oil. Recycled tank bottoms and sludges may be used as boiler fuel. In certain instances, non-hazardous petroleum wastes may be handled on-site or shipped off-site for treatment to minimize their toxicity. The appropriate legal requirements must be fulfilled while handling any non-hazardous petroleum waste. It is crucial to be informed of the exact restrictions in a certain location since these regulations may differ from state to state. In addition, any non-hazardous petroleum waste to be transported off-site for disposal must be labeled and packed according to current standards. The environmental effects of these wastes must be minimized by proper management.

1.5 Environmental Impacts

In addition to being very hazardous and toxic, petroleum waste has a significant environmental effect that causes serious concern for human health and the environment. Waste petroleum is generated in vast amounts throughout the petroleum sector, from drilling to refining to transport. Heavy metals, volatile organic compounds, polycyclic aromatic hydrocarbons, and other toxic substances are only some of the contaminants that may be present in such waste. Hydrocarbon emissions and soil, water, and groundwater pollution are the primary consequences of petroleum waste on the environment (Ossai et al. 2020). Air pollution is a significant concern in many parts of the world due to VOCs such as benzene, toluene, ethylbenzene, and xylene. Ground-level ozone generation, which these substances may contribute, is detrimental to plants, fish, and humans. Oil spills may be especially hazardous to ecosystems because of the widespread pollution of soil and groundwater. Soil particles may absorb petroleum pollutants, causing them to be more challenging to clean up. Groundwater contaminated with petroleum products is also a potential threat to human and animal health. Oil sludge is harmful to animals and the air and water they contaminate. Ingestion of petroleum products may lead to severe diseases or even animal deaths.

The environmental impacts of petroleum wastes are substantial since these materials are hazardous and may persist in the environment. Petroleum wastes may pollute soil, water, and air, resulting in several adverse ecological impacts (Ite and Ibok 2013). Petroleum adversely affects soil, air, groundwater, surface water, climate, wildlife, and humans; some significant impacts are depicted in Fig. 1.2. Hydrocarbons derived from petroleum, for instance, may accumulate in the food chain and cause long-term environmental damage. Furthermore, petroleum wastes may lead to the degradation of animal habitats and an increase in the polluting of nearby water sources.

Moreover, petroleum wastes in aquatic habitats might reduce the variety of species and disrupt food webs; hence, every individual should take appropriate steps to mitigate the adverse effects of petroleum waste on the environment. All individuals should use proper storage, waste management plans, and technology to lessen these wastes' environmental damage. Also, the government should enforce rules that limit the amount of oil waste that goes into the environment. Waste management strategies, technological solutions, and proper storage and disposal procedures are all examples of what may be done to lessen pollution in the natural world.

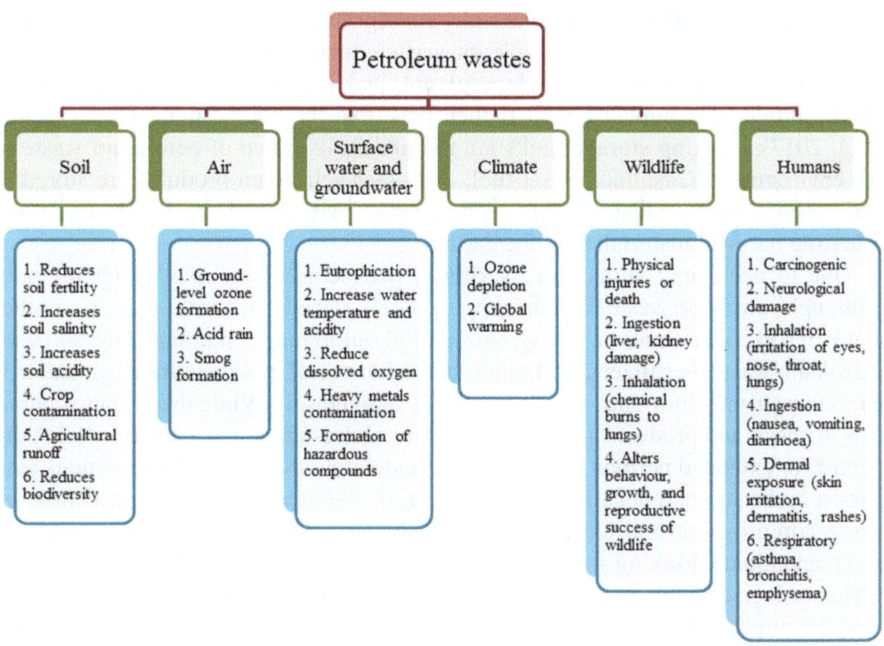

Fig. 1.2 Adverse effects of petroleum wastes on the ecosystem

1.5.1 Soil Pollution

As petroleum waste is released from various sources, it may often enter the ecosystem and have long-lasting effects on the soil. Soil is an essential natural resource since it is the basis of life on Earth (Lal 2012). Soil contamination is the most direct consequence of petroleum waste on the soil. Petroleum components may leak into the soil, contaminating it with hydrocarbons, heavy metals, and other contaminants. This contaminant may harm the soil and decrease its fertility.

Moreover, contaminated soil may include poisons that are hazardous to living creatures' health. Petroleum waste may also influence the soil's physical composition. The compaction of the earth caused by petroleum products makes it more difficult for plants to absorb nutrients and water, resulting in poor plant development and decreased production. In addition to physical and chemical damage, petroleum waste may bring plant-disease-causing bacteria into the soil. Microorganisms such as bacteria, fungi, and viruses can cause root rot and other plant diseases. These diseases may lower agricultural production and degrade crop quality.

Furthermore, petroleum waste may increase the soil's salinity, harming the soil and plants. Moreover, petroleum waste may increase soil acidity, making it harder for plants to absorb nutrients. These consequences may diminish the soil's productivity and reduce crop yields (Tang et al. 2012). Petroleum waste-caused soil contamination is a severe environmental hazard. It is caused by the introduction of petroleum products into the soil by sources like leaky underground tanks, accidental spills, landfills, and agricultural runoff. Petroleum waste contains several deadly substances, including benzene, toluene, ethylene glycol, and PAHs. These compounds harm plant, animal, and human health if they penetrate the soil, water, or air (Varjani et al. 2017). Leaking storage tanks are the principal source of petroleum waste in the environment. Gasoline, diesel fuel, and other petroleum products are stored in underground storage tanks (USTs). These tanks may corrode and leak over time, releasing harmful materials into the earth.

Due to accidental spills and overflows, petroleum waste may reach the soil. Although petroleum waste is often inappropriately placed in landfills, they are another cause of soil contamination. Lastly, agricultural runoff may contaminate the soil with petroleum waste. Fertilizers, pesticides, herbicide runoff, polluted drainage, and irrigation water may introduce petroleum waste into the soil. While the organic chemicals in petroleum products are resistant to natural degradation, petroleum waste may persist for extended periods in the soil, thus indicating that harmful compounds may persist in the soil and negatively affect the ecosystem and human health. Petroleum waste comprises several hazardous chemicals, which may leach into groundwater and contaminate drinking water sources when released into the soil, thus resulting in various adverse health impacts on people, including cancer and reproductive issues. Another effect of petroleum waste on soil is agricultural contamination (Varjani and Upasani 2019). Petroleum waste includes several chemicals that plants may absorb, contaminating the food supply. In addition to its direct effects on human health, soil contamination produced by petroleum waste may indirectly affect the ecosystem.

For instance, petroleum waste may be hazardous to organisms like worms and bacteria, resulting in decreased soil fertility and biodiversity. Preventing petroleum waste from entering the environment is the most effective method for reducing soil contamination from such waste. Proper monitoring and maintenance of USTs, storage of petroleum products in above-ground tanks, and environmentally safe disposal of waste products are all ways to achieve the desired goal. In addition, following to best management techniques, such as avoiding over-fertilization and applying pesticides and herbicides only when necessary, may prevent agricultural runoff. In conclusion, petroleum waste-caused soil contamination is a severe environmental hazard (Yuniati 2018). It may have several adverse effects on human health and the environment, including pollution of groundwater and crops, decreased soil fertility and decreased biodiversity. The most effective method for preventing this form of soil contamination is to prevent petroleum waste from entering the ecosystem.

1.5.2 Air Pollution

Air quality is significantly impacted by petroleum waste. Many processes, including oil and gas extraction, transportation, and refining, generate petroleum waste. Petroleum waste may be discharged into the environment and contribute to air pollution if not handled appropriately. Crude oil is a fossil fuel consisting of a complex combination of hydrocarbons from which petroleum products are formed. These hydrocarbons are used in various products, ranging from gasoline to plastics, and are essential to daily existence.

Nevertheless, when these products are improperly used or abandoned, petroleum waste may be released into the environment and cause air pollution. When petroleum waste is discharged into the atmosphere, it includes several pollutants, including particulate matter, volatile organic compounds, nitrogen oxides, and sulfur dioxide. Asthma, respiratory ailments, and cardiovascular disease are a few health issues these pollutants may cause. These pollutants have the potential to cause acid rain, ground-level ozone, and smog, all of which may be harmful to the environment and human health. Smog may irritate the eyes and throat and cause headaches and other respiratory problems. In addition to posing concerns to human health, petroleum waste may also harm the environment. Other petroleum waste products than emissions from burning petroleum may also pollute the air. Petroleum vapors may enter the atmosphere due to leaks from storage tanks, pipelines, and other sources. Volatile organic chemicals included in the gases can cause smog and other air pollutants (Cetin et al. 2003). Air quality may be significantly impacted by oil spills as well. Oil spills may release volatile organic chemicals and particle debris into the air. In various ways, petroleum waste may reach the atmosphere. It may be emitted directly into the atmosphere either deliberately during the manufacturing of petroleum products or unintentionally during spills and other incidents. It may also be indirectly emitted when petroleum products are used, such as when gasoline is spent in automobiles. The effects of these contaminants on air quality may be both immediate and long-term. Short-term health issues and reduced visibility may result from the pollutants. Long-term exposure to them may result in acid rain, harming aquatic, and terrestrial life. The incorrect disposal of petroleum waste may also cause air pollution. Inappropriate disposal of petroleum waste may result in its discharge into the atmosphere, which can contribute to smog and other air pollutants. The most efficient strategy to minimize air pollution caused by petroleum waste is to limit the number of petroleum products utilized and generated, to use and dispose of petroleum products properly, and to implement rules to reduce emissions. Petroleum product combustion is a significant cause of air pollution (Blake and Rowland 1995). Carbon dioxide, sulfur dioxide, and nitrogen oxide are emitted into the environment during burning. The greenhouse gas carbon dioxide contributes to global warming, whereas sulfur dioxide and nitrogen oxide contribute to acid rain.

Moreover, burning petroleum products creates soot and other particles that may be breathed and cause respiratory issues. The effects of these contaminants on air quality may be both immediate and long-term. Short-term health issues and reduced visibility may result from the pollutants. Long-term exposure to them may result in acid rain, harming aquatic and terrestrial life. The incorrect disposal of petroleum waste may also cause air pollution. Inappropriate disposal of petroleum waste may result in its discharge into the atmosphere, contributing to smog and other air pollutants.

Moreover, the disposal of petroleum waste in landfills may produce volatile organic compounds, which may worsen air pollution. Effective management and disposal may lessen the influence of petroleum waste on air quality. It should be appropriately disposed of to avoid discharging petroleum waste into the atmosphere. Moreover, utilising more effective combustion techniques and other fuels should help reduce emissions from burning petroleum. Hence, petroleum waste has a considerable effect on air quality. Although spills and leaks may produce petroleum odours and particle matter, burning petroleum emits various air pollutants. The incorrect disposal of petroleum waste may also cause air pollution. The management and disposal of petroleum waste should be done to minimize the adverse effects on air quality. Air pollution from petroleum waste is a significant environmental concern (Ragothaman and Anderson 2017). Reduced usage and manufacture of petroleum products are the most effective ways to reduce air pollution brought on by waste from the industry. Increasing efficiency and reducing petroleum consumption in industry, transportation, and heating and cooling may achieve this. Therefore, it is essential to utilise and dispose of petroleum products appropriately; this encompasses the proper storage and disposal of petroleum products and the proper upkeep of automobiles and other equipment. Finally, laws governing the industry to implement the best technology to reduce emissions and limit the number of pollutants that may be discharged should be enforced to reduce the amount of petroleum waste released into the atmosphere.

1.5.3 Surface Water and Groundwater Pollution

Petroleum wastes, including crude oil and refined petroleum products, may significantly impact surface water. The health of aquatic ecosystems may be adversely affected for a while by these toxins, which have the possibility of causing several environmental issues. An oil slick is the most apparent sign of petroleum pollution in surface water. These slicks are spurred on by spilling or leaking petroleum products like gasoline or diesel into bodies of water from storage tanks. Large portions of the water's surface may be covered by oil slicks, which restrict light from reaching aquatic plants and obstruct marine life's ability to breathe (Liu et al. 2016).

Moreover, oil slicks may lead to shoreline pollution, which is hard to remove and may have long-term effects on the ecosystem's health. Petroleum pollution has chemical and biological repercussions in addition to its apparent effects. Petroleum products may include several harmful substances, including polycyclic aromatic hydrocarbons, when they enter a body of water (PAHs). Aquatic species can absorb these substances, which may have various detrimental impacts on human health. Moreover, PAHs can adhere to sediments, pollute the whole food chain, and negatively influence ecosystem health. Eutrophication, or the excessive development of algae and other aquatic plants, is another effect of petroleum pollution, making the water hazy and decreasing the quantity of oxygen that is readily accessible, both of which can be fatal to aquatic life (Jonson et al. 2017).

Moreover, it may result in annoyance issues, including unpleasant tastes and smells. Last, petroleum pollution may raise water temperatures, lowering dissolved oxygen levels and making it harder for aquatic life to breathe. Fish and other marine species may have less reproductive success, which might result in population decreases. These effects may be challenging to clear up and may have long-lasting implications on the health of aquatic ecosystems. Thus, it is crucial to take action to stop petroleum from contaminating surface water to safeguard both the environment and public health.

Human activities rely primarily on groundwater, and its pollution may significantly impact the ecosystem. When discharged into the environment, petroleum waste is a significant contaminant that may affect groundwater. There are many ways that petroleum waste may pollute groundwater. The most frequent method of contamination is the environmental discharge of petroleum-based products like oil and gasoline due to intentional environmental pollution and unintentional spills and leaks of petroleum products. These products have the potential to contaminate groundwater when they are discharged because they may spread there. Petroleum waste may contaminate groundwater directly and indirectly (Logeshwaran et al. 2018). The discharge of petroleum products may alter the chemistry of the soil and rocks that make up the aquifer into the environment, which can change the groundwater quality. For instance, the quantity of oxygen accessible to aquatic species living in the aquifer may be decreased by the discharge of oil into the environment, which would lower the amount of oxygen dissolved in the groundwater, resulting in aquatic species dying, which would degrade the water's quality.

Moreover, the presence of petroleum waste in groundwater might result in the development of potentially harmful substances like benzene. Furthermore, petroleum waste may make the groundwater more acidic, which can corrode metal pipelines and other infrastructure (Ismail and El-Shamy 2009). Lastly, the accessibility of groundwater might be impacted by petroleum waste. Petroleum waste may become stuck in the aquifer's soil and rocks when discharged into the environment, which can decrease the quantity of water accessible for human use. As a result, less water could be available for drinking and other purposes. Petroleum waste may also result in the synthesis of potentially dangerous chemicals and the corroding of metal pipelines. Lastly, it may impact groundwater accessibility, lowering the quantity of water accessible for human usage.

The contamination of groundwater by petroleum waste is a major environmental threat. Many pollutants in petroleum waste may harm human health and the environment. Heavy metals, VOCs, and PAHs are some pollutants. Heavy metals are difficult to remove or degrade because they may bioaccumulate in the environment. Many health issues, including headaches, vertigo, and respiratory irritation, may be caused by VOCs. Human cancer is known to be driven by PAHs, which may persist in the environment for a very long time. Petroleum waste poisoning of groundwater may have detrimental effects on both the environment and human health. Many health issues, including nausea, respiratory discomfort, and skin irritability, may be caused by contaminated groundwater. Also, drinking toxic water might raise your chance of developing some cancers. As harmful substances may build up in the food chain, contaminated groundwater can also harm aquatic animals (Rosell et al. 2006). It's crucial to make sure that petroleum waste is correctly handled to avoid groundwater contamination caused by such waste. Routine maintenance and inspection of any petroleum-related equipment and the correct disposal of petroleum waste products can aid in attaining the desired result. In addition, precautions, including proper installation and frequent tank inspections, should be implemented to minimise the danger of underground storage tank leaks. Lastly, it is essential to regularly clean up spills to prevent them from contaminating the groundwater table. By managing petroleum waste properly, it is feasible to lessen the chance of groundwater pollution and contribute to environmental protection.

1.5.4 Climate Change

Petroleum waste includes any byproducts or substances from the manufacturing of petroleum that are dumped as garbage. Being highly hazardous and potentially impacting air quality when released into the atmosphere negatively, petroleum waste is a significant environmental problem. Chemicals like hydrocarbons, sulfur, nitrogen, and other volatile organic compounds are all present in petroleum products. When exposed to sunlight and other airborne contaminants, these substances combine to generate ozone. Ozone is a kind of air pollution that seriously affects the environment and human health. The stratosphere is a part of the Earth's atmosphere that contains the ozone layer. It is responsible for safeguarding the Earth's surface from the sun's damaging UV radiation and comprises three oxygen molecules. The ozone layer is deteriorating due to the emission of certain chemicals into the atmosphere, a process known as ozone depletion. These ozone-depleting compounds, such as chlorofluorocarbons (CFCs), hydrochlorofluorocarbons (HCFCs), halons, and carbon tetrachloride, are human-made. One of the primary sources of compounds that deplete the ozone layer is the release of petroleum waste into the atmosphere. Different VOCs found in petroleum waste are released into the atmosphere when burnt or evaporated. These VOCs generate ozone when they interact with sunlight and other air contaminants. If not handled appropriately, the ensuing ozone may be very hazardous and result in serious air quality issues. Other air pollutants like nitrogen oxides (NOx) and sulfur dioxide may worsen petroleum waste's effects on ozone (SO_2). The ground-level ozone produced by these contaminants and the atmospheric VOCs is much more harmful than the ozone in the higher atmosphere. Ground-level ozone has been linked to several health issues, such as respiratory ailments and eye discomfort. Petroleum waste may harm the environment and directly influence ozone, possibly due to the forming of tropospheric ozone due to the VOCs produced by petroleum waste. A greenhouse gas called tropospheric ozone absorbs heat in the atmosphere and contributes to global warming (Burnham et al. 2012).

The detrimental consequences of petroleum waste on the environment have been recognised for a long time. Still, their role in accelerating global warming is only being truly realised. Toxic byproducts from the petroleum industry contribute significantly to global warming. Global warming is exacerbated by burning petroleum byproducts, which releases more carbon dioxide into the atmosphere. Nevertheless, petroleum spills may have serious environmental consequences, including contaminating air and water supplies and removing harmful chemicals. Greenhouse gases are those that prevent heat from escaping from the atmosphere. CO_2 is released into the air when fossil fuels, like gasoline, are burned.

Consequently, the levels of greenhouse gases in the atmosphere increase, driving global warming because a higher amount of heat is trapped in the atmosphere due to the increased carbon dioxide levels. As a result, temperatures rise, and the chance of extreme weather events like floods, droughts, and heat waves rises. Oil spills can also cause a lot of damage to the environment because they release toxic chemicals into the air and water. These factors may render the air dirty, leading to breathing and other health problems. They can also cause water to become contaminated, making drinking unsafe and killing aquatic life. Methane is made from petroleum waste, which is another way that it changes the climate. Methane is a potent greenhouse gas when oil waste is not burned correctly. The burning process isn't finished, and the released gases aren't burned off, causing methane to be released into the air, which adds to the number of greenhouse gases and causes the Earth to warm. Changes in the climate are caused by oil waste in many ways. Burning petroleum waste increases the number of greenhouse gases in the atmosphere, warming the planet. Oil spills can pollute the air and water and release dangerous chemicals into the environment. Also, methane is released when petroleum waste is burned without being completely burned, a potent greenhouse gas. Lastly, as a major source of greenhouse gases, petroleum wastes can significantly impact global warming. About 25% of the world's carbon dioxide comes from petroleum products, which can contribute to the destruction of the ozone layer (Wei et al. 2014).

1.5.5 Wildlife

Petroleum waste has a significant and far-reaching influence on animals, impacting the surrounding and global ecology. When petroleum products are used and disposed of inappropriately, they may negatively affect the environment and animals. Petroleum products are manufactured from crude oil that is mined from the ground. The frequent dumping of petroleum waste into the atmosphere, ocean, and land may harm animals. When burnt, petroleum-based products like gasoline and diesel release environmental pollutants. These pollutants have the potential to adversely impact animals by causing acid rain, smog, and global warming. Acid rain may impede plant development if the soil has fewer nutrients.

Smog may hinder an organism's ability to photosynthesize by reducing the quantity of light and heat that reaches the ground. Climate changes may alter due to global warming, making it challenging for species to adapt to their new habitat. Over time, indirect consequences include modifications to the food chain, reproductive success, and species diversity. Petroleum wastes have extensive indirect effects on animals that may have long-term implications on the environment's quality of life. Petroleum byproducts may build up in the environment and affect the food chain. Petroleum wastes may be present in contaminated water, sediment, and soil. They may be ingested by plants and animals, resulting in a reduction in the number of species present and changes to the species diversity of the environment. Petroleum wastes could negatively impact the ability of animals to reproduce. Pollutants like PAHs may interfere with reproduction by altering hormone levels, lowering reproduction success. As a result, populations may become weaker and less diverse, and some species may even become extinct (Barron and Holder 2003).

Petroleum wastes have substantial direct impacts on animals and significantly damage the ecosystem. These wastes originate from several sources, including oil refineries, transportation, storage systems, and oil and gas production. Wildlife may be harmed directly or indirectly by petroleum waste. Direct consequences are those that come from pollution immediately and directly. They include bodily injury or death caused by contact with petroleum wastes or consuming contaminated food or water. Ingestion of contaminated food or water, physical harm or death brought on by contact with toxic substances, and inhalation of petroleum fumes are only a few examples of how petroleum wastes directly affect animals. Animals exposed to petroleum wastes in the environment may experience physical injury, including burns, skin irritation, and respiratory problems. Consuming contaminated food or drink may result in death or serious injury to internal organs like the liver and kidneys. Petroleum vapor inhalation may result in deadly chemical burns to the lungs. In addition to the potential for physical harm, coming into proximity with petroleum wastes also carries the risk of toxicological consequences. They include alterations in animal behavior, poor reproduction, and decreased survival rates. Animal development, behavior, and reproductive success may all be affected by exposure to petroleum waste due to biochemical and hormonal changes (Ludlow et al. 2015). These modifications may alter the ecosystem's species diversity and perhaps result in the extinction of certain species. Direct discharge into water sources or runoff from the land may cause petroleum pollution. When this happens, the water may become contaminated and unsafe for drinking, swimming, or fishing.

Moreover, oil spills may be particularly hazardous since the oil can film the water's top, blocking the passage of oxygen to the creatures below. Oxygen-dependent organisms may perish if aquatic environments are drastically altered. Petroleum waste may also be directly dumped onto land, contaminating the soil and preventing plant development (Ukaogo et al. 2020). Petroleum products may leach into the ground and pollute the water table when stored in land containers, rendering the water unsuitable for drinking, swimming, or fishing. The toxins in petroleum products may expose wildlife and harm or kill them. Large spills may also coat an animal's hair or feathers, making it difficult for them to regulate their body temperatures and causing hypothermia. Moreover, food supplies contaminated by oil spills are unfit for human consumption.

Petroleum wastes have destructive, far-reaching consequences on animals both directly and indirectly. Contamination of the air, water, or land may cause ecosystems to collapse, species to perish, and food supplies to become contaminated. These consequences can result in physical harm, behavioral changes, poor reproduction, and even the extinction of whole species. It is crucial to take action to minimize the number of petroleum waste emitted into the environment, including better legislation, enforcing already-existing rules, and using best management practices to lessen the number of petroleum waste discharged into the atmosphere. These actions are necessary to safeguard both the ecosystem and the living creatures.

1.5.6 Human Health

Petroleum waste is a significant pollutant associated with several adverse health effects. Many complex mixes of hundreds of molecules that constitute petroleum products are detrimental to human wellness. The consequences of exposure to these chemicals may vary from mild skin irritations to cancer and can occur by ingestion, inhalation, or dermal contact. Petroleum waste may have various health impacts depending on the kind of waste and the exposure route. Cancer, reproductive issues, and brain impairment may all result from prolonged exposure to petroleum pollution. Asthma, bronchitis, and emphysema are just a few respiratory conditions that may develop due to breathing in petroleum waste (Schraufnagel et al. 2019). Vomiting, diarrhea, and nausea may all result from ingesting petroleum waste. Dermatitis, rashes, and skin irritation may be brought on by dermal exposure. Petroleum waste might have indirect health impacts in addition to direct ones. For instance, petroleum waste may pollute drinking water sources, exposing people to harmful and carcinogenic compounds. Petroleum waste may also be dumped on land, which can pollute the environment, and the air and expose people if they eat contaminated food. Other environmental issues, such as climate change, may increase the health impacts of petroleum waste. Increased flooding may distribute petroleum waste across a broader region and raise the risk of human exposure when the frequency and intensity of severe weather events rise due to climate change (Ite and Ibok 2013).

Hazardous materials like petroleum wastes may significantly negatively influence the environment, human health, and other living things. Several processes produce these wastes, including oil and gas exploration, production, refining, and transportation. Petroleum wastes are complicated mixtures of organic and inorganic substances, including heavy metals, PAHs, VOCs, and other dangerous contaminants. Petroleum wastes may have severe adverse health impacts on humans. Several health issues, including skin and eye irritation, respiratory issues, and cancer, may be caused by exposure to petroleum pollutants. Moreover, petroleum products may include VOCs, resulting in symptoms including headaches and vertigo. Higher cancer risk has been associated with prolonged VOC exposure (Ite et al. 2013).

Reducing production rates and ensuring appropriate disposal is the key to mitigating the health risks associated with petroleum waste. Using best practices for the safe and secure disposal of waste is one way to ensure that petroleum waste is dealt with in an environmentally friendly and responsible manner. In conclusion, petroleum waste has the potential to affect human health in several ways negatively. Consequently, it's crucial to take measures to minimize garbage output and guarantee proper disposal.

In conclusion, petroleum waste may significantly affect the economy, the environment, and human health. These contaminants may pollute soil, water, and the atmosphere and are hazardous and persistent. Petroleum wastes may also contribute to climate change and the degradation of ecosystems. As a result, it is essential to effectively monitor and control the production and use of petroleum products. Regulating and reducing the discharge of petroleum waste into the atmosphere is critical to protect the ozone layer and mitigate the effects of climate change.

1.6 Environmental Remediation

Remediation of petroleum waste is crucial in safeguarding the environment from the consequences of oil spills, leaks, and other pollution from petroleum sources. Remediation is the process of decontaminating and restoring a site that has been polluted by petroleum waste. This approach aids in lowering the danger of adverse effects on human health, safeguarding natural resources, and minimizing long-term environmental harm brought on by petroleum waste. Treatment and remediation of contaminated soils, groundwater, and surface water caused by oil and other petroleum products constitute remediation of petroleum waste. Petroleum waste is a severe environmental risk because, if improperly managed, it may harm the ecosystem permanently. Effective remediation strategies are essential to halting future environmental contamination and protecting human health. Physical removal of polluted soil, sediment, and groundwater is a frequent remediation strategy. The two main types of remediation are Ex-situ and In-situ. The various ex-situ and in-situ remediation techniques are depicted in Fig. 1.3.

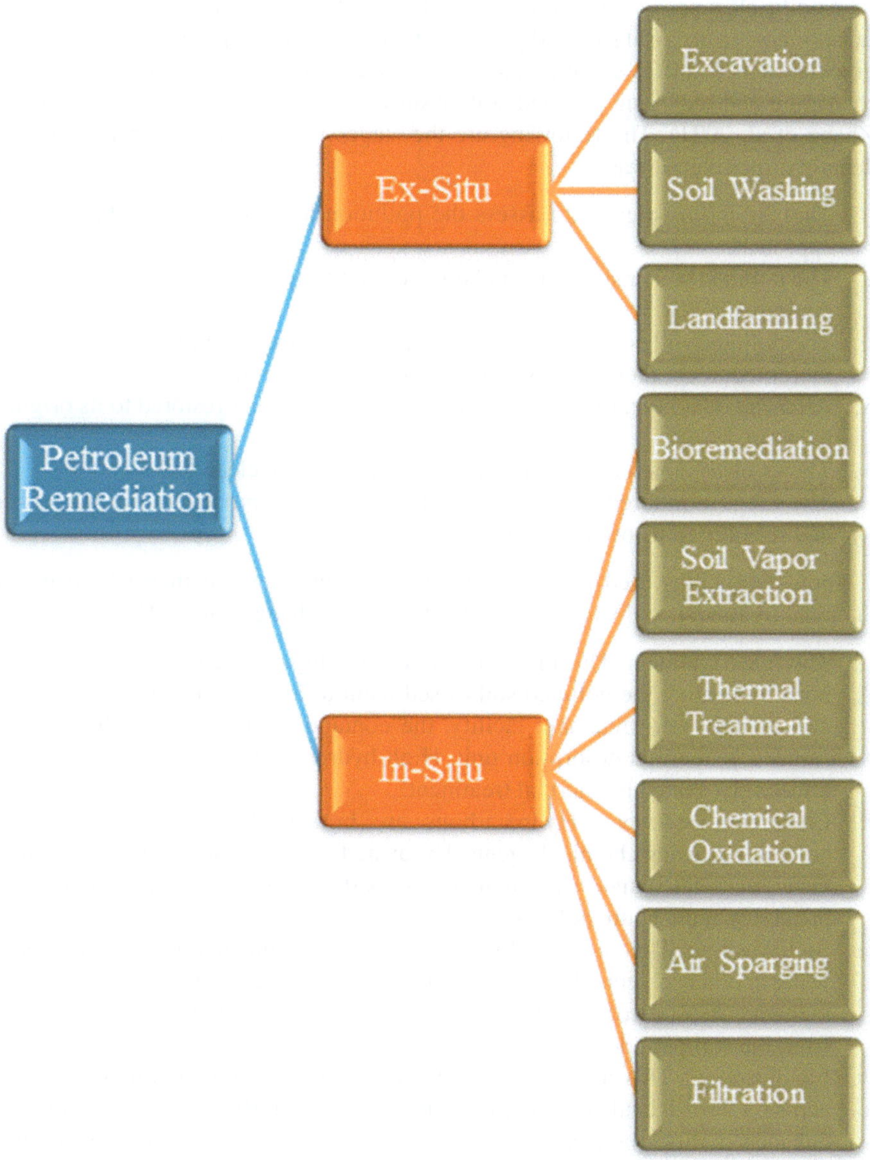

Fig. 1.3 Petroleum remediation techniques

1.6.1 Ex-Situ Remediation

Ex-situ remediation is a kind of remediation in which the contaminated material is excavated from its original site and treated elsewhere. It entails physically removing hazardous materials from the site and bringing them to a treatment facility. Excavation, treatment, disposal, and monitoring are some of the procedures involved in the ex-situ cleanup of petroleum waste. The contaminated material is excavated, transported to a disposal site, and properly disposed of in this procedure. Excavation remediation is a kind of ex-situ remediation used to treat petroleum waste. It entails excavating and removing polluted soil or silt from the site for treatment elsewhere (Zhang et al. 2021). The following are the stages involved in the excavation and cleanup of petroleum waste:

i. The site is investigated to assess the pollution level and the suitable cleanup approach.
ii. The polluted soil or silt is dug and transferred to a treatment facility using heavy equipment.
iii. The polluted soil or sediment is treated using a variety of approaches, including bioremediation, thermal treatment, and chemical treatment.
iv. After the treatment, the treated soil is either discarded or restored to its original location.
v. The excavated area is backfilled and compacted with clean soil.
vi. The site is monitored to confirm that the remediation procedure was successful and restored to its original state.

Some frequently used ex-situ remediation techniques are thermal treatment, chemical treatment, soil washing, and landfarming (Zhang et al. 2021).

i. Thermal Treatment: This method includes burning or volatilizing the hydrocarbons by heating the polluted soil or sediment to high temperatures. Excavated and brought to a treatment facility, the contaminated material is heated in a furnace or thermal desorption unit. High hydrocarbon concentrations may be effectively treated by thermal treatment.
ii. Chemical Treatment: Hydrocarbons are broken down into less harmful compounds using chemical agents. Excavated and brought to a treatment plant, the contaminated material is treated with oxidizing agents or other chemicals to break down the hydrocarbons.
iii. Soil Washing: This method physically separates pollutants from the soil using water or other solvents. The contaminated soil is excavated, transported to a treatment facility, and treated with water and other agents to remove the hydrocarbons from the soil particles.
iv. Landfarming: This approach includes distributing polluted soil or sediment across a broad land area and combining it with fertilizers and microbes to enhance hydrocarbon breakdown. Periodically, the contaminated material is tilled to improve oxygenation and mixing.

Nonetheless, the excavation procedure may be costly and detrimental to the environment. It is often used at smaller sites when contamination is confined and the amount of contaminated material is relatively low (Chen et al. 2003). On-site treatment of the contaminated material can lower its hazardous material concentration and enhance its worth as a resource (Kuppusamy et al. 2015).

1.6.2 In-Situ Remediation (ISR)

ISR is employed to clean polluted soils and groundwater without excavation. In the petroleum sector, it is often used to clean up spills, leaks, and other emissions. Besides treating the contamination in place, in-situ remediation refers to various remediation approaches to remediate environmental pollution at a site. This approach injects chemicals like surfactants, oxidants, and bioremediation agents into the contamination site. The chemicals react with petroleum pollutants and break them into less dangerous compounds. This procedure is often employed when pollution is extensive as an alternative to expensive and sometimes hazardous excavation. It is also possible to treat petroleum waste in-situ using various methods. Injecting surfactants, bacteria, and nutrients into a polluted region is a common part of these methods. These chemicals aid in the degradation of hydrocarbon-based molecules and lessen their toxicity (Camenzuli and Freidman 2015).

Petroleum hydrocarbons, such as gasoline, diesel, crude oil, motor oil, heavy metals, and other hazardous compounds, are among the pollutants that ISR is most often used to remediate. Depending on the kind of pollution and the particulars of the site, in-situ remediation techniques might differ significantly. Bioremediation, in which naturally existing microorganisms degrade the contaminants, and chemical oxidation, in which chemical oxidants (such as hydrogen peroxide or ozone) are added to the polluted site to degrade the pollutants, are the two most popular ISR techniques. Other ISR techniques include thermal treatments, air sparging, soil vapor extraction, and other physical treatments. Since it may be less expensive and time-consuming than digging and removing the contaminated material, ISR is often favored. Also, since ISR does not need extensive soil excavation and other materials, it often causes less environmental disruption. The nature and concentration of the pollutants, the site features, and the efficiency of the selected remediation approach are only a few of the variables that significantly impact the outcome of ISR. To be sure that the toxins have been successfully eliminated, in-situ remediation also requires careful monitoring and assessment of the site. ISR is a crucial tool for managing environmental pollution and has the potential to be a successful and affordable solution for cleaning up oil spills, leaks, and other discharges. ISR may effectively address environmental contamination while having negligible adverse effects on the ecosystem. The removal of polluted soils, groundwater, and surface water at the contamination site is accomplished by ISR (Song et al. 2017). The advantages of ISR are listed as follows:

- ISR minimizes the need for extensive excavation, which is often costly and disruptive. Removing polluted soil or water, which may be expensive for large-scale cleanups, is optional for the in-situ remediation method. Instead, the toxins are handled on the spot, enabling a more focused remediation method.
- ISR may capture, break down, or eliminate pollutants without excavation or relocation of soil or water. By cleaning the location and halting additional pollution migration, the procedure may also lower the danger of contamination spreading to adjacent places.
- ISR is often less costly than conventional, especially for large-scale cleanups. ISR treatments are often cheaper than excavation and transportation of contaminated materials. The procedure may also be finished quickly, enabling speedier remediation and a quicker return to regular operations.
- ISR is an efficient technique for removing various contaminants, including hydrocarbons derived from petroleum. Oil, gasoline, diesel fuel, and other petroleum-based products are among the petroleum pollutants the method may collect and treat. Moreover, the technique may minimize or completely stop the leakage of petroleum pollutants into the environment.
- Ultimately, in-situ remediation may be more sustainably accomplished by cleaning up polluted soils, groundwater, and surface water in the petroleum sector. The technique may lessen or eliminate the need for further excavation and transportation of contaminated materials by treating pollutants where they are. As a result, the environmental effect of the cleaning process is lessened, and the area may be kept clean and safe for future generations.

Although ISR is a cost-effective and efficient technique for removing petroleum hydrocarbons, it has a few drawbacks. Some of the disadvantages of ISR are:

- The process might take months or even years to complete, depending on the extent of the region that must be treated. Moreover, weather factors like torrential rain or high winds might impact the operation, slowing it down and raising costs.
- ISR also requires using potentially hazardous chemicals for both people and the environment. Solvents and surfactants, which must be used in the procedure, may be dangerous and pose health concerns if not handled carefully. These compounds may also be difficult to degrade and can persist in the environment for a very long time, both of which can harm the ecosystem.
- The success of the in-situ remediation procedure is still being determined. The polluted region is not guaranteed to be entirely petroleum-free, even if the treatment is effective. Also, the process can redistribute the petroleum, which might create additional pollution in other places.

Some of the most used ISR techniques are.

1.6.2.1 Bioremediation

Bioremediation is the application of naturally existing microorganisms to degrade contaminants. Microorganisms such as bacteria, fungi, and algae are introduced to the contaminated region, where petroleum pollutants are metabolized, transforming them into less hazardous chemicals. Since the microbes utilize hydrocarbons as an energy source, this approach successfully treats petroleum hydrocarbons. This technique uses natural organisms to decompose dangerous substances into harmless byproducts. Adding oxygen or nutrients to the environment may boost this process by stimulating the development of microorganisms. Both soil and groundwater pollution produced by petroleum waste may be cleaned up via bioremediation (Labianca et al. 2020).

Bioremediation is a technique used by the petroleum industry to purify polluted soil and water. This procedure employs naturally existing microbes to decompose and convert dangerous contaminants, such as petroleum hydrocarbons, into innocuous byproducts, such as carbon dioxide and water. This procedure is considered one of the safest and most cost-effective ways to restore polluted sites. Since the 1970s, bioremediation has been employed to remediate petroleum spills and other kinds of pollution. It has grown in popularity owing to its efficacy and relative affordability. Introducing strains of microbes that have evolved to degrade and metabolize hydrocarbons enables the process to function. These bacteria are often found in the environment, but their quantities may be insufficient to remediate a spill adequately. The approach starts with the identification and evaluation of the polluted region. The cleaning crew then uses the appropriate bioremediation chemical to enhance the microbial population. Typically, the product contains nutrients, oxygen, and other chemicals that promote the development and activity of microorganisms. As microorganisms are established, they begin to degrade contaminants. Depending on the magnitude and complexity of the contamination, this procedure might take anywhere from a few weeks to a few months. When the microorganisms break down the hydrocarbons, they emit carbon dioxide and water as harmless byproducts (Labianca et al. 2020).

Phytoremediation is a bioremediation that uses plants to remove many forms of environmental contaminants, including petroleum. It is a bioremediation in which plants degrade, metabolize, and immobilize toxins such as heavy metals, petroleum hydrocarbons, and other pollutants. It is a cost-effective and environmentally friendly method for remediating contaminated sites since it does not need the disposal or transportation of hazardous material. The phytoremediation process starts with identifying suitable plant species for the contaminant. The selected plants must withstand the pollution and the site's environmental conditions. The plants are then planted at the spot containing the pollutant. The plant roots absorb the contaminant from the soil, which is then digested or immobilized by the plant (Dietz and Schnoor 2001). The phytoremediation procedure may be divided into two major categories: rhizoremediation and phytoextraction.

i. Rhizoremediation uses plants to break down or immobilize soil or groundwater contaminants; this is achieved via the plant's roots, which offer a substrate for developing microorganisms to digest contaminants (Otte and Donna 2006).
ii. Phytoextraction employs plants to remove contaminants from the soil and store them in the plant's tissues. Typically, metal-tolerant plants that can absorb and retain metal pollutants in their roots are used (Ali et al. 2013).

1.6.2.2 Soil Vapor Extraction

Soil vapor extraction (SVE) is a remediation method that employs a vacuum system to extract VOCs from the soil. The system consists of a series of wells linked to a vacuum pump that generates a subsurface vacuum. This vacuum pulls volatile organic compounds from the soil into the pump, where they are collected and handled. SVE is a technique to remove volatile organic compounds VOCs from polluted soil and groundwater (Labianca et al. 2020). The petroleum industry often uses this method to decontaminate locations polluted with petroleum hydrocarbons such as gasoline and diesel fuel. SVE is a passive vapor extraction system that operates without external energy. Instead, it harnesses the pollutants' inherent vapor pressure to extract them from the soil and groundwater. A series of wells are placed inside the polluted region as part of the SVE system. These wells facilitate the escape of contaminated vapors into the atmosphere. Installation of extraction wells is the first phase of the SVE process. Wells are typically installed at regular intervals along the perimeter of the polluted region. When the wells have been installed, a vacuum is delivered to the soil and groundwater to create a pressure differential between the polluted region and the atmosphere. This pressure differential draws volatile organic chemicals from the soil into wells (Rathfelder et al. 1995). After the polluted vapors have been extracted from the soil and groundwater, they are treated using a system such as an activated carbon filter. The activated carbon filter prevents the pollutants from entering the atmosphere by absorbing them. After treatment, the vapours are discharged into the environment.

1.6.2.3 Chemical Oxidation

Chemical oxidation is a prevalent approach for remediating petroleum waste. This method employs chemical agents to convert dangerous substances into harmless byproducts. This approach is often used with other remediation methods to lower the concentration of dangerous substances and the risk of environmental contamination. Chemical oxidation remediation is a method used by the petroleum industry to decontaminate polluted soils and groundwater. This method includes adding a chemical oxidant, such as hydrogen peroxide and ozone, to polluted media. The oxidant interacts with the pollutants, decomposing them into innocuous chemicals (Labianca et al. 2020). Chemical oxidation often removes petroleum hydrocarbons from contaminated soils and groundwater, such as benzene, toluene, ethylbenzene, and xylene (BTEX). The technique eliminates BTEX chemicals since they are more readily oxidised than other petroleum compounds. The process begins by injecting an oxidant into polluted soil or groundwater. The oxidant interacts with the pollutants, breaking them into harmless carbon dioxide and other chemicals (Yuniati 2018).

1.6.2.4 Thermal Treatment

Thermal treatment is a remediation method that uses heat to degrade petroleum pollutants. The contaminated region is heated in this method, causing the petroleum pollutants to evaporate. The collected vapours are typically handled by adsorption or air sparging. As a successful method for removing petroleum toxins from the subsurface, this method is often employed to remediate polluted groundwater. Thermal treatment remediation is an efficient technique used by the petroleum sector to clean up contaminated soil, groundwater, and air. This technique utilises heat to convert organic pollutants, such as petroleum hydrocarbons, into less dangerous compounds. The thermal treatment may remove soil, groundwater, and organic air pollutants (Vidonish et al. 2016). Thermal treatment remediation injects hot air or steam into a polluted region and raises the temperature of the contaminated material, causing the organic pollutants to degrade into less harmful chemicals. This technique may be used to remove VOCs from the air and to convert hydrocarbons into carbon dioxide and water vapor. Inorganic pollutants in the soil, such as heavy metals, may also be eliminated by thermal treatment. The heat produced during the thermal treatment procedure may help oxidize and decrease the pollutants, making them less harmful (Khaitan et al. 2006). Thermal treatment remediation is often used with other remediation procedures, including bioremediation and chemical oxidation.

1.6.2.5 Air Sparging

Air sparging is a method for treating contaminated groundwater. This method involves injecting air into the subsurface to create a bubble that drives impurities to the surface. After the contaminants have been brought to the surface, various remediation methods, such as soil vapour extraction and bioremediation, may be used to remediate them. Air sparging is a remediation technique the petroleum industry uses to remove pollutants from the soil and groundwater. The process pumps pressurised air into the subsurface to generate air bubbles that rise through the soil and groundwater. When the air bubbles ascend, they generate a vacuum in their wake, causing pollutants to be sucked from the subsurface and into the air bubbles (Nadim et al. 2000). This procedure is called vapour extraction. It efficiently removes volatile chemicals such as petroleum hydrocarbons, chlorinated solvents, and other organic compounds. The process of air sparging involves delivering air into the subsurface via a network of pipelines or other injection devices. The injection of air at a pressure lower than the soil and groundwater allows air bubbles to rise and extract pollutants from the subsurface. The air bubbles also produce a vacuum, which allows pollutants to be pulled out of the soil and into the air bubbles. After the air bubbles have risen to the surface, the pollutants are released into the atmosphere, where they are treated by additional means based on the removed substance (Benner et al. 2002). Air sparging is often used with bioremediation, chemical oxidation, and soil vapor extraction to enhance the efficiency of the remediation procedure.

1.6.2.6 Filtration

Filtration (also known as filtration extraction and filtration remediation) removes petroleum pollutants from water. This method eliminates contaminants from the water by passing it through a filter. This technique is often used to remediate contaminated surfaces and groundwater. The petroleum industry has used this filtration technique for years to remove hydrocarbons and other environmental toxins. The first step in filter remediation is extracting polluted material from the affected region. This substance is subsequently put through a filter to eliminate impurities. The filters may be constructed from sand, gravel, activated carbon, or other substances. The choice of filter material is determined by the pollutant to be eliminated. When the contaminated material has been filtered, it is disposed of appropriately. It may be placed in a landfill, incinerated, processed to lower the concentration of toxins, or otherwise reused. Many variables influence the success of filter remediation. They include the nature of the filter utilized, the size of the filter, the type of pollutant, the concentration of the contaminant, and the filtration period (Mohammadi et al. 2020).

Bioremediation, soil vapor extraction, chemical oxidation, thermal treatment, air sparging, and filtration are some of the most common forms of in-situ remediation. Their relative benefits and drawbacks are compared, contrasted, and summarized in Table 1.2.

The choice of remediation approach will rely on the kind of contamination, its extent, and the available resources for cleaning up the contaminated region. Any of these methods mitigate the dangers presented by petroleum waste and benefit the environment and human health. Each treatment has benefits and drawbacks; choosing the appropriate procedure for the contamination is essential. Also, assessing the possible dangers connected with each design is vital since some may generate more pollution than they remove. Correctly managed remediation of petroleum waste may contribute to the protection of the environment and human health, as well as to the preservation of a safe environment for future generations.

1.7 Conclusion and Future Perspective

Petroleum is an essential energy source for contemporary civilization and is utilized in various processes. However, the manufacturing and utilization of petroleum products generate significant waste that might harm the environment and the general public's health. Petroleum wastes provide severe environmental problems, such as

Table 1.2 Advantages and disadvantages of various remediation techniques

Remediation method	Advantages	Disadvantages
Bioremediation	• It requires minimal energy • It doesn't need any expensive equipment or materials • It is relatively non-invasive • It doesn't release any harmful by-products • No chemicals are required	• Quite slow • Requires the addition of air or oxygen to the contaminated site, which can be difficult while dealing with hydrocarbons • Some hazardous substances, such as PAHs, are challenging to break down using microorganisms
Soil vapor extraction	• Compared to other remediation techniques, it is cost-effective • It requires less labor and time • It can be used for various soils, making it versatile for various contamination scenarios • Employs vacuum for the removal of VOCs, thus considered to be a safe process	• Unable to effectively remediate some pollutants, such as heavy metals and pesticides • Low permeability sites may not be suitable for this method because the suction may not reach deep enough into the soil to remove VOCs • Site conditions with shallow groundwater or high-water tables exclude this as a feasible choice
Chemical oxidation	• Faster than other remediation methods • Quite versatile for removing petroleum hydrocarbons, pesticides, VOCs, and heavy metals • Effective in treating soil and groundwater	• If the process is not correctly regulated, it may produce pollutants like nitrogen oxides that are dangerous to human health • These oxidants have the potential to interact with other environmental components and develop further harmful consequences
Thermal treatment	• An effective way to clean polluted soils and sediments is because it can rapidly and efficiently break down petroleum hydrocarbons • It doesn't employ any potentially harmful chemicals or solvents • It can treat pollutants such as oil, grease, contaminated soil, and other sediments	• Specialized equipment is needed for the process, which often has substantial energy expenditures • The process may release hazardous fumes or other substances into the environment, endangering people, and animals • It necessitates the removal of a significant volume of soil, which might cause soil erosion and the eviction of species

(continued)

Table 1.2 (continued)

Remediation method	Advantages	Disadvantages
Air sparging	• Relatively cost-effective • Very efficient in eliminating petroleum-based pollutants from soil that are located below the soil's surface • This restoration technique does not involve using harmful chemicals or other potentially dangerous materials, making it safe and ecologically friendly	• Certain pollutants may be too heavy to be lifted by the air bubbles and removed from the water, rendering this method ineffective • It may also cause air bubbles to get trapped in the soil, resulting in changes to soil porosity and a reduction in the soil's capacity to retain water • It requires a significantly larger area for the equipment installation
Filtration	• It is safer for the environment since it does not involve dangerous chemicals • It enables the speedy removal of hydrocarbons from the environment • Capable of removing dissolved and particulate contaminants • It can treat contaminated surface water and groundwater, allowing for a more comprehensive remediation effort	• Using specialized filters, pumps, and other equipment makes the method expensive • It requires routine maintenance and replacement, which can add to the cost • Pollutants could persist in the environment even if the filtering procedure is effective but at lower concentrations • The procedure may be time-consuming and ineffective when dealing with significant amounts of contaminants

Note VOCs—volatile organic compounds

air, water, and soil pollution, contaminating animal and aquatic resources and posing threats to human health. Managing petroleum waste is vital for mitigating its environmental implications. Recycling, reduction, reuse, recovery, treatment, and disposal may improve petroleum waste management. Implementing standards and guidelines to ensure the waste is handled safely and responsibly is necessary to manage petroleum waste properly. Governments and industry stakeholders must work to create petroleum waste management guidelines and best practices.

Moreover, public education and awareness initiatives may contribute to promoting sustainable behaviors and reducing petroleum waste production. The handling of petroleum waste provides economic prospects, such as the production of new technology and waste management sector employment. Recycling and recovery of valuable materials may generate income and decrease disposal expenses. Moreover, sustainable waste management procedures boost a company's brand and attract environmentally sensitive consumers.

The petroleum industry's vision for the future is complicated and multidimensional, with various possibilities and difficulties. On the one hand, the petroleum industry is anticipated to increase due to the ongoing need for energy, especially in emerging nations with rising populations and burgeoning economies. New reserves may be found, and unconventional resources, such as shale oil and gas, may be extracted due to technological and exploratory advancements. On the other hand, the petroleum industry will be significantly impacted by the shift to a low-carbon economy and growing concerns about climate change. The demand for oil and gas may decrease as nations turn to renewable energy sources and lessen their dependence on fossil fuels. Lower pricing and worse profitability for petroleum firms might result from this. Petroleum waste management will remain a significant environmental and financial challenge. The quantity of waste produced throughout exploration, production, refining, transportation, storage, and consumption will rise if the world depends on oil and gas for energy. One potential future perspective is creating and accepting more sustainable practices in the petroleum sector, such as lowering waste output, boosting recycling and reusing, and supporting more environmentally friendly manufacturing and distribution techniques. Another perspective is the increasing use of cutting-edge technology for the remediation and disposal of petroleum wastes, such as bioremediation, soil vapor extraction, chemical oxidation, thermal treatment, air sparging, and filtration. By ensuring safer and more efficient waste disposal, these technologies may help lessen the adverse effects of petroleum waste on the environment. In conclusion, the petroleum sector, the government, and society must collaborate to create creative methods for handling petroleum wastes sustainably and ecologically consciously.

References

Aalsalem MY, Khan WZ, Gharibi W, Khan MK, Arshad Q (2018) Wireless Sensor Networks in oil and gas industry: recent advances, taxonomy, requirements, and open challenges. J Netw Comput Appl 113:87–97. https://doi.org/10.1016/j.jnca.2018.04.004

Ali H, Khan E, Sajad MA (2013) Phytoremediation of heavy metals-concepts and applications. Chemosphere 91(7):869–881. https://doi.org/10.1016/j.chemosphere.2013.01.075

Asim N, Badiei M, Torkashvand M, Mohammad M, Alghoul MA, Gasaymeh SS, Sopian K (2021) Wastes from the petroleum industries as sustainable resource materials in construction sectors: opportunities, limitations, and directions. J Clean Prod 284:125459. https://doi.org/10.1016/j.jclepro.2020.125459

Barron MG, Holder E (2003) Are exposure and ecological risks of PAHs underestimated at petroleum contaminated sites? Hum Ecol Risk Assess 9(6):1533–1545. https://doi.org/10.1080/10807030390251029

Benner ML, Mohtar RH, Lee LS (2002) Factors affecting air sparging remediation systems using field data and numerical simulations. J Hazard Mater 95(3):305–329. https://doi.org/10.1016/S0304-3894(02)00144-9

Blake DR, Rowland FS (1995) Urban leakage of liquefied petroleum gas and its impact on Mexico City air quality. Science 269(5226):953–956. https://doi.org/10.1126/science.269.5226.953

Burnham A, Han J, Clark CE, Wang M, Dunn JB, Palou-Rivera I (2012) Life-cycle greenhouse gas emissions of shale gas, natural gas, coal, and petroleum. Environ Sci Technol 46(2):619–627. https://doi.org/10.1021/es201942m

Camenzuli D, Freidman BL (2015) On-site and in situ remediation technologies applicable to petroleum hydrocarbon contaminated sites in the Antarctic and Arctic. Polar Res 34(1):24492. https://doi.org/10.3402/polar.v34.24492
Cetin E, Odabasi M, Seyfioglu R (2003) Ambient volatile organic compound (VOC) concentrations around a petrochemical complex and a petroleum refinery. Sci Total Environ 312(1–3):103–112. https://doi.org/10.1016/S0048-9697(03)00197-9
Charles M, Sayle S (2010) Offshore drill cuttings treatment technology evaluation. In: Paper presented at the SPE international conference on health, safety and environment in oil and gas exploration and production, vol 1, pp 405–415. https://doi.org/10.2118/126333-MS
Chen Z, Huang GH, Chan CW, Geng LQ, Xia J (2003) Development of an expert system for the remediation of petroleum-contaminated sites. Environ Model Assess 8(4):323–334. https://doi.org/10.1023/B:ENMO.0000004584.53505.15
Demirbas A, Bafail A, Nizami AS (2016) Heavy oil upgrading: unlocking the future fuel supply. Pet Sci Technol 34(4):303–308. https://doi.org/10.1080/10916466.2015.1136949
Deville JP, Fritz B, Jarrett M (2011) Development of water-based drilling fluids customized for shale reservoirs. SPE Drill Complet 26(04):484–491. https://doi.org/10.2118/140868-PA
Dietz AC, Schnoor JL (2001) Advances in phytoremediation. Environ Health Perspect 109(SUPPL. 1):163–168. https://doi.org/10.1289/ehp.01109s1163
Fu H, Chen Y, Liu T, Zhu X, Yang Y, Song H (2021) Research on hazardous waste removal management: identification of the hazardous characteristics of fluid catalytic cracking spent catalysts. Molecules 26(8):2289. https://doi.org/10.3390/molecules26082289
Guerin TF (2002) Heavy equipment maintenance wastes and environmental management in the mining industry. J Environ Manag 66(2):185–199. https://doi.org/10.1006/jema.2002.0583
Hu G, Li J, Zeng G (2013) Recent development in the treatment of oily sludge from petroleum industry: a review. J Hazard Mater 261:470–490. https://doi.org/10.1016/j.jhazmat.2013.07.069
Hu G, Liu H, Chen C, Hou H, Li J, Hewage K, Sadiq R (2021) Low-temperature thermal desorption and secure landfill for oil-based drill cuttings management: pollution control, human health risk, and probabilistic cost assessment. J Hazard Mater 410:124570. https://doi.org/10.1016/j.jhazmat.2020.124570
Ismail AIM, El-Shamy AM (2009) Engineering behaviour of soil materials on the corrosion of mild steel. Appl Clay Sci 42(3–4):356–362. https://doi.org/10.1016/j.clay.2008.03.003
Ite AE, Ibok UJ (2013) Gas flaring and venting associated with petroleum exploration and production in the Nigeria's Niger Delta. Am J Environ Prot 1(4):70–77. https://doi.org/10.12691/env-1-4-1
Ite AE, Ibok UJ Ite MU, Petters SW (2013) Petroleum exploration and production: past and present environmental issues in the Nigeria's Niger Delta. Am J Environ Prot 1(4):78–90. https://doi.org/10.12691/env-1-4-2
Jia C, Zheng M, Zhang Y (2012) Unconventional hydrocarbon resources in China and the prospect of exploration and development. Pet Explor Dev 39(2):139–146. https://doi.org/10.1016/S1876-3804(12)60026-3
Jonson JE, Borken-Kleefeld J, Simpson D, Nyíri A, Posch M, Heyes C (2017) Impact of excess NO x emissions from diesel cars on air quality, public health and eutrophication in Europe. Environ Res Lett 12(9):094017. https://doi.org/10.1088/1748-9326/aa8850
Kardena E, Helmy Q (2015) Petroleum oil and gas industry waste treatment; common practice in Indonesia. J Pet & Environ Biotechnol 06(05). https://doi.org/10.4172/2157-7463.1000241
Khaitan S, Kalainesan S, Erickson LE, Kulakow P, Martin S, Karthikeyan R, Hutchinson SLL, Davis LC, Illangasekare TH, Ng'oma C (2006) Remediation of sites contaminated by oil refinery operations. Environ Prog 25(1):20–31. https://doi.org/10.1002/ep.10083
Kokal S (2005) Crude-oil emulsions: a state-of-the-art review. SPE Prod Facil 20(1):5–12. https://doi.org/10.2118/77497-pa
Kuppusamy S, Palanisami T, Megharaj M, Venkateswarlu K, Naidu R (2015) In-situ remediation approaches for the management of contaminated sites: a comprehensive overview. In: de Voogt P (ed) Reviews of environmental contamination and toxicology, vol 236. Springer. https://doi.org/10.1007/978-3-319-20013-2

Labianca C, De Gisi S, Picardi F, Todaro F, Notarnicola M (2020) Remediation of a petroleum hydrocarbon-contaminated site by soil vapor extraction: a full-scale case study. Appl Sci 10(12):4261. https://doi.org/10.3390/app10124261

Lal R (2012) Climate change and soil degradation mitigation by sustainable management of soils and other natural resources. Agric Res 1(3):199–212. https://doi.org/10.1007/s40003-012-0031-9

Leemann JE (1988) Hazardous waste minimization: part V. Waste minimization in the petroleum industry. J Air Pollut Control Assoc 38(6):814–823. https://doi.org/10.1080/08940630.1988.10466423

Li J, Lin F, Li K, Zheng F, Yan B, Che L, Tian W, Chen G, Yoshikawa K (2021) A critical review on energy recovery and non-hazardous disposal of oily sludge from petroleum industry by pyrolysis. J Hazard Mater 406:124706. https://doi.org/10.1016/j.jhazmat.2020.124706

Liu Z, Liu J, Gardner WS, Shank GC, Ostrom NE (2016) The impact of deepwater horizon oil spill on petroleum hydrocarbons in surface waters of the northern Gulf of Mexico. Deep-Sea Res Part II: Top Stud Oceanogr 129:292–300. https://doi.org/10.1016/j.dsr2.2014.01.013

Logeshwaran P, Megharaj M, Chadalavada S, Bowman M, Naidu R (2018) Petroleum hydrocarbons (PH) in groundwater aquifers: an overview of environmental fate, toxicity, microbial degradation and risk-based remediation approaches. Environ Technol Innov 10:175–193. https://doi.org/10.1016/j.eti.2018.02.001

Ludlow SM, Brigham RM, Davis SK (2015) Oil and natural gas development has mixed effects on the density and reproductive success of grassland songbirds. Condor 117(1):64–75. https://doi.org/10.1650/CONDOR-14-79.1

Mohammadi L, Rahdar A, Bazrafshan E, Dahmardeh H, Susan MABH, Kyzas GZ (2020) Petroleum hydrocarbon removal from wastewaters: a review. Processes 8(4):1–34. https://doi.org/10.3390/PR8040447

Nadim F, Hoag GE, Liu S, Carley RJ, Zack P (2000) Detection and remediation of soil and aquifer systems contaminated with petroleum products: an overview. J Petrol Sci Eng 26(1–4):169–178. https://doi.org/10.1016/S0920-4105(00)00031-0

Nas S, Toralde JS, Wuest C (2009) Offshore managed pressure drilling experiences in Asia Pacific. All Days 2:1099–1111. https://doi.org/10.2118/119875-MS

Nichols WJ (2001) The U.S. environmental protection agency: national oil and hazardous substances pollution contingency plan, subpart j product schedule (40 CFR 300.900). In: International oil spill conference proceedings, vol 2, pp 1479–1483. https://doi.org/10.7901/2169-3358-2001-2-1479

Nowak P, Kucharska K, Kamiński M (2019) Ecological and health effects of lubricant oils emitted into the environment. Int J Environ Res Public Health 16(16):1–13. https://doi.org/10.3390/ijerph16163002

Ossai IC, Ahmed A, Hassan A, Hamid FS (2020) Remediation of soil and water contaminated with petroleum hydrocarbon: a review. Environ Technol Innov 17:100526. https://doi.org/10.1016/j.eti.2019.100526

Otte M, Donna J (2006) Constructed wetlands for phytoremediation. In: Phytoremediation rhizoremediation, pp 57–67. https://doi.org/10.1007/978-1-4020-4999-4_5

Ragothaman A, Anderson W (2017) Air quality impacts of petroleum refining and petrochemical industries. Environments 4(3):66. https://doi.org/10.3390/environments4030066

Ramírez-García R, Gohil N, Singh V (2019) Recent advances, challenges, and opportunities in bioremediation of hazardous materials. In: Phytomanagement of polluted sites. Elsevier, pp 517–568. https://doi.org/10.1016/B978-0-12-813912-7.00021-1

Rana S (2008) Facts and data on environmental risks-oil and gas drilling operations. In: SPE Asia Pacific oil and gas conference and exhibition 2008 - "gas now: delivering on expectations," vol 1, pp 448–468. https://doi.org/10.2118/114993-ms

Rathfelder K, Lang JR, Abriola LM (1995) Soil vapor extraction and bioventing: applications, limitations, and future research directions. Rev Geophys 33(S2):1067–1081. https://doi.org/10.1029/95RG00402

Rosell M, Lacorte S, Barceló D (2006) Analysis, occurrence and fate of MTBE in the aquatic environment over the past decade. TrAC - Trends Anal Chem 25(10):1016–1029. https://doi.org/10.1016/j.trac.2006.06.011
Santella N, Steinberg LJ, Sengul H (2010) Petroleum and hazardous material releases from industrial facilities associated with hurricane Katrina. Risk Anal 30(4):635–649. https://doi.org/10.1111/j.1539-6924.2010.01390.x
Schraufnagel DE, Balmes JR, Cowl CT, De Matteis S, Jung SH, Mortimer K, Perez-Padilla R, Rice MB, Riojas-Rodriguez H, Sood A, Thurston GD, To T, Vanker A, Wuebbles DJ (2019) Air pollution and noncommunicable diseases. Chest 155(2):417–426. https://doi.org/10.1016/j.chest.2018.10.041
Shah NK, Li Z, Ierapetritou MG (2011) Petroleum refining operations: key issues, advances, and opportunities. Ind Eng Chem Res 50(3):1161–1170. https://doi.org/10.1021/ie1010004
Singh S, Shikha (2019) Treatment and recycling of wastewater from oil refinery/petroleum industry. In: Ram Lakhan Singh RPS (ed) Advances in biological treatment of industrial waste water and their recycling for a sustainable future. Springer, Singapore, pp 303–332. https://doi.org/10.1007/978-981-13-1468-1_10
Song B, Zeng G, Gong J, Liang J, Xu P, Liu Z, Zhang Y, Zhang C, Cheng M, Liu Y, Ye S, Yi H, Ren X (2017) Evaluation methods for assessing effectiveness of in situ remediation of soil and sediment contaminated with organic pollutants and heavy metals. Environ Int 105:43–55. https://doi.org/10.1016/j.envint.2017.05.001
Tang J, Lu X, Sun Q, Zhu W (2012) Aging effect of petroleum hydrocarbons in soil under different attenuation conditions. Agr Ecosyst Environ 149:109–117. https://doi.org/10.1016/j.agee.2011.12.020
Ukaogo PO, Ewuzie U, Onwuka CV (2020) Environmental pollution: causes, effects, and the remedies. In: Microorganisms for sustainable environment and health. INC. https://doi.org/10.1016/B978-0-12-819001-2.00021-8
Varjani S, Upasani VN (2019) Influence of abiotic factors, natural attenuation, bioaugmentation and nutrient supplementation on bioremediation of petroleum crude contaminated agricultural soil. J Environ Manag 245(May):358–366. https://doi.org/10.1016/j.jenvman.2019.05.070
Varjani SJ, Gnansounou E, Pandey A (2017) Comprehensive review on toxicity of persistent organic pollutants from petroleum refinery waste and their degradation by microorganisms. Chemosphere 188:280–291. https://doi.org/10.1016/j.chemosphere.2017.09.005
Vidonish JE, Zygourakis K, Masiello CA, Sabadell G, Alvarez PJJ (2016) Thermal treatment of hydrocarbon-impacted soils: a review of technology innovation for sustainable remediation. Engineering 2(4):426–437. https://doi.org/10.1016/J.ENG.2016.04.005
Wei W, Cheng S, Li G, Wang G, Wang H (2014) Characteristics of ozone and ozone precursors (VOCs and NOx) around a petroleum refinery in Beijing, China. J Environ Sci (china) 26(2):332–342. https://doi.org/10.1016/S1001-0742(13)60412-X
Yap J, Leong Y-K, Liu J (2011) Structural recovery behavior of barite-loaded bentonite drilling muds. J Petrol Sci Eng 78(2):552–558. https://doi.org/10.1016/j.petrol.2011.06.010
Yu L, Han M, He F (2017) A review of treating oily wastewater. Arab J Chem 10:S1913–S1922. https://doi.org/10.1016/j.arabjc.2013.07.020
Yuniati MD (2018) Bioremediation of petroleum-contaminated soil: a review. IOP Conf Ser: Earth Environ Sci 118(1):012063. https://doi.org/10.1088/1755-1315/118/1/012063
Zhang B, Matchinski EJ, Chen B, Ye X, Jing L, Lee K (2019) Marine oil spills—oil pollution, sources and effects. In: World seas: an environmental evaluation, 2nd edn. Elsevier, pp 391–406. https://doi.org/10.1016/B978-0-12-805052-1.00024-3
Zhang Y, Labianca C, Chen L, De Gisi S, Notarnicola M, Guo B, Sun J, Ding S, Wang L (2021) Sustainable ex-situ remediation of contaminated sediment: a review. Environ Pollut 287:117333. https://doi.org/10.1016/j.envpol.2021.117333

Chapter 2
Toxicity of Persistent Hydrocarbon Pollutants, Sources and Sustainable Remediation Process

Jaydeep Kanungo, Teyaswini Sahoo, Laxmi Priya Swain, and Ipsita Dipamitra Behera

Abstract Hydrocarbon pollution from the petroleum industry is one of the world's most critical environmental problems. Polluting the environment by inadvertent release of petroleum products is a serious concern. Hydrocarbon components have long been recognized as organic pollutants with well-documented carcinogenic and neurological effects. There are three major categories of hydrocarbons in petroleum: pyrogenic, petrogenic, and biological. Numerous petroleum waste and organic contaminants are consistently produced throughout oil exploration and processing. Leaks of petroleum oil, sewage sludge, and tarry or creosote waste are additional environmentally hazardous sources of petroleum hydrocarbons. The most harmful hydrocarbons evaporate quickly following oil spills, causing damage to the ecology in this way. Aromatic compounds, nitrogen, aliphatic, amines, and oxides are all part of the petroleum wastes. However, the composition may change from place to place due to differences in the local biological and geologic factors. As the demand for petroleum products increase, this ultimately increases environmental concern for its dumping. As there is no suitable dumping method, this causes various impacts on the ecosystem, affecting humans, soil, aquatic life, plants, etc. In plants, it leads to inhibition in seed germination enzymatic dysfunction and affects the process of photosynthesis by causing changes in chlorophyll concentration. In humans, it affects the respiratory system, such as lung (pneumonia) or lung damage, skin and eye irritation, dizziness, headache, etc. Similarly, it affects aquatic life by affecting reproductive process, lowering the production of eggs, and decreasing oxygen content, which affects flora, fauna, and aquatic animals. Several physicochemical and biological methods are used for the remediation of petroleum hydrocarbon wastes. Physiochemical methods are generally not environment-friendly and cost-effective; hence, we go for various biological methods such as phytoremediation, bioaugmentation, and biostimulation.

J. Kanungo · T. Sahoo · L. P. Swain · I. D. Behera (✉)
Department of Chemical Engineering, Indira Gandhi Institute of Technology, Sarang 759146, Odisha, India
e-mail: ipsitadbehera@gmail.com

I. D. Behera and A. P. Das (eds.), *Impact of Petroleum Waste on Environmental Pollution and its Sustainable Management Through Circular Economy*, Environmental Science and Engineering, https://doi.org/10.1007/978-3-031-48220-5_2

Keywords Petroleum hydrocarbon · Crude oil · Petroleum refinery · Oily sludge · Sustainable bioremediation

2.1 Introduction

The predominant proportion of hydrocarbons on Earth are naturally derived and primarily found in crude oil. These hydrocarbons serve as a significant source of energy in the contemporary world. With the increase in usage of various petroleum products, there is a high demand for crude oil in all sectors of industries and refineries worldwide. Huge expansion and development of crude oil have also seen a wide distribution of hydrocarbon-containing products such as diesel, natural gas, lubricating oil, kerosene, paraffin, naphthene, fuel oil, etc. The growth has resulted in the generation of petroleum waste products due to improper disposal methods. The exploration, extraction, and production process of all the hydrocarbon products give rise to pollutants such as solid waste, wastewater, filter clay, and spent caustic containing high consistency of heavy metals and also leading to the emission of harmful gases such as oxides of nitrogen, carbon dioxide, carbon monoxide, hydrogen sulfide, etc. The primary sources of pollution are oil exploration, storage, transportation, and oil processing in refineries. Oil spills in oceans where oil is released into the ocean accidentally or intentionally, transportation of petroleum hydrocarbons from the production site to the distribution sites, leakages in pipelines carrying gas exploration activities and crude oil from manufactured gas production sites, retail service stations, and accidents during storage of petroleum and its byproducts, drilling and production refining (Behera et al. 2022).

Hydrocarbons are compounds made of carbons and hydrogen, which can be straight or branched, saturated or unsaturated, aliphatic, alicyclic, aromatic, and polyaromatic compounds. The nonhydrocarbon compounds comprise phenol, naphthenic acid, metalloporphyrin, thiol, heterocyclic nitrogen, asphaltene sulfur compounds, etc. The chemical and physical qualities of petroleum extracted from various sources exhibit significant variations (Rosner and Markowitz 2013). Petroleum hydrocarbons can be classified into four distinct fractions based on their chemical composition. The first fraction is saturating, also known as alkanes. The second fraction is aromatics, which includes compounds like benzene, toluene, ethylbenzene, and xylenes (collectively referred to as BTEX) and polyaromatic hydrocarbons (PAHs). The third fraction consists of resins, which are compounds containing nitrogen, sulfur, and oxygen that are dissolved in oil. Lastly, the fourth fraction is asphaltenes, which are large and complex molecules that are dispersed in crude oil in a colloidal manner (Hu et al. 2013).

Crude oil spills are a frequent source of pollution. Porphyrins, which have a pyrrolic structure, are strongly correlated with the amount of metals in crude oil. A wide range (thousands) of organic chemicals, mostly hydrocarbons, with a smattering of oxygen-, nitrogen-, and sulphur-containing organic compounds and some inorganic components, i.e., metals, make up crude oil. Several different types of petroleum fractions are produced throughout the refining process. These include lube oil waxes, light oil, asphaltenes, naphtha, diesel, kerosene, etc. Light ends refer to the at-atmosphere-pressure-distilled light fractions, whereas heavy ends refer to the high-pressure-distilled heavy fractions such as asphaltenes and lubricating oil. Light ends vary from heavy ends in their hydrocarbon makeup, with the former consisting of saturated and unsaturated hydrocarbons of lower molecular weight and a lesser concentration of aromatic molecules (Srivastava et al. 2019).

Hydrocarbon pollution or contamination generally occurs due to organic substances that cause toxicity and are mostly harmful to the environment as well as other living organisms. The aliphatic and aromatic hydrocarbons can cause serious human and environmental health hazards. The hydrocarbons have less surface tension as well as viscosity; hence, they penetrate the lungs in no time, causing diseases such as pneumonia and affecting the surfactant, airway, alveoli, capillaries, etc., leading to inflammation, atelectasis, and fever. Hydrocarbons not only affect human health but, at the same time, they affect soil fertility and microbial activity (NCBI). The effects of hydrocarbon pollution on marine life, terrestrial life, and the atmosphere are poisonous, lethal, and catastrophic (Behera et al. 2021). Hydrocarbon products, especially oils, adversely affect fishes' fertility and zooplankton's growth. Oil spills and leakages pose a global hazard to aquatic or marine life, yet around 35 million barrels of oil are transported annually across the seas. Industrial waste causes soil contamination. One of the leading causes of groundwater pollution and overall decreased productivity of agricultural land is the disposal of toxic wastes from refineries (Varjani and Upasani 2017).

Due to the hazardous environmental impact, we have to reduce the long-term effect that leads to concern for the ecosystem. The contamination level and number of pollutants can be reduced by both physiochemical and biological methods. The first approach involves the use of physiochemical methods, which are often employed in traditional engineering practices. The high expenses associated with physicochemical decontamination procedures may be attributed to the substantial costs of excavating and transporting significant volumes of contaminated materials for treatment. Some important physiochemical methods include centrifugation, solvent extraction, pyrolysis, microwave radiation, electrokinetic method, froth flotation, etc. The second method is a biological method, where we use bacteria to break down hazardous and toxic substances into less hazardous ones. Some of the biological methods are bioaugmentation, phytoremediation, zero valent ion (ZVI) combined with /the biological methods, soil fauna in bioremediation, flocculation bio treatment, etc. The physiochemical method has too many disadvantages as compared to the biological method, such as the generation of secondary pollutants, high chemical consumption, and high energy consumption, whereas the biological method is eco-friendly and cost-effective, we are using the biological method over the physiochemical method as the efficient one.

With the potential threat that petroleum hydrocarbons pose to humans and the environment, it is important to test and characterize the biodegradation and biotransformation processes of hydrocarbons in contaminated soil to develop bioremediation techniques for cleaning such soils to levels that ensure their safe disposal or reuse

(Gao et al. 2022). Biodegradation is a kind of metabolism in which microorganisms convert pollutants into less dangerous and non-hazardous compounds that may then be incorporated into natural biogeochemical cycles. Several elements, including nutrients, oxygen, PH value, composition, concentration, and soil chemical and physical features, influence how quickly petroleum degrades in soil (Varjani and Upasani 2016).

The bioremediation method is another biological method that is a non-destructive, cost-effective, and eco-friendly process that increases the rate of biodegradation process of contaminants or pollutants through optimization of limiting conditions. The best method is chosen according to its biodegradation efficiency.

2.2 Composition of Hydrocarbon Pollutants

Crude oil is a combination of volatile hydrocarbons (compounds mostly consisting of hydrogen and carbon), also contains some amount of nitrogen, sulfur, and oxygen, and is, therefore, one of the largest sources of hydrocarbon pollutants. There is no simple way to identify these substances because of their intricate chemical architectures. Petroleum's elemental makeup is well-defined, yet the ratios of organic compounds within it vary widely. By weight, carbon accounts for 83–87%, hydrogen for 10–14%, nitrogen for 1–2%, oxygen for 0.05–0.6%, sulfur for 0.05–6.0%, and other metals for less than 0.1%. Iron, nickel, copper, vanadium, etc., are among the most often encountered metals. Crude oil contains hydrocarbons, which may be found in almost all three stages. Lighter hydrocarbons, such as methane, ethane, propane, and butane, exist as gases at surface pressure and temperature, whereas heavier hydrocarbons exist as both liquids and gases.

The pollutants consisting of hydrocarbons are divided into different categories such as (i) Saturates (aliphatic), (ii) Aromatics (ringed hydrocarbons), (iii) Resins, and (iv) Asphaltenes.

i. Saturates (aliphatic): Single-bond hydrocarbons, such as alkanes (paraffin), cycloalkane (naphthene), isoalkanes, terpenes, and steranes, make up the saturates. The saturates make up the bulk of the components in crude oil. First, there are the gaseous n-alkanes; second, the low molecular weight (C8–C16) alkanes; third, the medium molecular weight (C17–C28) alkanes; and fourth, the high molecular weight (>C28) alkanes.
ii. Aromatics (ringed hydrocarbons): An essential class of hydrocarbons, aromatics may be found in almost all petroleum blends across the globe. Aromatics are hydrocarbons having alternating double bonds and a cyclic structure, yet they are unsaturated. Arenes are a class of hydrocarbons with a benzene ring, so-called because they contain six carbon atoms in a configuration similar to that of benzene or the phenol group of aromatic chemicals. Aromatics are a vital component of petrol blends since they provide the much sought-after octane. Refiners often aim to increase the aromatics content of petrol up to a level

established to address environmental concerns, even though some of them are hazardous to people and the atmosphere.

iii. Resins: Hydrocarbon resins are synthetic polymers made by polymerizing hydrocarbons and are most often used as tackifiers. These are available in both aromatic and aliphatic forms. There are three classes of hydrocarbon resins, including C5, C9, and DCPD. C5 resins are yellow granules of aliphatic hydrocarbon resin. While DCPD is classified as a cycloaliphatic resin, C9 resins, often known as aromatic hydrocarbon resins, are marketed in brown granule form and have a greater melt viscosity than C5 hydrocarbon resins. Hydrocarbon resins come in a wide range of melting temperatures and chemical compositions and may be made in a variety of methods. The coloration and thermal stability may be improved by hydrogenating them.

iv. Asphaltenes: When compared to other hydrocarbons, their molecular weight is larger. These components cannot dissolve in either pentane or heptane. Resin and asphaltenes are non-hydrocarbon polar chemicals, unlike the saturated and aromatic portions. Many nitrogen, sulfur, and oxygen atoms are added to the already complicated carbon structures of resins and asphaltenes (Chandra et al. 2013; Harayama et al. 2004). The biodegradability of a given material depends on the specific chemical behavior of its components.

2.3 Occurrences and Sources of Hydrocarbon Pollutants

Petroleum hydrocarbon waste is a common contamination in several environmental mediums, including soil, water, and air. Petroleum hydrocarbons are often seen in many locations, including refineries, where they serve as feedstocks for processes such as plastic manufacturing, as well as at facilities involved in the production of petrol and retail service stations (Uddin et al. 2021). Petroleum hydrocarbons are chemical compounds that may be found in soil due to human activities, including accidents, controlled spills, or as unintentional consequences of industrial, commercial, or personal operations (Negrete-Bolagay et al. 2021; Varjani et al. 2017). This category encompasses seepages originating from oil deposits as well as the breakdown of organic substances. The soil contains petroleum hydrocarbon waste as a result of the leaking from underground storage tanks (USTs) that were formerly used at fuel stations. The issue of pollution is a matter of significant concern due to its extensive impact on many living forms (Tsai and Kao 2009). Several significant sources of hydrocarbons include petrochemical companies, oil spills, automobile garages, incomplete combustion of fuels, petroleum imports, forest, and grass fires, as well as the biosynthesis of hydrocarbons by marine or terrestrial species. Various sources of petroleum hydrocarbon have been shown in Fig. 2.1.

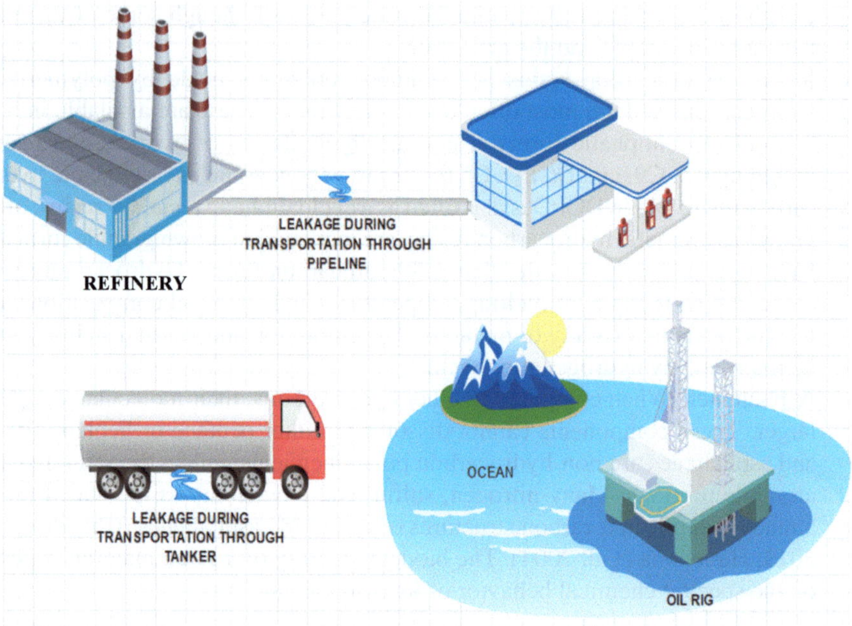

Fig. 2.1 Various sources through which petroleum hydrocarbon infiltrates the ecosystem

2.3.1 Petrochemical Industries

The petrochemical sector has significant importance in the growth and development of the Indian economy. The rise in energy consumption leads to a corresponding escalation in the formation of hazardous waste throughout the processes of oil exploration, refining, and production. The hydrocarbon pollutants produced by petrochemical companies mostly include intricate polycyclic aromatic hydrocarbons (PAHs). These substances possess mutagenic and carcinogenic properties, exhibit persistence in the environment, and have detrimental impacts on both the environment and human health. The water solubility of primary polycyclic aromatic hydrocarbons (PAHs) and BTEX chemicals (including benzene, toluene, ethylbenzene, and xylene) is rather high, which facilitates their contamination of water bodies (Souza et al. 2014). This industry is also a significant producer of large amounts of solid oily waste, which are difficult to recycle. The petrochemical sector generates wastewater, including hydrocarbons, which are complex organic compounds with hazardous properties (Jain et al. 2017).

2.3.2 Oil Spills

Oil spill refers to the discharge of a liquid petroleum hydrocarbon into the surrounding environment, which has detrimental effects on marine ecosystems, terrestrial bodies, and human populations. Oil spills may occur as a result of several factors, including the discharge of crude oil from tankers, offshore platforms, and drilling rigs. Additionally, spills of petroleum products like petrol, diesel, and heavier fuels used by big ships, such as bunker fuel, can contribute to oil leak incidents. Furthermore, the release of any oily debris or waste oil can also lead to oil spills (Pinedo et al. 2013). The composition of crude oil mostly comprises hydrocarbons, including various dangerous substances such as benzenes, toluene, poly-aromatic hydrocarbons, and oxygenated polycyclic aromatic hydrocarbons. The inhalation of these substances might potentially lead to detrimental health consequences for individuals. Spills oil can affect birds and mammals by penetrating the body and reducing the term of survival with vulnerable temperature fluctuations (Ismail and Lewis 2006; Jernelöv 2010).

2.3.3 Automobile Exhaust Emissions

Automobile exhaust emissions are generated as a result of the combustion engine's air–fuel mixture burning, leading to the release of carbon dioxide into the environment. The rise in the use of cars and automobiles has resulted in an increase in the consumption of automotive oil, contributing significantly to the polluting of water bodies with hydrocarbons. This kind of pollution arises when automotive oil is released into the ground and then seeps into the surrounding environment, potentially being transported into water bodies via the process of runoff. Diesel-powered, gasoline-powered, and hybrid cars all emit exhaust pollutants due to the diverse composition of hydrocarbons included in their respective fuels, which combust at varying rates and manners. Emissions of non-methane hydrocarbons are one of the largest anthropogenic sources of hydrocarbons. Nitrogen oxide and carbon monoxide, which are formed primarily in bulk gases, and hydrocarbons, which are formed in quench areas, are the major pollutants. Some of the major hydrocarbon pollutants from non-methane hydrocarbons are composed of ethylene, toluene, acetylene, *m*, *p*-Xylenes, benzene, propylene, and *i*-pentane.

2.3.4 Anthropogenic Sources

2.3.4.1 Stacking of Oily Slag and Sludge

Oily sludge is a byproduct of the petroleum industry that consists of a combination of sand, water, metals, and a significant concentration of hydrocarbons (HCs). Oily sludge is a kind of harmful organic sludge that petroleum oil refineries produce. It is classified as a priority pollution. The primary source of its generation may be attributed to the sequential activities of drilling, extraction, transportation, refining, and storage of crude oil. The oily sludge leftovers typically consist of water (15–50%), hydrocarbons (30–80%), particles (5–40%), and metals (Hu et al. 2013). A significant quantity of oil sludge comprises many perilous substances, including petroleum hydrocarbons (PHCs) and heavy metals (HMs). When deposited in soils, the poisonous constituents included in oil sludge might potentially lead to nutrient insufficiency or impede the development of seeds and plants. The oil sludge was stored outdoors, resulting in a gradual reduction of volatile components, such as light oil components, and moisture content. This process led to the formation of aged oil sludge (AOS). The chemical is classified as very toxic, carcinogenic, and mutagenic due to its higher concentrations of dangerous compounds, including resins, phenol, benzene, asphaltenes, polycyclic aromatic hydrocarbons, and heavy metals (Mishra et al. 2021).

2.3.4.2 Partial Burning of Fuel

When a fuel component is only partially oxidized during combustion, it is said to have undergone incomplete or partial combustion. Variation in atmospheric composition is caused by the release of different gaseous pollutants produced by incomplete combustion of fuel. Hydrocarbon fuels typically consist of hydrogen and carbon, either in their elemental or compound forms. Carbon monoxide and other partly reacted flue gas components (gases and liquid or solid aerosols) may also occur in trace amounts at sites where flue gas is produced. Sulfur-based fossil fuel (coal and heavy oils) combustion produces nitrogen oxides and carbon monoxide. Most common fuels also include non-combustible components such as mineral matter (ash), water, and inert gases, as well as trace quantities of sulfur, which is oxidized to sulfur dioxide (SO_2) or sulfur trioxide (SO_3) during burning.

2.3.4.3 Volcanic Eruption

Volcanic oil and gas are known to collect inside volcanic rocks, as well as in hydrogen reservoirs associated with volcanic rocks, which are sealed or form traps. The manner in which hydrocarbon accumulations are formed and distributed in volcanic rocks differs from that in non-volcanic (silica) clastic rocks. The possible risks associated with volcanic gases are of significant concern. These gases come from lava flow as well as the volcano that erupts violently. Carbon dioxide, water vapor, sulfur dioxide, hydrogen chloride, hydrogen fluoride, etc., are the most common volcanic gases. Sulfur dioxide can cause breathing and respiratory problems such as asthma. All these pollutants lead to acid rain and air pollution, which is severely hazardous to the environment.

2.3.4.4 Pesticides

Pesticides, including insecticides, fungicides, and herbicides, are sources of hydrocarbon contamination in both water as well as soil. Only a small proportion leaves in the soil, and the rest is eroded by water streams. Among all types of pesticides, herbicides are considered to be the most harmful due to their direct application into the soil for weed control. This characteristic renders them susceptible to being quickly carried away by water streams. Compounds with chlorine, carbon, and hydrogen are referred to as chlorinated hydrocarbons (CHC). Organochlorine pesticides like lindane and dichlorodiphenyltrichloroethane fall under this category, as do industrial chemicals like polychlorinated biphenyls (PCB) and chlorine waste products like dioxins and furans. Insecticides based on chlorinated hydrocarbons may be absorbed through fat. They have the potential for prolonged environmental persistence and clinical harm.

2.4 Impacts of Hydrocarbon Pollutants

The concentration of hydrocarbon pollutants in the environment has seen an escalation due to the utilization of diverse petrochemicals and the discharge of associated industrial waste products and effluents. There have been reports indicating that they have the potential to contaminate soil, hence posing a growing concern for environmental well-being. The majority of petroleum hydrocarbons are introduced into the environment by biological processes. Hydrocarbon pollutants are one of the persistent organic pollutants that cause massive damage to the environment and ecosystem due to biomagnification (Chandra et al. 2013). The widespread discharge of hydrocarbon pollutants from various sources such as petrochemical industries, oil refinery sites, leakages, and exhaust gas from automobiles causes soil, groundwater as well as ocean contamination (Saeki et al. 2009; Janbandhu and Fulekar 2011; Prince et al. 2013). Several components found in petroleum crude are known to be refractory and extremely hazardous owing to the existence of hemotoxic, carcinogenic, and teratogenic compounds, such as BTEX (benzene, toluene, ethylbenzene, and xylene) and PAHs (Zhang et al. 2011; Costa et al. 2012; Meckenstock et al. 2016).

Mostly, the wastewater or petroleum sludge consisting of a large proportion of hydrocarbons hurts human beings, plants, and animals. Due to a lack of oxygen and the absence of certain nutrients, fertility decreases due to the presence of oil content, which ultimately leads to a decrease in crop yield and the unavailability of proper food because of soil contamination by hydrocarbons. Oxygen shortage is also another concern due to hydrocarbon contamination (Behera et al. 2020). Oil spills from different sources inhibit root penetration and lead to blockage of pores of the soil, which causes death to the plants Hydrocarbon contamination also restricts the penetration of light into the water, thus affecting aquatic plants by making them incapable of the process of photosynthesis the food chain. Pollutants in polluted water are absorbed by plants and then transferred up the food chain to humans and other animals. Discharge of wastewater consisting of persistent amounts of hydrocarbons into the water bodies is a serious risk to the sea birds as well as they spend most of their time in the water, causing indigestion, absorption, and breathing problems. Scavengers consume their prey, the contaminated fish from the water streams. The pollutants are also consumed by the fish through gills, which accumulate in various parts such as gall bladder, liver, stomach, etc. making it unhealthy. There is a loss of breeding capacity; fertility decreases; hence the production of eggs also declines, which leads to retardation of its growth.

2.4.1 Soil

The presence of petroleum hydrocarbons in soils has the potential to impact soil health adversely. Furthermore, it is noteworthy that the aforementioned substance exhibits its impact on human health at much higher concentrations in comparison to its impacts on other aspects. The use of these substances has the potential to negatively impact soil microorganisms by diminishing their population and impairing their metabolic functions. Soil microorganisms play a crucial role in facilitating nutrient availability for plants, although the presence of pollutants has a detrimental impact on these microorganisms, leading to a decline in their population and functional capacity (Ambaye et al. 2022).

The impact of petroleum hydrocarbons on the environment and soil is contingent upon their chemical and physical attributes, as well as the diverse hydrocarbon constituents. These hydrocarbons predominantly influence soil properties through adsorption, biodegradation, and leaching processes. These organisms can potentially become part of the ecological food web and induce adverse health impacts on human populations. The significance of agricultural soil in the provision of food and the maintenance of ecosystem equilibrium is of utmost importance (Turner and Renegar 2017). Nevertheless, the pollution of soil has detrimental effects on its fertility as well as its physical and chemical qualities. The introduction of petroleum hydrocarbons into the uppermost layer of soil, known as topsoil, as well as the layer underneath it, referred to as subsoil, has been shown to result in detrimental effects such as the deterioration of soil texture and structure, as well as a reduction in pore spaces and saturated hydraulic conductivity. Furthermore, the impact of this phenomenon extends to the biological features of the soil, particularly the population of soil microorganisms and enzymatic activities. Consequently, this has an indirect influence on the availability of nutrients for plants.

The petroleum hydrocarbon-contaminated soils may create nutrient deficiencies because seed germination is inhibited, resulting in restricted growth or death upon contact with plants. The permanent nature of Petroleum hydrocarbon Components in soil pores or their adsorption onto the surface might be attributed to their high viscosity. The mineral elements of soil contribute to the formation of a continuous surface cover (Adeniyi and Afolabi 2002). These factors would result in a decrease in the hydraulic conductivity, hygroscopic moisture, and water retention capacity of soils. Specifically, the constituents with greater molecular mass present in sludge and their resultant breakdown products may accumulate in close proximity to the upper layer of soil, resulting in the formation of hydrophobic crusts. These crusts subsequently reduce the availability of water and restrict the flow of water and air.

2.4.2 Water and Marine Bodies

Wildlife is impacted by petroleum hydrocarbon because it forms a thick film on their bodies. Many petroleum hydrocarbons, because of weathering, grow progressively stickier with time and hence attach to animals much more. Many marine creatures and birds may be poisoned by petroleum hydrocarbons because they float on the water's surface. The presence of a petroleum hydrocarbon is not always avoided by birds and marine animals. There have been sightings of marine animals, including seals and dolphins, feeding and swimming in or around a petroleum hydrocarbon. Petroleum hydrocarbon may entice fish because it resembles food floating in the water.

The presence of oil slicks poses a threat to marine birds since they are drawn to aggregations of fish and may inadvertently traverse these oil-contaminated areas in pursuit of their prey. Hypothermia in avian species may lead to a deterioration or impairment of the insulating and hydrophobic characteristics of their plumage. Birds are rendered vulnerable to predation because of the adverse effects of petroleum hydrocarbon on their feathers, resulting in reduced flight capability (Feng et al. 2021). The enduring consequences include developmental anomalies seen in marine organisms, such as diminished jaw size, absence of pigmentation, and incomplete cranial fusion (Van Meter et al. 2006; Alonso-Alvarez et al. 2007; Varjani 2014). These impacts induce alterations in the population of a species or community, thereby leading to modifications in the whole of an ecosystem (Walker et al. 2006).

Fur seal offspring are at risk of drowning if petroleum hydrocarbon substances adhere to their flippers, leading to an inability to effectively navigate aquatic environments. Dugongs may have inflammation or illness as a result of petroleum hydrocarbon adhering to the sensory hairs around their lips, leading to feeding challenges. The ingestion of petroleum hydrocarbons may result in internal damage to the bodies of animals and birds, such as the development of ulcers or bleeding in their stomachs, as a consequence of unintentional exposure. Marine animals and turtles may have adverse effects on their airways and lungs, including congestion, pneumonia, emphysema, and potential fatality, as a result of inhalation of petroleum hydrocarbon droplets. The impact on marine flora is hazardous. The adverse impact of petroleum hydrocarbon on avian species includes the impairment of adrenal tissue functionality, resulting in compromised regulation of blood pressure and fluid concentration inside the bird's body (The Impact of Petroleum Hydrocarbon on Wildlife).

2.4.3 Human Beings

Exposure to sources that have a higher concentration of heavy metals has been shown to result in significant health complications. Direct exposures include two primary routes: inhalation of polluted air, namely the volatile components that are exhaled as gases, and direct contact with the skin, often occurring when individuals traverse through contaminated locations. Indirect oil exposures may occur as a result of individuals bathing in water that has been polluted with oil, as well as consuming food that has been similarly contaminated. Health issues include a range of conditions, such as skin and ocular irritation, respiratory and neurological complications, as well as psychological distress. Thermal Power Plants (TPHs) exert a significant influence on mental well-being and elicit physical and physiological responses, thereby posing toxicity risks to genetic, immunological, and endocrine systems. Despite the limited understanding of the long-term effects of TPHs in humans, it is evident that some symptoms may last throughout the postexposure period for many years. Therefore, the issue of safeguarding the health of those exposed to TPHs is a subject of significant importance. Health risk assessments play a crucial role in facilitating the identification of possible adverse consequences associated with exposure, whether they occur immediately or persist over an extended period. This chapter offers a detailed analysis of the impact of TPHs on human health, aiming to enhance our knowledge in this area.

Due to its low surface tension and viscosity, hydrocarbon has the ability to effectively infiltrate the pulmonary system. Consequently, this results in the development of severe necrotizing pneumonia. The compounds have the potential to cause damage to surfactant, airway epithelium, alveolar septum, and pulmonary capillaries, resulting in inflammation, atelectasis, and fever. The act of inhaling or aspirating substances might potentially result in the development of a reactive airway syndrome resembling asthma, as well as chemical pneumonitis. The symptoms often manifest as a cough and/or dyspnea. Certain effects may also arise as a result of the metabolic conversion of the hydrocarbon into a neurotoxic. Extended periods of exposure, as seen in occupational settings, may also lead to neuropathy, diminishment in brain volume, and encephalopathy. The following list comprises many cardiovascular disorders. Contact of the skin with some substances may result in moderate irritation, while persistent contact can lead to chemical burns that vary in severity from superficial to full-thickness burns. Full-thickness burns have the potential to result in systemic manifestations. The dermatological condition characterized by skin irritation in the perioral region is often referred to as "glue sniffer's rash" in the vernacular. Skin lesions may manifest as bullae or blistering. Additional skin signs may include jaundice and/or inflammation of the mucous membranes. The diseases and defects in the human body caused by the presence of petroleum hydrocarbons are mentioned in Table 2.1 and Fig. 2.2.

Table 2.1 Different PHPs and their impact on lifeforms

Sl. no	Pollutants	Impact	References
1	Toluene	Neurological effects	Walser et al. (2014)
2	Xylene	Hearing problems	Zhang et al. (2020)
3	Benzene	Irregular heartbeat, myeloid leukemia	Galbraith et al. (2010)
4	Styrene	CNS disorder, liver and kidney toxicity	Lee et al. (2006)
5	Pentane	Skin irritation	Morita et al. (1986)
6	Ethyl benzene	Developmental toxicity	Davidson et al. (2021)

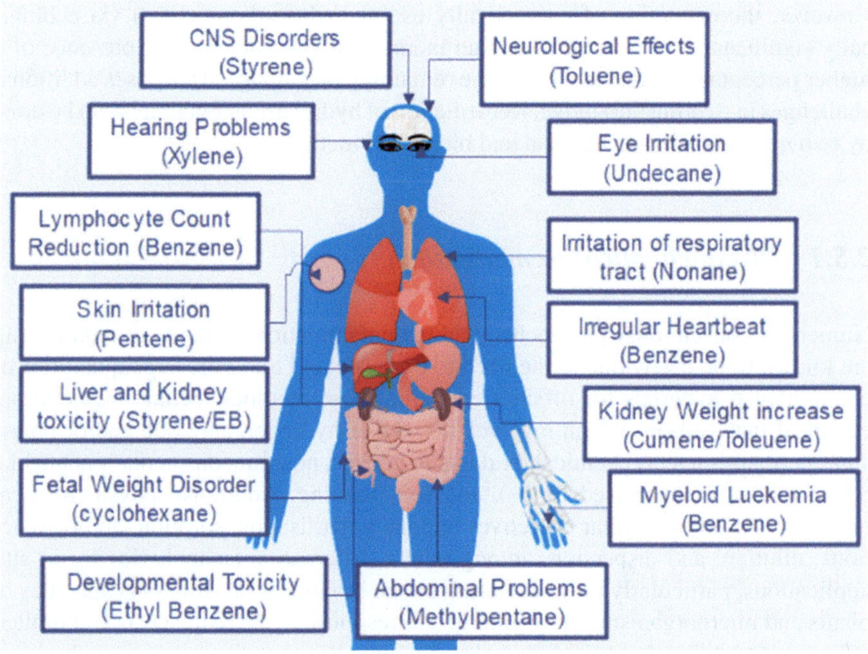

Fig. 2.2 The diseases and defects in the human body are caused by the presence of petroleum hydrocarbons

2.5 Remediation of Hydrocarbons

Hydrocarbon pollutants are of great concern to the environment as well as human health. Hence, it requires remediation, which is not easy as the chemical substances present have different classes of compounds. In order to eradicate this problem, efforts should be made both nationally and internationally. The objective of remediation is to eliminate or mitigate the presence of pollutants that are polluting the soil or groundwater, rendering them non-hazardous. Remediation is done after knowing the oil content, composition, and physiochemical properties. The existence of hazardous compounds such as benzene, PAHs, noxious by-products, and metallic elements inside the sludge raises apprehensions about the well-being of both humans and the surrounding ecosystem (Barraza et al. 2018; Maurice et al. 2019). Due to the negative environmental and economic impacts of oily sludge, numerous strategies have been introduced in recent years to lessen its production, extract any oil with marketable value, and treat it before disposal (Das et al. 2018; Johnson and Affam 2019). However, the extraction of energetically useful hydrocarbons (HCs) via ecologically significant processes results in an increase in viscosity and the presence of a higher percentage of heavier HCs in the remaining oily sludge. This poses additional challenges in treating the sludge. Remediation of hydrocarbon pollutants can be done by two methods: physiochemical and biological method.

2.5.1 Physiochemical Remediation

Numerous conventional physiochemical decontamination methods in engineering are known to be costly due to the need to excavate and transport large quantities of contaminated materials for offsite treatment. These methods include incineration, chemical inactivation (which involves the use of hydrogen peroxide and/or potassium permanganate as chemical oxidants to convert non-aqueous contaminants like petroleum into mineralised forms), and soil washing. Additional physiochemical procedures used for similar objectives include volatilisation, sorption, abiotic reactions, dilution, and dispersion, among others. Alternative technologies for in-situ applications, particularly those centering on the biological remediation capacities of plants and microorganisms, have emerged in response to the rising costs and limited efficacy of traditional physiochemical treatments. Bioremediation refers to the practice of using microorganisms and other forms of life to aid in the breakdown and elimination of pollutants. This process is a highly efficient, cost-effective, adaptable, and ecologically sustainable approach. Centrifugal separation, freezing, distillation, and filtering are all examples of physical processes. When strong electric fields are applied to an emulsion, the charged droplets migrate to the electrodes, discharge, and mix. Naggar et al. (2010) and Taiwo and Otolorin et al. (2009) reported that solvent extraction has been extensively used as a technique for the removal of semi-volatile and non-volatile organic molecules from matrices of soil and water. The process involves the combination of oily wastes and solvents in certain ratios to achieve full miscibility while simultaneously removing water, solid particles, and carbonaceous pollutants via the extraction solvent. Subsequently, the solvent/oil combination is subjected to a distillation process to achieve the separation of oil and solvent. Da Silva et al. 2012 reported that centrifugation has been extensively used for the treatment of large-scale oily sludge in several fields despite the limited number of scientific publications on this topic in recent times. A specialized high-speed rotating apparatus is used to provide a significant centrifugal force capable of rapidly separating components of varying densities (e.g., solids, oil, water, and pasty mixes found in oily sludge). Reducing the viscosity of oily sludge by pretreatment procedures is critical for optimizing centrifugation performance and minimizing energy usage. Organic solvents, demulsifying agents, steam injection, and direct heating are all examples of such processes (Siddiqi et al. 2022). Heat pretreatment reduces the viscosity of petroleum sludge. This lower viscosity allowed the sludge to be successfully treated using a disc/bowl centrifuge. During the first phase of centrifugation, more than 80% of the waste volume was converted into liquid effluent. After mixing the centrifuge residue with hot water, a second centrifugation was performed on the resulting solids. Two centrifuges were run, and the liquid that resulted was combined and then delivered to a refinery. Cambiella et al. (year) found that the water–oil separation process may be significantly improved by adding a coagulant salt, namely calcium chloride ($CaCl_2$), in the concentration range of 0.01–0.5 M. The inclusion of this coagulant salt led to a significant increase in oil separation efficiency, with values ranging from 92 to 96% being achieved. Liu et al. (2009) and Qin et al. (2015) reported that pyrolysis is the process of thermally decomposing organic compounds at high temperatures (500–1000 °C) inside an environment devoid of reactive substances. Gasification is a distinct process that involves the conversion of organic molecules into a flammable gas or syngas, often in the presence of oxygen levels ranging from 20 to 40%. The pyrolysis process results in the formation of hydrocarbons with reduced molecular weight, which may exist in either a condensed state or as non-condensable gases. Furthermore, it generates a significant secondary product referred to as char.

2.5.1.1 Disadvantages of Physiochemical Methods

There are a lot of disadvantages of physicochemical methods as compared to biological methods, and these are as follows.

2.6 Generation of Secondary Pollutants

The production of secondary pollutants during the physicochemical treatment procedure of heavy, oily sludge might result in substantial consequences for the environment and human health. Physicochemical techniques are often used for the remediation of sludge; nevertheless, it is important to note that these processes may give rise to the creation of other chemicals that carry dangers extending beyond the initial contaminants found in the sludge. Physicochemical phenomena, such as oxidation, chemical precipitation, and coagulation/flocculation, may induce the conversion of main contaminants found in dense, oily sludge into novel compounds (Johnson and Affam 2019). The aforementioned by-products possess the potential to be similar to or even more deleterious than the initial contaminants. As an example, the process of oxidizing hydrocarbons has the potential to generate intermediate molecules that possess toxicological properties or have the ability to induce carcinogenesis. Secondary pollutants may have increased toxicity and persistence in comparison to the primary pollutants due to physicochemical reactions (Hu et al. 2013). This phenomenon has the potential to result in enduring effects on ecosystems and human health since these chemicals may exhibit greater resistance to natural degradation mechanisms. The introduction of secondary pollutants into the atmosphere or aquatic systems as a result of physicochemical treatment procedures may contribute to the degradation of air and water quality. These contaminants have the potential to disseminate over expansive regions and have an impact on a diverse array of creatures, including human populations as well. Inhalation of airborne secondary pollutants poses a risk, while waterborne pollutants have the potential to pollute water bodies and have adverse effects on aquatic organisms. The impact of secondary pollutants on ecosystems may be seen by their influence on many trophic levels. Accumulation of these substances inside organisms may occur, resulting in the processes of bioaccumulation and biomagnification throughout the food chain (Jasmine and Mukherji 2019). The aforementioned phenomenon may lead to detrimental consequences for the variety of species, the success of reproduction, and the general health of the ecosystem.

Also, exposure to secondary pollutants has the potential to result in various health hazards for both human beings and animals. The potential health effects of these substances vary depending on their characteristics, including respiratory issues, skin irritations, neurological abnormalities, and perhaps leading to chronic ailments such as cancer. The production of secondary pollutants might provide challenges to the implementation of regulatory measures. Environmental laws sometimes prioritize the control of certain pollutants, potentially leading to inadequate coverage of newly emerging chemicals under current regulatory frameworks. The inadvertent production of secondary pollutants has the potential to compromise the overall efficiency of the physicochemical treatment procedure. In the event that the resultant by-products are not well handled, there is a possibility that they might undermine the advantages derived from the treatment of the initial contaminants.

2.7 High Chemical Consumption

The excessive use of chemicals in the physicochemical treatment procedure for heavy oily sludge might result in many environmental, economic, and operational consequences. Although the use of these substances is often essential for facilitating the separation, destabilization, and elimination of impurities from the sludge, their overuse may lead to various unfavorable outcomes. The overutilization of chemicals may lead to the presence of leftover chemicals in the treated sludge or effluent, which may subsequently be discharged into the surrounding environment. The presence of these leftover compounds has the potential to have adverse impacts on several aspects of the environment, including aquatic organisms, soil composition, and water reservoirs (Yang et al. 2005). The introduction of chemically treated sludge effluent into water bodies has the potential to introduce contaminants that may have detrimental effects on aquatic ecosystems and degrade the overall quality of water. Certain treatment chemicals have the potential to exhibit toxicity towards aquatic creatures, hence leading to disturbances within local ecosystems and resulting in adverse effects on aquatic life. The escalation in chemical usage results in an elevation in operational expenses. The financial implications of procuring, storing, and managing substantial volumes of chemicals may exert significant pressure on project budgets. The production of increased quantities of chemically processed trash might lead to elevated expenses associated with waste disposal, particularly in cases where the waste is classified as hazardous. The excessive use of chemicals might impede the efficacy of the therapy procedure. Excessive exposure to chemicals might result in the occurrence of flocculation or coagulation issues, which can compromise the effectiveness of separation processes and need the implementation of supplementary treatments. The management and regulation of doses for a substantial volume of chemicals might introduce heightened intricacy to the treatment procedure, necessitating the use of more advanced equipment and individuals with specialized experience. The use of an excessive amount of chemicals may result in heightened levels of equipment degradation, necessitating more frequent maintenance interventions and possibly resulting in periods of operational inactivity. The escalation in the quantity of chemicals being managed amplifies the likelihood of workers being exposed to perilous compounds, giving rise to dangers pertaining to health and safety. The probability of mishaps or spills occurring during the storage, transportation, and handling of chemicals is heightened by larger quantities, hence presenting potential hazards to both the workforce and the surrounding ecosystem (Virkutyte et al. 2002). The manufacturing of chemicals used in treatment procedures often necessitates substantial energy consumption and raw material utilization, hence contributing to the depletion of resources and the emission of carbon. The carbon footprint of a facility may be influenced by the energy-intensive manufacturing process of treatment chemicals, as well as the potential emissions resulting from their use.

2.8 High Energy Consumption

The excessive energy consumption linked to physicochemical methods used in the treatment of heavy oily sludge might result in various adverse effects on the environment and the overall sustainability of the remediation process. Frequent high levels of energy consumption are often associated with elevated levels of greenhouse gas emissions, making a significant contribution to the phenomena of climate change and air pollution. The combustion of fossil fuels for energy generation results in the emission of carbon dioxide (CO_2) and several other pollutants into the Earth's atmosphere. The use of equipment that requires high levels of energy may result in the emission of diverse air pollutants, including sulphur dioxide (SO_2), nitrogen oxides (NOx), and particulate matter (Tan et al. 2007). These pollutants have adverse impacts on both air quality and human health. Energy-intensive processes have the potential to exacerbate the exhaustion of non-renewable resources, such as fossil fuels, which are already under significant pressure as a result of escalating global demand. Certain energy production technologies need substantial water use, which has the potential to worsen water shortage concerns in locations that are already experiencing water stress. Elevated energy consumption results in heightened operating expenditures, exerting financial pressure on project budgets and diminishing the economic feasibility of remediation efforts, particularly in the context of long-term initiatives. The dependence on excessive energy usage renders the remediation process susceptible to variations in energy costs, hence impacting project cost estimations and financial stability. The issue of high energy usage poses a significant obstacle to the long-term sustainability of the restoration process. With the depletion of fossil fuel reserves, there is a potential for a decline in the accessibility of cost-effective energy sources. In areas characterized by restricted availability of energy resources, the practicality of deploying energy-intensive technology for the treatment of heavy oily sludge may be hindered. Emissions resulting from high-energy processes have the potential to give rise to public health concerns, including respiratory ailments and other health issues associated with pollution (Appleton et al. 2005). Energy-intensive operations have the potential to cause disturbances within communities as a result of noise pollution, heightened traffic levels, and other adverse effects linked to the production and utilization of energy. Compliance with environmental standards and acquisition of requisite licenses may pose increased difficulties as a result of the environmental consequences linked to elevated energy usage.

As physiochemical methods have so many disadvantages as compared to biological methods, we will opt for the biological methods of treatment of hydrocarbon pollutants. In biological methods, bioaugmentation and biostimulation are the two most efficient tactics for bio-cleaning of a contaminated location. Consortium applications using many strains tend to provide better results than those involving a single strain. Among all the biological methods, we choose bioaugmentation as one of the best methods for the removal of hydrocarbon pollutants. The increasing costs and limited efficiency of the physiochemical process have developed bioremediation techniques for the treatment of hydrocarbon pollutants. Green technologies are used to facilitate the bioremediation of petroleum-contaminated sites via biological processes (Varjani et al. 2015). Bioremediation refers to the use of living organisms to break down and detoxify contaminants (Dua et al. 2002; Ron and Rosenberg 2014; Sajna et al. 2015; Varjani and Upasani 2016). This method is ecologically friendly, efficient, cost-effective, and adaptable (Farhadian et al. 2008) reported on physicochemical methods are summarized in Table 2.2.

Table 2.2 Comparison and summary of physicochemical method of petroleum hydrocarbon treatment

Methodology	Advantages	Limitations	References
Solvent extraction	Efficient and easy method	High cost, use of large organic solvents, not eco-friendly	Taiwo and Otolorin (2009), Naggar et al. (2010)
Centrifugation	Efficient and easy method, with no chemicals	High maintenance cost, noise pollution, and pretreatment required to reduce viscosity	Da Silva et al. (2012)
Pyrolysis	Efficient and easy method, large treatment capacity	High maintenance cost, high energy consumption, not suitable for high moisture content sludge	Liu et al. (2009), Qin et al. (2015)
Freeze/Thrawn	Easy method, applicable for cold regions, less time required	High cost, low efficiency	Chen and He (2003)
Surfactant EOR	Easy method, less time required	High cost, use of chemical surfactants	Lima et al. (2011), Long et al. (2013)
Microwave irradiation	Fast process, no chemicals used	High equipment cost, high energy consumption	Tan et al. (2007), Appleton et al. (2005)
Electrokinetic method	Fast process, no chemicals used	Difficulty in operation, low capacity	Yang et al. (2005), Virkutyte et al. (2002)
Ultrasonic irradiation	Fast process, no use of chemicals	High equipment cost, less capacity	Pilli et al. (2011)
Froth flotation	Generation of large amounts of wastewater	Less energy consumption	Al-Otoom et al. (2010)

2.8.1 Biological Methods of Treatment

The biological treatment of hydrocarbons involves the breakdown of petroleum hydrocarbons, resulting in the transformation of diverse organic pollutants into simpler and non-toxic chemicals. This process minimizes the environmental impact over an extended duration. One of the important advantages of biological treatment is that it is eco-friendly and doesn't disturb the environment on a long-term basis. The major disadvantage of this treatment is that the process takes an extended period as it is a slow process ranging from months to several years for the successful removal of contaminants (Mohanta et al. 2023). The viability of biological treatment technology mainly depends on limiting factors and the location of the contaminants. Treatment of the pollutants varies in soil, surface water, groundwater, etc. If the treatment is on-site, then it's in situ, but if the treatment is offsite, then it is ex-situ. Hydrocarbon pollution removal using bioremediation is an important and long-term technique. Oil spills may be cleaned up and the ecosystem restored through bioremediation, which involves the conversion of hazardous compounds into less harmful ones, including carbon dioxide, water, and fatty acids (Kanungo et al. 2023). However, the biodegradability and bioavailability of the pollutants, as well as the presence of hazardous substances, may hinder the effectiveness of this form of therapy. Bioaugmentation and biostimulation are the most widely used bioremediation techniques, and they have been put into practice both in the lab and in the field (Ayotamuno et al. 2010; Ouyang et al. 2005; Cerqueira et al. 2011). Cerqueira et al. (2011) reported that the isolated pure bacterial strains such as *Bacillus megaterium*, *Stenotrophomonas acidaminiphila*, and *Bacillus cibi* from petrochemical oily sludge and another two strains, Pseudomonas aeruginosa and Bacillus cereus from petrochemical waste polluted soil showed the efficiency of aliphatic hydrocarbon degradation in the range 86–91% and aromatic hydrocarbon degradation in the range 33–64% when bio augmented in pure culture. On the other hand, the consortium of these five bacterial strains showed the efficiency of degradation towards aliphatic fractions and aromatic fractions at 90.7 and 51.8%, respectively, in an incubation period of 40 days. Varjani and Upasani (2016) stated that the isolated pure bacterial strain *Pseudomonas aeruginosa* from petroleum-polluted soil showed the efficacy of hydrocarbon (C8-C36+) degradation up to 61% in 60 days in a liquid medium. The liquid medium was supplemented with 3% (v/v), and the operating conditions were maintained at 37 °C, 180 rpm for an incubation period of 60 days. Abena et al. (2019) reported that bioaugmentation of five bacterial consortia showed 48% TPH degradation in 40 days. The strains were isolated from crude oil-contaminated soil and identified as *Raoultella sp., Serratia sp., Bacillus sp., Acinetobacter sp.,* and *Acinetobacter sp.* Reports on biodegradation TPH by bioaugmentation strategy have been summarized in Table 2.3.

2.8.2 Bioaugmentation

Bioaugmentation is the process of reducing or eliminating pollution by using microorganisms to break down hazardous hydrocarbons. Here, contaminated water is injected alongside microorganisms capable of degrading hydrocarbons. Genetically engineered microbes are added to the contaminated sample to reduce the harmful pollutants. It is the process of inoculating efficient exogenous microorganisms to the contaminated site to maintain the microbial activities for effective degradation of pollutants. It can be achieved by inoculation of pure strain or by a group of the culture as a consortium. To date, several studies have been documented through single bacterial strains and by the bacterial consortium for the effective degradation of petroleum hydrocarbons. Bioaugmentation is a very effective initiative towards

Table 2.3 Degradation of petroleum hydrocarbon by using pure bacterial culture

Bacterial culture	Experiment type and hydrocarbon nature	Incubation period (day)	Degradation efficiency (%)	References
Pseudomonas sp. NCIM 5514	Microcosm study Crude oil contaminated soil, Ankleshwar, India	60	60	Varjani and Upasani (2016)
Arthrobacter citreus and *Rhodococcus jostii*	Liquid media Oily tank sludge, Kerman province	35	75	Hamidi et al. (2021)
Acinetobacter sp. *Pseudomonas* sp.	Microcosm study, crude oil contaminated soil	70	82	Suja et al. (2014)
Pseudomonas sp., *Achromobacter* sp., and *Ochrobactrum* sp.	Microcosm study Diesel oil contaminated agricultural soil	32	32.5	Colla et al. (2014)
Acinetobacter SZ-1 KF453955	Microcosm study Crude oil contaminated soil, Shaanxi province, China	49	34	Wu et al. (2016)
Bacillus sp. GR1-01, 02, 05, 13, 14, 15, 21, and E1	Liquid medium Refinery sludge, IOCL, Guwahati, India	28	19–48	Roy et al. (2018)

(continued)

Table 2.3 (continued)

Bacterial culture	Experiment type and hydrocarbon nature	Incubation period (day)	Degradation efficiency (%)	References
Indigenous bacterial culture	Crude oil contaminated soil, Liaohe Oilfield, China	84	27	Xu and Lu (2010)
Ochrobactrum NCIM 5515, *Stenotrophomonas* sp. BAB 3482 and NCIM 5530, *Pseudomonas* sp. BAB 3483, BAB 3485 and NCIM 5514	Microcosm study Liquid media Crude oil contaminated soil, Ankleshwar, India	56 75	82 83	Varjani et al. (2015, 2020)

the sustainability of the environment and especially superior substrate. This process is a sustainable approach with superior substrate specificity.

2.9 Conclusion

The worldwide environmental problem of hydrocarbon pollution originating from the petrochemical sector has emerged as a widespread issue that needs immediate attention. The unintentional discharge of petroleum products presents a significant risk to both ecosystems and human welfare. The identification of hydrocarbons as organic pollutants has been well documented, highlighting their well-established potential for carcinogenic and neurological consequences. This emphasizes the urgent need for the development and implementation of efficient remediation solutions. The categorization of hydrocarbons into pyrogenic, petrogenic, and biological groups within the petroleum domain elucidates the intricate characteristics of these pollutants. During the process of crude oil extraction and processing, substantial amounts of petroleum waste and organic pollutants are produced, hence increasing the existing environmental concerns. The presence of petroleum hydrocarbons is exacerbated by the inclusion of supplementary sources, such as petroleum oil spills, sewage sludge, and tarry waste compounds. The rapid evaporation of highly detrimental hydrocarbons after oil spills imposes a significant ecological impact. The constituents of petroleum waste consist of aromatic chemicals, nitrogen compounds, aliphatic compounds, amines, and oxides, which exhibit variations caused by local biological and geological causes. The increasing need for petroleum compounds directly exacerbates the environmental hazards linked to inadequate disposal practices. The inadequate management of the environment leads to a diverse range of negative outcomes within ecosystems. The impacts include a spectrum of effects, including the inhibition of seed germination, the disruption of enzyme activities, and the alteration of chlorophyll concentrations in plants. The adverse effects of hydrocarbon pollution disproportionately impact human health, resulting in various respiratory disorders, as well as skin and eye irritation, among other significant health concerns. Aquatic organisms, which play a crucial role in ecosystems, have adverse effects such as reproductive impairments, decreased oxygen levels, and consequent declines in plant and animal populations. When considering this complex issue, several methods of remediation are used. Although physiochemical approaches are available, their lack of environmental friendliness and high expenses make it imperative to embrace biological alternatives. Phytoremediation, bioaugmentation, and biostimulation are notable methods that use the capabilities of living organisms to rectify imbalances in polluted settings. In light of the ongoing global challenges posed by hydrocarbon contamination, immediate and collaborative actions must be undertaken. The implementation of efficient regulatory measures aimed at reducing unintentional releases, the promotion of sustainable industrial practices, and the investment in novel and environmentally friendly remediation technologies will play a crucial role in limiting the extensive consequences of persistent hydrocarbon pollutants. By placing responsible and sustainable activities at the forefront, society has the potential to establish a trajectory towards a future characterized by improved cleanliness and enhanced well-being for both ecosystems and mankind.

References

Abena MTB, Li T, Shah MN, Zhong W (2019) Biodegradation of total petroleum hydrocarbons (TPH) in highly contaminated soils by natural attenuation and bioaugmentation. Chemosphere 234:864–874

Adeniyi AA, Afolabi JA (2002) Determination of total petroleum hydrocarbons and heavy metals in soils within the vicinity of facilities handling refined petroleum products in the Lagos metropolis. Environ Int 28(1–2):79–82. https://doi.org/10.1016/S0160-4120(02)00007-7

Alonso-Alvarez C, Pérez C, Velando A (2007) Effects of acute exposure to heavy fuel oil from the Prestige spill on a seabird. Aquat Toxicol 84(1):103–110

Al-Otoom A, Allawzi M, Al-Omari N, Al-Hsienat E (2010) Bitumen recovery from Jordanian oil sand by froth flotation using petroleum cycles oil cuts. Energy 35(10):4217–4225. https://doi.org/10.1016/j.energy.2010.07.008

Ambaye TG, Chebbi A, Formicola F, Prasad S, Gomez FH, Franzetti A, Vaccari M (2022) Remediation of soil polluted with petroleum hydrocarbons and its reuse for agriculture: recent progress, challenges, and perspectives. Chemosphere 293:133572. https://doi.org/10.1016/j.chemosphere.2022.133572

Appleton TJ, Colder RI, Kingman SW, Lowndes IS, Read AG (2005) Microwave technology for energy-efficient processing of waste. Appl Energy 81(1):85–113. https://doi.org/10.1016/j.apenergy.2004.07.002

Ayotamuno JM, Kogbara RB, Agele EA, Agoro OS (2010) Composting and phytoremediation treatment of petroleum sludge. Soil Sediment Contam 19(6):686–695

Barraza F, Maurice L, Uzu G, Becerra S, López F, Ochoa-Herrera V, Ruales J, Schreck E (2018) Distribution, contents and health risk assessment of metal(loid)s in small-scale farms in the Ecuadorian Amazon: an insight into impacts of oil activities. Sci Total Environ 622:106–120

Behera ID, Nayak M, Biswas S, Meikap BC, Sen R (2021) Enhanced biodegradation of total petroleum hydrocarbons by implementing a novel two-step bioaugmentation strategy using indigenous bacterial consortium. J Environ Manag 292. https://doi.org/10.1016/j.jenvman.2021.112746

Behera ID, Basak G, Kumar RR, Sen R, Meikap BC (2020) Treatment of petroleum refinery sludge by petroleum degrading bacterium Stenotrophomonas pavanii IRB19 as an efficient novel technology. J Environ Sci Health Part A 56(2):226–239. https://doi.org/10.1080/10934529.2020.1866924

Behera ID, Nayak M, Mishra A, Meikap BC, Sen R (2022) Strategic implementation of integrated bioaugmentation and biostimulation for efficient mitigation of petroleum hydrocarbon pollutants from terrestrial and aquatic environments. Mar Pollut Bull 177:113492. https://doi.org/10.1016/j.marpolbul.2022.113492

Cerqueira VS, Hollenbach EB, Maboni F, Vainstein M, Camargo F, Do-Carmo RPM, Bento FM (2011) Biodegradation potential of oily sludge by pure and mixed bacterial cultures. Bioresour Technol 102(23):11003–11010. https://doi.org/10.1016/j.biortech.2011.09.074

Chandra S, Sharma R Singh K et al (2013) Application of bioremediation technology in the environment contaminated with petroleum hydrocarbon. Ann Microbiol 63:417–431. https://doi.org/10.1007/s13213-012-0543-3

Chen G, He G (2003) Separation of water and oil from water-in-oil emulsion by freeze/thaw method. Sep Purif Technol 31(1):83–89. https://doi.org/10.1016/S1383-5866(02)00156-9

Colla TS, Andreazza R, Bücker F, de Souza MM, Tramontini L, Prado GR, Frazzon APG, de Oliveira Camargo FA, Bento FM (2014) Bioremediation assessment of diesel-biodiesel-contaminated soil using an alternative bioaugmentation strategy. Environ Sci Pollut Res 21(4):2592–2602. https://doi.org/10.1007/s11356-013-2139-2

Da Silva LJ, Alves FC, De França FP (2012) A review of the technological solutions for the treatment of oily sludges from petroleum refineries. Waste Manag Res 30(10):1016–1030. https://doi.org/10.1177/0734242X12448517

Das S, Kuppanan N, Channashettar VA, Lal B (2018) Remediation of oily sludge- and oil-contaminated soil from petroleum industry: recent developments and future prospects. In: Advances in soil microbiology: recent trends and future prospects: volume 1: soil-microbe interaction, pp 165–177

Davidson CJ, Hannigan JH, Bowen SE (2021) Effects of inhaled combined Benzene, Toluene, Ethylbenzene, and Xylenes (BTEX): Toward an environmental exposure model. Environ Toxicol Pharmacol 81:103518. https://doi.org/10.1016/j.etap.2020.103518

Dua M, Singh A, Sethunathan N, Johri A (2002) Biotechnology and bioremediation: successes and limitations. Appl Microbiol Biotechnol 59:143–152

Farhadian M, Vachelard C, Duchez D, Larroche C (2008) In situ bioremediation of monoaromatic pollutants in groundwater: a review. Bioresour Technol 9:5296–5308

Feng L, Jiang X, Huang Y, Wen D, Fu T, Fu R (2021) Petroleum hydrocarbon-contaminated soil bioremediation assisted by isolated bacterial consortium and sophorolipid. Environ Pollut 273:116476. https://doi.org/10.1016/j.envpol.2021.116476

Galbraith D, Gross SA, Paustenbach D (2010) Benzene and human health: a historical review and appraisal of associations with various diseases. Crit Rev Toxicol 40(sup2):1–46. https://doi.org/10.3109/10408444.2010.508162

Gao H, Wu M, Liu H, Xu Y, Liu Z (2022) Effect of petroleum hydrocarbon pollution levels on the soil microecosystem and ecological function. Environ Pollut 293:118511. https://doi.org/10.1016/j.envpol.2021.118511

Hamidi Y, Ataei SA, Sarrafi A (2021) Biodegradation of total petroleum hydrocarbons in oily sludge: a comparative study of biostimulation, bioaugmentation, and combination of methods. J Chem Technol Biotechnol 96(5):1302–1307. https://doi.org/10.1002/jctb.6646

Harayama S, Kasai Y, Hara A (2004) Microbial communities in oil-contaminated seawater. Curr Opin Biotechnol 15(3):205–214
Hu G, Li J, Zeng G (2013) Recent development in the treatment of oily sludge from petroleum industry: a review. J Hazard Mater 261:470–490. https://doi.org/10.1016/j.jhazmat.2013.07.069
Ismail K, Lewis G (2006) Multi-symptom illnesses, unexplained illness and Gulf War Syndrome. Philos Trans R Soc B Biol Sci 361(1468):543–551
Jain K, Desai C, Madamwar D (2017) Bacterial interactions with chromium and strategies for the remediation of hexavalent chromium pollution. In: Handbook of metal-microbe interactions and bioremediation. CRC Press, Boca Raton, pp 719–729
Janbandhu A, Fulekar MH (2011) Biodegradation of phenanthrene using adapted microbial consortium isolated from petrochemical contaminated environment. J Hazard Mater 187(1–3):333–340
Jasmine J, Mukherji S (2019) Impact of bioremediation strategies on slurry phase treatment of aged oily sludge from a refinery. J Environ Manage 246:625–635
Jernelöv A (2010) The threats from oil spills: now, then, and in the future. Ambio 39(5–6):353–366
Johnson OA, Affam AC (2019) Petroleum sludge treatment and disposal: a review. Environ Eng Res 24(2):191–201
Kanungo J, Sahoo S, Bal M, Behera ID (2023) Performance of bioremediation strategy in waste lubricating oil pollutants: a review. Geomicrobiol J. https://doi.org/10.1080/01490451.2023.2245395
Lee CW, Dai YT, Chien CH, Hsu DJ (2006) Characteristics and health impacts of volatile organic compounds in photocopy centers. Environ Res 100(2):139–149. https://doi.org/10.1016/j.envres.2005.05.003
Liu J, Jiang X, Zhou L, Han X, Cui Z (2009) Pyrolysis treatment of oil sludge and model-free kinetics analysis. J Hazard Mater 161(2–3):1208–1215. https://doi.org/10.1016/j.jhazmat.2008.04.072
Long X, Zhang G, Shen C, Sun G, Wang R, Yin L, Meng Q (2013) Application of rhamnolipid as a novel biodemulsifier for destabilizing waste crude oil. Biores Technol 131:1–5. https://doi.org/10.1016/j.biortech.2012.12.128
Lima TMS, Fonseca AF, Leão BA, Mounteer AH (2011). Oil recovery from fuel oil storage tank sludge using biosurfactants. J Bioremediation Biodegrad 02(04). https://doi.org/10.4172/2155-6199.1000125
Maurice L, López F, Becerra S, Jamhoury H, Le Menach K, Dévier MH, Budzinski H, Prunier J, Juteau-Martineau G, Ochoa-Herrera V, Quiroga D (2019) Drinking water quality in areas impacted by oil activities in Ecuador: associated health risks and social perception of human exposure. Sci Total Environ 690:1203–1217
Mishra A, Siddiqi H, Kumari U, Behera ID, Mukherjee S, Meikap BC (2021) Pyrolysis of waste lubricating oil/waste motor oil to generate high-grade fuel oil: a comprehensive review. Renew Sustain Energy Rev 150:111446. https://doi.org/10.1016/j.rser.2021.111446
Mohanta S, Pradhan B, Behera ID (2023) Impact and remediation of petroleum hydrocarbon pollutants on agricultural land: a review. Geomicrobiol J 1–15. https://doi.org/10.1080/01490451.2023.2243925
Morita S, Snider MT, Inada Y (1986) Increased N-pentane excretion in humans: a consequence of pulmonary oxygen exposure. Anesthesiology 64(6):730–733. https://doi.org/10.1097/00000542-198606000-00009
Elnaggar AY, Saad EA, Kandil AT, Elmoher HO (2010) Petroleum cuts as solvent extractor for oil recovery from petroleum sludge. J Pet Technol Altern Fuels 1(1):10–19
Negrete-Bolagay D, Zamora-Ledezma C, Chuya-Sumba C, De Sousa FB, Whitehead D, Alexis F, Guerrero VH (2021) Persistent organic pollutants: the trade-off between potential risks and sustainable remediation methods. J Environ Manag 300:113737. https://doi.org/10.1016/j.jenvman.2021.113737
Ouyang W, Liu H, Murygina V, Yu Y, Xiu Z, Kalyuzhnyi S (2005) Comparison of bio-augmentation and composting for remediation of oily sludge: a field-scale study in China. Process Biochem 40(12):3763–3768

Pilli S, Bhunia P, Yan S, LeBlanc RJ, Tyagi RD, Surampalli RY (2011) Ultrasonic pretreatment of sludge: a review. Ultrason Sonochem 18(1):1–18. https://doi.org/10.1016/j.ultsonch.2010.02.014

Pinedo J, Ibáñez R, Lijzen JPA, Irabien A (2013) Assessment of soil pollution based on total petroleum hydrocarbons and individual oil substances. J Environ Manag 130:72–79. https://doi.org/10.1016/j.jenvman.2013.08.048

Prince RC, McFarlin KM, Butler JD, Febbo EJ, Wang FC, Nedwed TJ (2013) The primary biodegradation of dispersed crude oil in the sea. Chemosphere 90:521–526. https://doi.org/10.1016/j.chemosphere.2012.08.020

Qin L, Han J, He X, Zhan Y, Yu F (2015) Recovery of energy and iron from oily sludge pyrolysis in a fluidized bed reactor. J Environ Manag 154:177–182. https://doi.org/10.1016/j.jenvman.2015.02.030

Ron EZ, Rosenberg E (2014) Enhanced bioremediation of oil spills in the sea. Curr Opin Biotechnol 27:191–194. https://doi.org/10.1016/j.copbio.2014.02.004

Rosner D, Markowitz G (2013) Persistent pollutants: a brief history of the discovery of the widespread toxicity of chlorinated hydrocarbons. Environ Res 120:126–133. https://doi.org/10.1016/j.envres.2012.08.011

Roy A, Dutta A, Pal S, Gupta A, Sarkar J, Chatterjee A, Saha A, Sarkar P, Sar P, Kazy SK (2018) Biostimulation and bioaugmentation of native microbial community accelerated bioremediation of oil refinery sludge. Biores Technol 253:22–32. https://doi.org/10.1016/j.biortech.2018.01.004

Saeki H, Sasaki M, Komatsu K, Miura A, Matsuda H (2009) Oil spill remediation by using the remediation agent JE1058BS that contains a biosurfactant produced by Gordonia sp. strain JE-1058. Bioresour Technol 100(2):572–577

Sajna KV, Sukumaran RK, Gottumukkala LD, Pandey A (2015) Crude oil biodegradation aided by biosurfactants from Pseudozyma sp. NII 08165 or its culture broth. Bioresour Technol 191:133–139. https://doi.org/10.1016/j.biortech.2015.04.126

Siddiqi H, Mishra A, Maiti P, Behera ID, Meikap BC (2022) In-situ and ex-situ co-pyrolysis studies of waste biomass with spent motor oil: elucidating the role of physical inhibition and mixing ratio to enhance fuel quality. Biores Technol 358:127364. https://doi.org/10.1016/j.biortech.2022.127364

Souza EC, Vessoni-Penna TC, Oliveira RPDS (2014) Biosurfactant-enhanced hydrocarbon bioremediation: an overview. Int Biodeterior Biodegrad 89:88–94. https://doi.org/10.1016/j.ibiod.2014.01.007

Srivastava M, Srivastava A, Yadav A, Rawat V (2019) Source and control of hydrocarbon pollution. Hydrocarb Pollut Effect Environ 1–21. https://doi.org/10.5772/intechopen.86487

Suja F, Rahim F, Taha MR, Hambali N, Rizal Razali M, Khalid A, Hamzah A (2014) Effects of local microbial bioaugmentation and biostimulation on the bioremediation of total petroleum hydrocarbons (TPH) in crude oil contaminated soil based on laboratory and field observations. Int Biodeterior Biodegrad 90:115–122. https://doi.org/10.1016/j.ibiod.2014.03.006

Taiwo EA, Otolorin JA (2009) Oil recovery from petroleum sludge by solvent extraction. Pet Sci Technol 27(8):836–844. https://doi.org/10.1080/10916460802455582

Tan W, Yang XG, Tan XF (2007) Study on demulsification of crude oil emulsions by microwave chemical method. Sep Sci Technol 42(6):1367–1377. https://doi.org/10.1080/01496390701193736

Tsai TT, Kao CM (2009) Treatment of petroleum-hydrocarbon contaminated soils using hydrogen peroxide oxidation catalyzed by waste basic oxygen furnace slag. J Hazard Mater 170(1):466–472. https://doi.org/10.1016/j.jhazmat.2009.04.073

Turner NR, Renegar DA (2017) Petroleum hydrocarbon toxicity to corals: A review. Mar Pollut Bull 119(2):1–16. https://doi.org/10.1016/j.marpolbul.2017.04.050

Uddin S, Fowler SW, Saeed T, Jupp B, Faizuddin M (2021) Petroleum hydrocarbon pollution in sediments from the Gulf and Omani waters: status and review. Mar Pollut Bull 173:112913. https://doi.org/10.1016/j.marpolbul.2021.112913

Van Meter RJ, Spotila JR, Avery HW (2006) Polycyclic aromatic hydrocarbons affect survival and development of common snapping turtle (Chelydra serpentina) embryos and hatchlings. Environ Pollut 142(3):466–475

Varjani SJ (2014) Hydrocarbon degrading and biosurfactants (bioemulsifiers) producing bacteria from petroleum oil wells. Kadi Sarva Vishwavidyalaya, Gandhinagar, India

Varjani SJ, Gnansounou E, Pandey A (2017) A comprehensive review on toxicity of persistent organic pollutants from petroleum refinery waste and their degradation by microorganisms. Chemosphere 188:280–291. https://doi.org/10.1016/j.chemosphere.2017.09.005

Varjani SJ, Upasani VN (2017) A new look at factors affecting microbial degradation of petroleum hydrocarbon pollutants. Int Biodeterior Biodegrad 120:71–83. https://doi.org/10.1016/j.ibiod.2017.02.006

Varjani SJ, Rana DP, Jain AK, Bateja S, Upasani VN (2015) Synergistic ex-situ biodegradation of crude oil by a halotolerant bacterial consortium of indigenous strains isolated from onshore sites of Gujarat, India. Int Biodeterior Biodegrad 103:116–124. https://doi.org/10.1016/j.ibiod.2015.03.030

Varjani S, Upasani V (2016) Biodegradation of petroleum hydrocarbons by oleophilic strain of Pseudomonas aeruginosa NCIM 5514. Biores Technol 222:195–201. https://doi.org/10.1016/j.biortech.2016.10.006

Varjani S, Pandey A, Upasani VN (2020) Oilfield waste treatment using novel hydrocarbon utilizing bacterial consortium—a microcosm approach. Sci Total Environ 745:141043. https://doi.org/10.1016/j.scitotenv.2020.141043

Virkutyte J, Sillanpää M, Latostenmaa P (2002) Electrokinetic soil remediation - critical overview. Sci Total Environ 289(1–3):97–121. https://doi.org/10.1016/S0048-9697(01)01027-0

Walker TR, Grant J, Archambault MC (2006) Accumulation of marine debris on an intertidal beach in an urban park (Halifax Harbour, Nova Scotia). Water Qual Res J 41(3):256–262. https://doi.org/10.2166/wqrj.2006.029

Walser T, Juraske R, Demou E, Hellweg S (2014) Indoor exposure to toluene from printed matter matters complementary views from life cycle assessment and risk assessment. Environ Sci Technol 48(1):689–697. https://doi.org/10.1021/es403804z

Wu M, Dick WA, Li W, Wang X, Yang Q, Wang T, Xu L, Zhang M, Chen L (2016) Bioaugmentation and biostimulation of hydrocarbon degradation and the microbial community in petroleum-contaminated soil. Int Biodeterior Biodegrad 107:158–164. https://doi.org/10.1016/j.ibiod.2015.11.019

Wu Y, Wei T, Yu B (2009) Demulsification of asphaltenes and resins stabilized emulsions via the freeze/thaw method. Energy Fuels 23(1):481–486. https://doi.org/10.1021/ef800600v

Xu Y, Lu M (2010) Bioremediation of crude oil-contaminated soil: comparison of different biostimulation and bioaugmentation treatments. J Hazard Mater 183(1–3):395–401. https://doi.org/10.1016/j.jhazmat.2010.07.038

Yang L, Nakhla G, Bassi A (2005) Electrokinetic dewatering of oily sludges. J Hazard Mater 125(1–3):130–140. https://doi.org/10.1016/j.jhazmat.2005.05.040

Zhang ZF, Zhang X, Zhang XM, Liu LY, Li YF, Sun W (2020) Indoor occurrence and health risk of formaldehyde, toluene, xylene, and total volatile organic compounds derived from an extensive monitoring campaign in Harbin, a megacity of China. Chemosphere 250:126324

Chapter 3
Challenges, Opportunities, and Strategies for Effective Petroleum Hydrocarbon Waste Management

Varsha Parashar and Chandrakant Thakur

Abstract The petroleum industry's rapid expansion has produced vast quantities of different wastes that need to be properly disposed of and valued. In terms of resources, refinery wastes are viewed as valuable assets with a high energy potential, so their management is crucial. These large amounts of refinery waste are a mixture of hydrocarbons, water, heavy metals, and fine particles. Numerous gases, high- and low-boiling-point constituents, wastewater, spent caustic, filter clay, and solid waste are among them. The sludge in the bottom of crude oil storage tanks is made up of water, asphaltenes, paraffin, hydrocarbons, and inorganic particles like sand, iron sulfides, and iron oxides. Petroleum sludge, which is created when changes in the environment alter the qualities of crude oil, primarily consists of hydrocarbons. Hydrocarbons are volatile organic compounds that are of major concern in refineries and are the most significant antecedents of fine particulate matter (PM2.5) and ground-level ozone production in the atmosphere. The basic waste management techniques include waste source reduction, reuse, and recycling, composting, incineration with or without energy recovery, fuel generation, and landfilling. Hydrocarbon emissions are reduced by defining equipment design standards, control systems, inspection requirements, and maintenance requirements. Fugitive emissions, wastewater treatment systems, storage tanks, and loading/unloading systems are the refineries' principal sources of volatile organic compounds. Possible treatment techniques like stabilization and solidification offer potential products with economic benefits while conserving resources and the environment. Physical treatments of petroleum wastes include in-line filtering, filter presses, centrifuges, hydrocyclones, and sludge dryers. The goal of final disposal is to store waste permanently in a location where it won't be transferred or expected to return (landfills, incineration, and deep well injection).

Keywords Petroleum · Hydrocarbons · Waste treatment · Waste disposal

V. Parashar (✉) · C. Thakur
Department of Chemical Engineering, National Institute of Technology, Raipur 492010, Chhattisgarh, India
e-mail: vparashar.phd2022.che@nitrr.ac.in

I. D. Behera and A. P. Das (eds.), *Impact of Petroleum Waste on Environmental Pollution and its Sustainable Management Through Circular Economy*, Environmental Science and Engineering, https://doi.org/10.1007/978-3-031-48220-5_3

3.1 Introduction

The petroleum industry is an essential part of modern society, providing fuel for transportation, heating, and electricity generation. However, the process of extracting, refining, and distributing petroleum products generates various types of waste, including solids, liquids, and gases. These wastes can be harmful to human health and the environment if not properly managed. The refining process in the petroleum industry involves the conversion of crude oil into various useful products, such as gasoline, diesel fuel, and lubricants, through a series of physical and chemical processes. This process generates various types of waste, including wastewater, spent catalysts, and sludge. The waste is composed of hydrocarbons, heavy metals, and other pollutants. To manage these wastes, various treatment techniques have been developed. These include physical, chemical, and biological methods, such as incineration, adsorption, bioremediation, and solidification/stabilization. Each technique has its advantages and disadvantages, and the selection of the most appropriate method depends on factors such as the type and quantity of waste, cost, and environmental considerations.

Numerous research papers have investigated the generation and management of waste in the petroleum industry. Ayinla et al. (2019a, b) reviewed various waste management approaches, including source reduction, recycling, and disposal methods such as incineration and landfilling. Huang et al. (2015a, b) analyzed the characteristics of refinery waste and discussed the benefits and limitations of various treatment methods. Rezaei et al. (2017a, b) reviewed the available treatment techniques for petroleum refinery waste and evaluated their effectiveness and feasibility. Wu et al. (2019a, b) discussed the treatment of petroleum refinery waste sludge using stabilization/solidification techniques. Shafaei et al. (2021a, b) reviewed current treatment technologies for petroleum sludge, including thermal, mechanical, and biological methods. The petroleum industry generates various types of waste that require careful management to prevent harm to human health and the environment. Several treatment techniques are available, and the selection of the most appropriate method depends on various factors. Ongoing research aims to improve waste management in the petroleum industry and reduce its impact on the environment.

3.1.1 Petroleum Hydrocarbon

The environmental impact of petroleum hydrocarbons involves various processes that occur when these substances are released into the environment. These processes (Fig. 3.1) can be categorized into physical abiotic reactions, biological interactions with microorganisms, and metabolic pathways. In terms of physical processes, one of the fastest is the evaporation of lighter compounds. The rate and effectiveness of evaporation can be influenced by environmental factors such as temperature, wind speed, water turbulence, and surface characteristics.

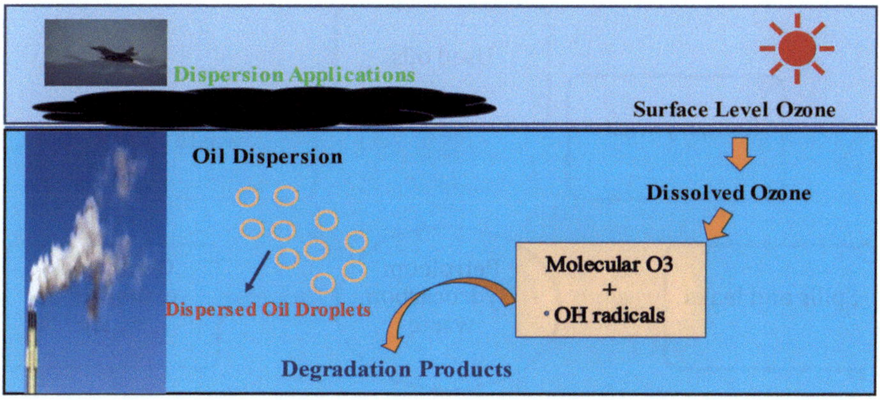

Fig. 3.1 Environmental impact of petroleum hydrocarbons through various processes

Petroleum hydrocarbons are organic compounds consisting mainly of carbon and hydrogen atoms. They are derived from fossilized organic materials and serve as vital energy sources and raw materials for various industries. Crude oil, the primary form of petroleum hydrocarbons, is extracted from underground reservoirs and undergoes refining processes to separate it into different fractions (Payne et al. 1991). Refining involves fractional distillation, where crude oil is separated based on boiling points. This process yields fractions like gasoline, diesel, jet fuel, and heating oil. Further refining techniques, such as cracking and reforming, are employed to convert heavier hydrocarbons into more valuable products.

Petroleum hydrocarbons possess advantageous properties that make them well-suited for diverse applications. They have a high energy content, making them an efficient fuel source for transportation, power generation, and industrial operations. Their lubricating properties are beneficial for machinery and engine applications. Additionally, their versatility allows for the creation of plastics, solvents, fertilizers, and synthetic materials. However, the utilization and management of petroleum hydrocarbons also present environmental challenges. Their extraction, production, and combustion contribute to air pollution and greenhouse gas emissions, exacerbating climate change. Accidental spills and leaks during transportation and storage can result in environmental contamination, negatively impacting ecosystems, wildlife, and human health (Mohammadi et al. 2020).

3.2 Petroleum and Hydrocarbon Waste

Petroleum hydrocarbon waste includes the waste materials, residues, or discarded substances that arise from the production, refining, distribution, and utilization of petroleum-based products. These wastes are in various states, such as solids, liquids,

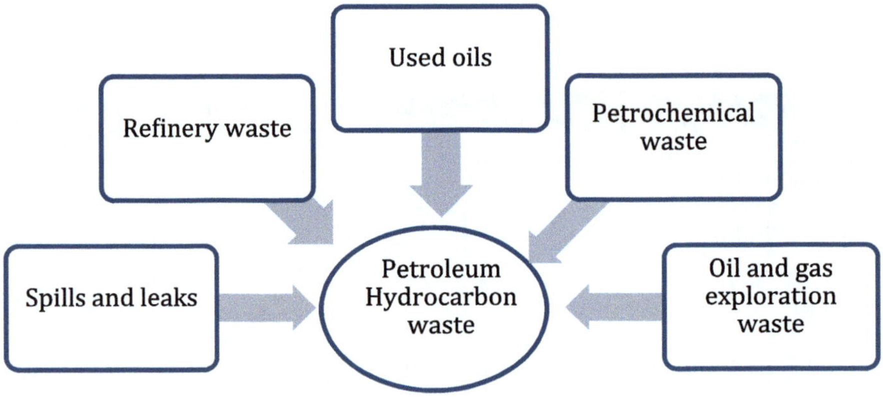

Fig. 3.2 Different types of petroleum hydrocarbon waste (IOGP)

or gases, and they often contain a diverse array of hydrocarbon compounds. There are several distinct categories of petroleum hydrocarbon waste, including (Fig. 3.2).

In Spills and leaks, releases of petroleum products during the transportation, storage, or handling phases can lead to environmental contamination. Spills can transpire on land or in water bodies, causing immediate harm to ecosystems, wildlife, and potentially human health. Refinery waste comes from refining processes that generate different waste streams, such as sludge, spent catalysts, and wastewater. These waste materials may contain residual hydrocarbons, heavy metals, and other contaminants, necessitating careful management and treatment to prevent environmental pollution (Index*ex* @ Www.Iogp.Org n.d.). Used waste oils, including used motor oil, lubricants, and hydraulic fluids, represent a substantial source of petroleum hydrocarbon waste. Insufficient disposal of these oils can result in soil and water contamination, posing risks to the environment and human well-being.

Petrochemical wastes, derived from petroleum hydrocarbons, generate waste products like off-spec or off-grade materials, process residues, and packaging waste. Proper disposal or recycling of these wastes is essential to averting environmental harm. Oil and gas exploration waste exploration activities, including drilling operations, yield waste materials such as drill cuttings, muds, and produced water. These waste substances can contain hydrocarbons, drilling additives, and naturally occurring radioactive materials (NORMs), necessitating appropriate management and disposal practices.

Effectively managing petroleum hydrocarbon waste is crucial to minimizing its adverse environmental impact. Regulatory frameworks and industry standards exist to ensure the safe handling, treatment, and disposal of these wastes. Waste management practices often involve containment, segregation, treatment, recycling, or the adoption of appropriate disposal methods, such as incineration, landfills, or specialized treatment facilities.

3.2.1 Refinery Waste Composition and Characteristics

Refinery waste is a complex mixture of various organic and inorganic substances, which makes its composition and characteristics highly variable. A study by Huang et al. (2015a, b) analyzed the composition of different refinery wastes and found that they typically contain hydrocarbons, heavy metals, and inorganic particles such as sand and clay. Hydrocarbons are the primary components of refinery waste and are of major concern due to their potential toxicity and impact on the environment. Additionally, refinery waste may contain other contaminants such as benzene, toluene, ethylbenzene, and xylene (BTEX), which are known to be harmful to human health and the environment. Asphaltenes and paraffin are other common constituents of refinery waste that contribute to the formation of sludge.

The characteristics of refinery waste are also affected by various factors, such as the type of crude oil used, the refining processes, and the type of equipment used in the refinery. A study by Ayinla et al. (2019a, b) discussed the variability of refinery waste characteristics due to these factors and highlighted the need for proper management and disposal of these wastes. Moreover, the presence of different types of waste streams, such as liquid, solid, and gaseous waste, further complicates the management of refinery waste. Each waste stream has unique characteristics and requires different treatment methods to ensure safe disposal. Therefore, understanding the composition and characteristics of different waste streams is crucial to developing effective waste management strategies.

In summary, refinery waste is a complex mixture of various organic and inorganic substances with variable composition and characteristics. The presence of different waste streams adds to the complexity of managing these wastes. Proper characterization of refinery waste is essential to developing effective waste management strategies.

3.2.2 Environmental Concerns Associated with Refinery Waste

The petroleum industry generates a large amount of waste, which can pose significant environmental concerns if not properly managed. Refinery waste can contain hazardous chemicals, such as heavy metals and volatile organic compounds, which can contaminate soil and groundwater if they are not managed correctly. The release of greenhouse gases during the refining process can also contribute to climate change. Refinery waste is generated during the production of petroleum products and is known to contain various hazardous components. The environmental concerns associated with refinery waste are primarily related to their impact on air, water, and soil.

Air pollution is a major concern due to the release of greenhouse gases, such as carbon dioxide and methane, during the processing of refinery waste. These gases contribute to global warming and climate change, which can have significant impacts on human health and the environment. Additionally, refinery waste may also release various toxic air pollutants, such as nitrogen oxides, sulfur oxides, and volatile organic compounds, which can cause respiratory problems and other health issues in humans and wildlife.

Water pollution is another major concern associated with refinery waste. The waste may contain various heavy metals, such as lead, arsenic, and mercury, as well as organic compounds, such as benzene, toluene, and xylene. These substances can leach into groundwater and surface water sources, leading to contamination and potential health hazards for humans and wildlife. Additionally, wastewater from refineries can also contribute to the eutrophication of water bodies, which can lead to algal blooms and the death of aquatic life.

Soil contamination is also a concern associated with refinery waste. The waste may contain various heavy metals and other toxic substances that can leach into the soil and groundwater, leading to contamination and potential health hazards for humans and wildlife. Additionally, the disposal of refinery waste in landfills and other storage facilities can contribute to the degradation of soil quality and the loss of natural habitat.

In addition to the environmental concerns associated with refinery waste, there are also health concerns for workers and communities surrounding refineries. Exposure to pollutants such as benzene, sulfur dioxide, and particulate matter can lead to respiratory problems, cancer, and other health issues. Therefore, it is important to ensure proper management of refinery waste to minimize its impact on the environment and human health. Several research papers have focused on the environmental concerns associated with refinery waste. For example, a study by Chibueze et al. (2020) examined the levels of heavy metals in soil samples taken from an oil refinery in Nigeria. The researchers found that the soil samples contained elevated levels of heavy metals, which could lead to soil contamination and potential health hazards. Almuktar et al. (2020) assessed the environmental impact of the discharge of treated wastewater from a refinery in Iraq. The researchers found that the discharge of the treated wastewater had a significant impact on the quality of the receiving water bodies, including increased levels of organic matter, oil and grease, and heavy metals.

To address these environmental concerns, various regulatory bodies have implemented policies and regulations to govern the management of refinery waste. For instance, the Environmental Protection Agency in the United States has established the Resource Conservation and Recovery Act (U.S. Environmental Protection Agency 2021), which regulates the treatment, storage, and disposal of hazardous waste generated by refineries. The environmental concerns associated with refinery waste highlight the importance of proper waste management practices in the petroleum industry. Research into new technologies and processes for managing refinery waste in an environmentally sustainable manner is ongoing (Dhanapal et al. 2019). The environmental concerns associated with refinery waste underscore the importance of implementing sustainable waste management practices in the petroleum industry. By reducing waste generation, implementing more efficient processes, and using innovative waste management techniques, it is possible to minimize the environmental impact of refinery waste.

3.2.3 *Current Trends of Transforming Petroleum Waste into Useful Biofuels in India*

India has been exploring the potential of transforming petroleum waste into useful biofuels. The current trend involves the use of technologies such as pyrolysis, gasification, and liquefaction to convert waste into valuable products. The Indian government has launched initiatives to promote the use of biofuels and encourage the adoption of waste-to-energy technologies. One such initiative is the National Biofuels Policy, which was launched in 2018 to promote the use of biofuels and create a market for them. The policy aims to achieve a 20% blending of biofuels with gasoline and diesel by 2030. Another initiative is the Bioenergy Awards for Cutting Edge Research (B-ACER), which was launched in 2019 to encourage research on bioenergy and waste-to-energy technologies.

Several research studies in India have explored the potential of converting petroleum waste into biofuels. Kaur et al. (2021) explored the use of pyrolysis to convert refinery sludge into bio-oil, which can be used as a substitute for diesel. Kumar et al. (2020) investigated the potential of gasification to convert petroleum waste into hydrogen-rich syngas, which can be used for power generation and other applications. In recent years, there has been a growing interest in transforming petroleum waste into useful biofuels in India. This trend can be attributed to the increasing demand for alternative sources of energy, the need to reduce dependence on fossil fuels, and the growing concern over environmental pollution. One of the most promising technologies for converting petroleum waste into biofuels is pyrolysis. Pyrolysis is a thermochemical process that breaks down organic materials into smaller molecules in the absence of oxygen. The resulting liquid product, known as bio-oil, can be further refined into fuels such as gasoline, diesel, and jet fuel. In addition, the solid and gaseous byproducts of pyrolysis can be used for electricity generation or as a feedstock for other chemical processes.

Several ongoing projects in India are focused on the conversion of petroleum waste to biofuels. For instance, the Indian Oil Corporation has developed a technology for converting waste plastics and tire oil into diesel fuel, which is being implemented on a commercial scale. Another example is the joint venture between Hindustan Petroleum Corporation and LanzaTech, which aims to convert refinery waste gases into ethanol. However, there are also challenges to the development of biofuels from petroleum waste in India, including the lack of a clear regulatory framework, the limited availability of financing, and the need for technological innovation. Nevertheless, with the Indian government's focus on reducing fossil fuel dependence and promoting sustainable development, there is a growing opportunity for the growth of the biofuels industry in India.

It is important to note that while India has made significant progress in transforming petroleum waste into useful biofuels, there are still challenges that need to be addressed. One of the major challenges is the high cost of production of biofuels from petroleum waste, which makes them less competitive than fossil fuels. However, the government of India has been implementing various policies and incentives to encourage the production of biofuels from waste, such as the National Biofuel Policy, which provides financial support and other incentives to biofuel producers. Another challenge is the lack of infrastructure and distribution facilities for biofuels, which limits their market penetration. In response, the government of India has initiated several projects aimed at building biofuel infrastructure and promoting its use, such as the National Biofuel Program, which aims to set up biofuel production plants across the country. Additionally, research and development efforts are ongoing to find ways to improve the efficiency and cost-effectiveness of biofuel production from petroleum waste. For instance, researchers at the Indian Institute of Technology in Bombay have developed a process that uses catalysts to convert petroleum waste into biodiesel at a lower cost and with higher efficiency. Overall, while there are challenges to transforming petroleum waste into useful biofuels in India, the country has made significant progress in this area and is continuing to invest in policies, infrastructure, and research to promote the development of this industry.

3.2.4 Statistical Data on Petroleum Waste

India is a major player in the petroleum industry and generates a significant amount of petroleum waste (Ministry of Petroleum and Natural Gas, Government of India 2020). In the years 2020–2021, the total amount of hazardous waste generated by the petroleum industry in India was 1,22,547.67 metric tons, with the majority (1,21,682.52 MT) generated by oil refineries. The CPCB reports that 97.8% of hazardous waste generated by the petroleum industry was recycled, and 2.2% was sent for disposal in the same period. However, the concentration of pollutants in the water bodies around the refineries was found to exceed the prescribed limits, indicating the need for better wastewater treatment and management practices (CPCB 2019–2020). The majority of hazardous waste generated by the petroleum industry is in the form of sludge and solid waste. The petroleum industry in India generates around 35–40 million metric tons of solid waste every year, which has a high calorific value and can be utilized for energy generation through waste-to-energy technologies (Fig. 3.3).

The consumption of petroleum products in India was 194.2 million metric tons in 2019–20, indicating that the generation of petroleum waste in India will also increase, highlighting the need for effective waste management strategies (Petroleum Planning and Analysis Cell, 2019–20). In response to growing concerns, the government has introduced policies and regulations for the management of petroleum waste, including the establishment of hazardous waste treatment facilities (National Green Tribunal n.d.). Overall, the statistical data on petroleum waste in India suggests that

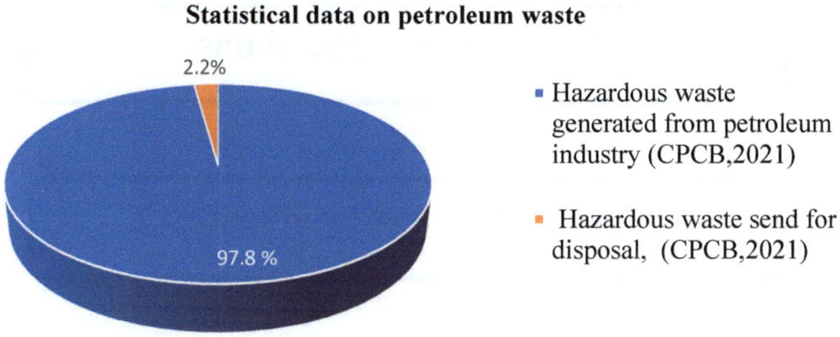

Fig. 3.3 Statistical data on petroleum waste (CPCB)

the industry is taking steps towards sustainable waste management practices, with a focus on recycling and the recovery of hazardous waste. However, there is still room for improvement in wastewater treatment and management practices to ensure minimal impact on the environment (Pandey et al., 2018).

3.3 Challenges in Petroleum and Hydrocarbon Waste Management

Petroleum hydrocarbon management encompasses various activities such as exploration, production, transportation, storage, and utilization of hydrocarbon resources like oil and natural gas. While these resources are vital for meeting global energy needs, their management presents numerous challenges. These challenges arise from diverse factors, including technical complexities, environmental concerns, economic considerations, and regulatory requirements (Fig. 3.4).

One primary challenge is the exploration and production of petroleum hydrocarbons. The search for new reserves involves advanced technologies and significant investments. As easily accessible reserves decline, locating and extracting new reserves becomes increasingly difficult, necessitating more complex and costly extraction methods (International Association of Oil and Gas Producers (IOGP).

Transportation and storage pose additional challenges. Petroleum hydrocarbons must be transported safely and efficiently over long distances, often through pipelines, tankers, or trucks. Ensuring the integrity of these transportation systems is crucial to preventing accidents, spills, and leaks that can cause substantial environmental damage and pose risks to human health and safety. Adequate storage facilities are also necessary to securely contain these hazardous substances. Spills and leaks can have severe environmental consequences. Accidents during transportation, handling, or storage can lead to the release of hydrocarbons, resulting in contamination of land and water bodies and disruption of ecosystems. Cleanup and mitigation of the environmental impact require significant resources and specialized expertise.

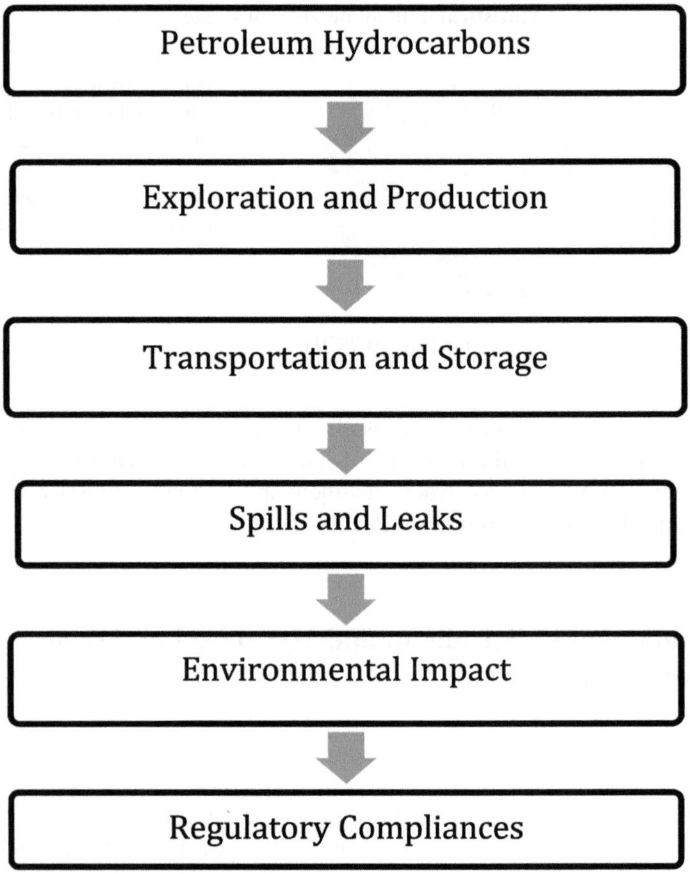

Fig. 3.4 Outline of the challenges involved in petroleum hydrocarbon waste management (IOGP, API, EPA)

The environmental impact of petroleum hydrocarbon management is a major concern. The extraction, processing, and utilization of these resources generate greenhouse gas emissions, contributing to climate change. Furthermore, exploration and production activities can disrupt fragile ecosystems and habitats, causing biodiversity loss and ecological imbalances. Effective management strategies should address these environmental concerns to ensure sustainable practices and minimize the industry's ecological footprint. Compliance with regulations presents another significant challenge in petroleum hydrocarbon management. Governments and international bodies have established standards and regulations to ensure the safe and responsible handling of these resources. Meeting these requirements, which encompass safety, environmental protection, and labor practices, demands extensive monitoring, reporting, and adherence to stringent guidelines American Petroleum Institute (API).

Addressing these challenges necessitates collaboration among industry stakeholders, governments, and environmental organizations. Technological advancements, innovation in extraction and production methods, stricter regulations, and increased awareness of environmental and social responsibilities are crucial in managing petroleum hydrocarbons in a sustainable and responsible manner. Overcoming these challenges allows the industry to strive for a balance between energy demands, environmental protection, and long-term sustainability.

3.3.1 Reduction of Hydrocarbon Emissions

The reduction of hydrocarbon emissions is a critical aspect of environmental protection and public health. Hydrocarbon emissions are released from various sources, including transportation, industrial processes, and natural sources such as wildfires. Hydrocarbons can react with other pollutants in the atmosphere to form ground-level ozone, which can lead to respiratory problems and other health issues. To reduce hydrocarbon emissions, various strategies can be implemented, including regulatory measures, technological advancements, and changes in human behavior.

One approach to reducing hydrocarbon emissions is through the use of emissions control technologies in vehicles and industrial processes. The automotive industry, for example, has implemented a range of technologies to reduce emissions from vehicles, including catalytic converters, exhaust gas recirculation systems, and gasoline direct injection (Lopez-Sotelo et al. 2018). These technologies can significantly reduce hydrocarbon emissions, with some studies reporting reductions of up to 90% (Li et al. 2021). Similarly, industrial processes can use technologies such as flare gas recovery systems and vapor recovery units to capture and recycle hydrocarbon emissions (Chakravarty et al. 2020). Another approach to reducing hydrocarbon emissions is through the implementation of regulatory measures. Governments can implement regulations on emissions from various sources, such as vehicles and industrial processes. For example, the US Environmental Protection Agency (EPA) has implemented regulations on the emissions of volatile organic compounds (VOCs) from various sources, including petroleum refineries and chemical manufacturing facilities (Adeosun et al. 2021). These regulations require companies to implement control technologies and processes to reduce their emissions. Further, changes in human behavior can also contribute to the reduction of hydrocarbon emissions. For example, individuals can reduce their transportation-related emissions by using public transportation or electric vehicles. Additionally, individuals can reduce their energy consumption at home and work by implementing energy-efficient practices such as turning off lights and using energy-efficient appliances.

Overall, a combination of strategies is needed to effectively reduce hydrocarbon emissions. The implementation of emissions control technologies, regulatory measures, and changes in human behavior can all contribute to the reduction of hydrocarbon emissions and the protection of public health and the environment. Hydrocarbon emissions are a major contributor to air pollution and climate change, and concerted efforts are needed to mitigate their impact on the environment and human health (Liu et al. 2021).

3.3.2 *Petroleum Waste Resources and Recovery*

Petroleum waste refers to the solid and liquid byproducts generated during the exploration, production, refining, and transportation of crude oil and its derivatives (Abdulredha et al. 2020; Li et al. 2020). Petroleum waste can be categorized into hazardous and non-hazardous waste, with examples including drilling mud, produced water, sludge, spent catalysts, and contaminated soil (Seshadri et al. 2020). To manage petroleum waste, various methods are used, including biological treatment and thermal treatment (Adeosun et al. 2021). Bioremediation involves using microorganisms to break down waste materials, and it can be carried out in situ or ex-situ (Adeosun et al. 2021). On the other hand, thermal treatment involves heating waste materials to high temperatures to break them down, and it is often used for hazardous waste (Chakravarty et al. 2020). Recovery of petroleum waste can be achieved through recycling, reuse, and repurposing (Yang et al. 2021). Spent catalysts from refining processes can be regenerated and reused, and used oil can be re-refined into new lubricants (Yang et al. 2021). Petroleum coke, a byproduct of oil refining, can also be repurposed as a fuel source or as a raw material for the production of carbon-based products such as electrodes and anodes (Adeosun et al. 2021).

In conclusion, petroleum waste management and recovery are critical aspects of sustainable oil and gas production (Abdulredha et al. 2020). Proper management and recovery of petroleum waste can reduce the negative impact on the environment and public health while providing valuable resources for further use (Seshadri et al. 2020). The oil and gas industry is constantly seeking new and innovative ways to manage and recover petroleum waste, and continued efforts in this area are essential for a sustainable future.

3.4 Opportunities from Petroleum and Hydrocarbon Waste

Opportunities in the field of transforming petroleum waste into useful biofuels in India is the third largest energy consumer in the world after China and the United States, and approximately 80% of India's primary energy needs are met by fossil fuels, primarily crude oil, and natural gas (International Energy Agency 2021). The demand for energy in India is expected to grow significantly in the coming years due to population growth, urbanization, and industrialization. However, India is also home to a large amount of petroleum waste, which can have serious environmental consequences if not managed properly. The transformation of petroleum waste into

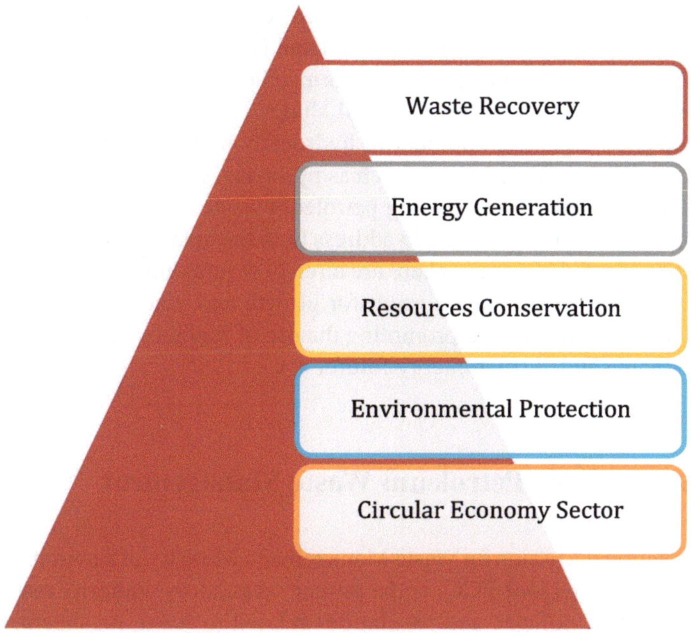

Fig. 3.5 Opportunities from petroleum hydrocarbon waste

useful biofuels presents a significant opportunity to address both India's energy needs and its waste management challenges (Fig. 3.5).

Petroleum waste is generated in India from a variety of sources, including oil refineries, petrochemical plants, and industrial processes (Singh et al. 2019). This waste typically contains a mixture of organic and inorganic compounds, including hydrocarbons, heavy metals, and other contaminants (Nasrullah et al. 2020). The disposal of petroleum waste can lead to soil and water contamination, which can have negative impacts on human health and the environment. This issue involves the transformation of petroleum waste into biofuels. Biofuels are renewable energy sources that are derived from organic matter, such as crops, waste, and other biomass sources (Chen et al. 2021). Biofuels have the potential to reduce greenhouse gas emissions, improve energy security, and promote rural development (International Energy Agency 2018).

In India, there are already many initiatives underway to transform petroleum waste into biofuels. For example, the Indian Oil Corporation (IOC) has established a pilot plant for the conversion of petroleum sludge into biodiesel (Singh et al. 2019). The plant uses a process known as pyrolysis to convert the waste into liquid fuel. The plant has a capacity of 100 kg per hour and has been successfully tested at the IOC's refinery in Panipat. Another initiative is the conversion of waste cooking oil into biodiesel. This process has the potential to reduce waste and provide a source of renewable energy. The Indian government has launched a program to encourage the use of biodiesel in the transport sector and has set a target of blending 5% biodiesel with diesel by 2030 (Ministry of Petroleum and Natural Gas 2019).

There are also opportunities to transform other types of petroleum waste into biofuels, such as refinery effluent and oil sludge (Singh et al. 2019). These waste streams contain a significant number of hydrocarbons, which can be converted into biofuels through various processes, such as pyrolysis, gasification, and fermentation (Nasrullah et al. 2020). Transforming petroleum waste into useful biofuels presents a significant opportunity for India to address both the country's energy needs and its waste management challenges. There are already several initiatives underway in this area, but there is still significant room for growth and innovation. By investing in research and development and promoting the use of biofuels, India can help create a more sustainable and cleaner energy future.

3.5 Strategies for Petroleum Waste Management

Rules and National Mission on Waste Management, Ministry of Environment, Forest, and Climate Change (MoEFCC) is the primary regulatory authority responsible for the formulation of policies and regulations related to waste management in India (Ministry of Environment, Forest, and Climate Change). The key legislation that governs the management of hazardous wastes, including petroleum wastes, in India, is the Hazardous and Other Wastes (Management and Transboundary Movement) Rules, 2016 (Ministry of Environment, Forest, and Climate Change 2016). This legislation specifies the procedures for the handling, storage, transportation, and disposal of hazardous wastes, including the registration and authorization of waste generators and treatment, storage, and disposal facilities.

India has also launched various national missions and initiatives to address waste management issues, including those related to petroleum waste. The Swachh Bharat Abhiyan (Clean India Mission) launched in 2014 aims to create awareness about cleanliness, encourage waste segregation, and promote scientific waste management practices across the country (Government of India n.d.). Another initiative is the National Mission for Clean Ganga (NMCG), which aims to rejuvenate the Ganga River by reducing pollution loads in the river, including industrial waste such as petroleum waste generated along the river (National Mission for Clean Ganga n.d.).

3.5.1 Petroleum Waste Management Techniques

The petroleum industry generates various types of waste during the production, refining, and distribution of petroleum products. These wastes may include solid, liquid, and gaseous materials that can cause significant environmental and health problems if not managed properly. To minimize these impacts, various waste management techniques have been developed and employed by the industry. Here we discuss some of the most common techniques used for petroleum waste management.

3.5.1.1 Source Reduction

Source reduction, also known as waste minimization, is a waste management technique aimed at reducing the amount of waste generated at its source (U.S. Environmental Protection Agency 2022; Williams 1994). The technique involves the identification of waste sources, modification of production processes, and substitution of hazardous or wasteful materials with less harmful alternatives (U.S. Environmental Protection Agency 2022). One of the main advantages of source reduction is that it eliminates the need for waste treatment and disposal, which reduces the environmental impacts and costs associated with waste management (Williams 1994). Additionally, source reduction can improve operational efficiency, reduce raw material and energy consumption, and enhance the overall sustainability of the production process (U.S. Environmental Protection Agency 2022). Despite the potential benefits, source reduction also has some disadvantages. It may require significant changes in production processes and infrastructure, which can be costly and time-consuming (Williams 1994). Moreover, some companies may be resistant to implementing source reduction measures, as they may perceive them as a threat to their profitability (U.S. Environmental Protection Agency 2022).

Opportunities for source reduction exist across various sectors, including the petroleum industry (Odeyinka et al. 2016). For instance, petroleum refineries can reduce waste generation by implementing process modifications, such as optimizing production scheduling, improving material handling practices, and recycling process streams (Odeyinka et al. 2016). In addition, the use of cleaner production technologies and the substitution of hazardous chemicals with less harmful alternatives can also contribute to waste reduction (Odeyinka et al. 2016).

Source reduction has a long history, dating back to the 1970s when the concept of waste minimization was first introduced (Williams 1994). Since then, the technique has evolved, and various strategies and tools have been developed to support its implementation, such as the development of pollution prevention plans and the use of life cycle assessments (U.S. Environmental Protection Agency 2022). Source reduction is a waste management technique that aims to reduce waste generation at its source by modifying production processes, substituting hazardous materials with less harmful alternatives, and optimizing material handling practices (Williams 1994). While it has many advantages, such as reducing environmental impacts and costs associated with waste management, it may also require significant changes in production processes and infrastructure (Williams 1994). Nonetheless, opportunities for source reduction exist across various sectors, including the petroleum industry (Odeyinka et al. 2016), and their implementation can contribute to enhancing the overall sustainability of production processes (U.S. Environmental Protection Agency 2022).

3.5.1.2 Reuse and Recycling

The petroleum industry has been implementing reuse and recycling techniques for several decades to reduce waste generation and minimize environmental impacts. One of the advantages of these techniques is that they help conserve energy and natural resources, which is essential for sustainable development. In addition, reuse and recycling can result in cost savings for the industry by reducing the need for disposal and raw material acquisition. Saleem et al. (2014) conducted a study on the recycling of waste lubricating oil and found that the reuse of waste oil as a lubricant resulted in better engine performance and lower environmental impacts compared to the use of virgin oil. The authors also highlighted the importance of proper treatment and processing of waste oil to ensure that it meets the required quality standards for reuse.

The U.S. Environmental Protection Agency (EPA) has also recognized the importance of reuse and recycling in the petroleum industry and has developed guidelines and recommendations for waste minimization (EPA 1995). The agency encourages the industry to adopt reuse and recycling practices to reduce waste generation and promote sustainability. Abdallah et al. (2015) provided an overview of the recycling of petroleum waste and discussed various techniques that can be used to recycle waste materials such as spent catalysts, sludges, and lubricating oils. The authors highlighted the importance of proper waste management practices, including segregation, storage, and transportation, to ensure that waste materials can be recycled safely and effectively. While there are many benefits to reuse and recycling in the petroleum industry, there are also some challenges that need to be addressed. For example, the quality of recycled materials may not always be as high as that of virgin materials, and some waste materials may not be suitable for reuse or recycling. In addition, the costs associated with implementing reuse and recycling programs can be high, and regulatory compliance can be complex.

In conclusion, reuse and recycling are important waste management techniques that can help the petroleum industry reduce waste generation and minimize environmental impacts. The industry has a long history of implementing these techniques, and there are numerous opportunities for further implementation. However, to ensure the success of these programs, proper waste management practices must be followed, and the industry must continue to invest in research and development to improve the quality and efficiency of recycling processes.

3.5.1.3 Composting

Petroleum waste is a major environmental concern due to its potential impact on soil, water, and air quality (Lu et al. 2018; Wong et al. 2017). However, composting is a sustainable solution for managing petroleum waste (Mukherjee et al. 2013). Composting is a natural process in which organic materials are decomposed and transformed into a stable product, called compost. This process can also be utilized for managing petroleum waste, including sludge and contaminated soil. Composting petroleum waste involves the addition of organic materials, such as yard waste or food waste, to the waste material. The mixture is then subjected to controlled aerobic conditions that facilitate the biodegradation of petroleum hydrocarbons by microorganisms (Lu et al. 2018). The resulting compost can then be used as a soil amendment or fertilizer. Composting offers several advantages over traditional waste management methods, including reducing waste volume, producing a valuable product, and reducing environmental impacts. Also, composting can reduce the volume and weight of petroleum waste by up to 50% and 80%, respectively, and can also reduce the concentration of petroleum hydrocarbons in the waste material by up to 90% (Mukherjee et al. 2013). The resulting compost can be used to improve soil quality and increase crop yields, making it a valuable resource for agriculture and horticulture. Composting is also a low-cost and environmentally friendly method of waste management.

However, there are some disadvantages to composting. Composting requires a large space to accommodate the waste material and the composting process. It can take several months to complete and requires continuous monitoring and management to ensure optimal conditions for decomposition (Wong et al. 2017). Additionally, composting may not be suitable for all types of petroleum waste, such as waste with high levels of heavy metals or other contaminants. Despite these challenges, composting can be integrated into existing waste management systems to provide a sustainable solution for petroleum waste management. It can also be used in conjunction with other waste treatment technologies, such as bioremediation or thermal treatment, to achieve a more comprehensive waste management solution. Composting can create new economic opportunities, such as the production of high-quality compost for agricultural or horticultural use. The use of composting for managing petroleum waste is a relatively new field, but it has gained interest in recent years due to the increasing need for sustainable waste management practices. Studies have shown that composting can effectively reduce the concentration of petroleum hydrocarbons in waste materials and produce a stable, valuable product (Lu et al. 2018; Mukherjee et al. 2013). Research is ongoing to optimize composting processes for petroleum waste management.

In conclusion, composting is a sustainable and effective solution for managing petroleum waste. It offers numerous advantages over traditional waste management methods, including reducing waste volume, producing a valuable product, and reducing environmental impacts. However, it is important to carefully consider the type of waste being composted and to monitor the composting process to ensure optimal conditions for decomposition. Further research is needed to optimize composting processes for petroleum waste management and explore new economic opportunities associated with compost production.

3.5.1.4 Incineration

Petroleum waste poses a significant environmental threat due to its potential impact on soil, water, and air quality (Kumar et al. 2017). Incineration is one of the thermal treatment methods that can be used to manage petroleum waste (Nguyen et al. 2021). This chapter will discuss the advantages and disadvantages of incineration, its potential use for the treatment of certain types of petroleum waste, and recent research related to incineration for petroleum waste management. Incineration has several advantages over traditional waste management methods. It can destroy hazardous waste and reduce its volume, making it easier to dispose of (Wong et al. 2017). Additionally, incineration can produce energy that can be used to power industrial processes or generate electricity (Kumar et al. 2017). This makes incineration a potentially cost-effective and sustainable waste management option.

Despite its advantages, incineration also has some significant disadvantages. One major concern is the potential release of pollutants into the air during the incineration process. This can include dioxins and heavy metals, which can have harmful effects on human health and the environment (Zhang et al. 2020). The cost of operating and maintaining an incinerator can also be high (Wong et al. 2017). Incineration can be a useful method for the treatment of certain types of petroleum waste, such as oily sludge and contaminated soil (Nguyen et al. 2021). It can also be used to treat waste generated during the cleanup of oil spills (Kumar et al. 2017). In addition, advances in technology have made incineration more efficient and effective at treating waste (Lu et al. 2018). One study published in the Journal of Hazardous Materials evaluated the use of incineration for the treatment of petroleum sludge. The results showed that incineration was effective at reducing the volume and toxicity of the sludge and that the treated waste met regulatory requirements for disposal (Zhang et al. 2020). However, the study also noted that careful monitoring of the incineration process is necessary to ensure that pollutants are not released into the air.

Incineration can be a useful method for the treatment of certain types of petroleum waste, but its use should be carefully evaluated based on the specific waste stream and environmental regulations in place. While advances in technology have made incineration more efficient and effective, concerns about air pollution and the cost of operating and maintaining an incinerator remain. Future research should focus on optimizing incineration processes for petroleum waste management and exploring new opportunities for energy recovery.

3.5.1.5 Landfilling

Landfilling is a widely used method for the disposal of non-hazardous solid wastes, including petroleum waste, that cannot be reused, recycled, or incinerated. This technique involves burying the waste in a specially designed landfill site, where it undergoes natural biodegradation. The landfill is typically lined with impermeable barriers to prevent contamination of the surrounding soil and water (Shafaei et al. 2021a, b). The advantages of landfilling include its cost-effectiveness, low energy requirements, and ability to handle large volumes of waste over a long period. Additionally, landfills can capture and use methane gas produced by the decomposing waste to generate electricity (Rahman et al. 2018).

However, there are also several disadvantages associated with landfilling. Landfills require a significant amount of land, which can be a problem in densely populated areas. If not properly managed, landfills can potentially pollute the surrounding soil and water. Some petroleum waste may take a long time to decompose in landfills, which means the waste will remain in the landfill for a long period. Finally, decomposing waste can produce unpleasant odors that can be a nuisance to nearby residents (Olanipekun et al. 2019). There are opportunities to improve landfilling as a petroleum waste management technique. Advances in technology can lead to better management of landfill sites to reduce the potential for environmental pollution. Additionally, recycling and source reduction can reduce the volume of waste that needs to be disposed of in landfills (Rahman et al. 2018).

Landfills have been used for waste disposal for centuries, with the first modern sanitary landfill established in the United States in 1937. Since then, landfills have become a common method of waste disposal around the world (Shafaei et al. 2021a, b). Landfilling is an effective method for the disposal of non-hazardous solid wastes, including petroleum waste, but it has both advantages and disadvantages. With advancements in technology and an emphasis on recycling and source reduction, landfilling can continue to be a valuable tool in petroleum waste management.

3.5.1.6 Stabilization and Solidification

Stabilization and solidification are techniques commonly used to treat petroleum waste by immobilizing contaminants within a solid matrix (Wu et al. 2019a, b). The process involves adding binders or reagents to the waste to stabilize the contaminants, reduce their leachability, and then solidify the waste into a stable and durable material (Rezaei et al. 2017a, b). Stabilization and solidification can be done in situ, where the waste is treated in place, or ex-situ, where the waste is removed and treated offsite.

One of the advantages of Stabilization and solidification is the reduced leachability of contaminants, which is achieved by stabilizing the waste and reducing the mobility of contaminants (Khan et al. 2004). In addition, the Stabilization and solidification processes improve the physical and chemical stability of the waste, resulting in a reduction in waste volume. The final product can be used as a construction material or as a barrier layer in landfills. However, the process has some disadvantages, such as a high cost compared to other treatment methods, a limited ability to treat volatile contaminants, and the fact that the treated waste may not be suitable for all types of reuse (Rezaei et al. 2017a, b). The long-term performance and durability of the solidified waste also need to be evaluated.

Stabilization and solidification have been found to be effective in treating various types of petroleum wastes, including oily sludge, drilling waste, and refinery waste. The process has been shown to reduce the leachability of contaminants such as heavy metals, polycyclic aromatic hydrocarbons (PAHs), and volatile organic compounds (VOCs) from the waste (Rezaei et al. 2017a, b). The use of industrial by-products as binders, such as fly ash or cement kiln dust, can reduce costs and improve the sustainability of the Stabilization and solidification processes (Wu et al. 2019a, b).

Additionally, the use of nanomaterials as binders may increase the effectiveness of the process. Stabilization and solidification have been used for petroleum waste management since the 1980s, and the technique has been continuously developed and improved since then (Khan et al. 2004). Recent studies have focused on improving the performance and durability of the solidified waste and evaluating the long-term environmental impact of the process.

3.5.1.7 Deep Well Injection

Deep well injection involves injecting waste materials deep into the ground, where they are stored in natural geological formations. This technique is commonly used for hazardous liquid wastes that cannot be treated by other methods. Deep well injection (DWI) is a technique for disposing of liquid waste in underground formations. This technique involves the injection of waste into deep geological formations, such as rock strata or porous zones, that are located below the groundwater table. The waste is injected through a borehole into a designated zone and is then allowed to disperse and mix with the surrounding rock or soil, where it is expected to be naturally degraded over time.

Advantages of DWI include its ability to safely and permanently dispose of large volumes of liquid waste, as well as its ability to prevent the release of hazardous chemicals into the environment. It also requires a relatively small land area and is generally considered to be a cost-effective method of waste disposal. However, there are also several disadvantages associated with DWI, including the potential for contamination of groundwater and surface water resources, as well as the risk of induced seismic activity. There is also the risk of failure of the injection well casing or the formation itself, leading to the escape of waste fluids into the environment. Additionally, there is public opposition to DWI because of the potential for harm to the environment and human health.

Despite these drawbacks, DWI remains a widely used method for the disposal of liquid waste, including petroleum waste. The majority of petroleum waste generated in the United States is disposed of by deep well injection. DWI is also used for the disposal of produced water generated from oil and gas extraction, as well as for the disposal of other hazardous wastes (Hoda et al. 2017).

In conclusion, deep well injection is a technique that is widely used for the disposal of petroleum waste and other liquid wastes. While it has several advantages, including its ability to dispose of large volumes of waste safely and permanently, there are also several environmental and health concerns associated with its use. Various research papers have been published on petroleum waste management techniques. For instance, the paper by Ayinla et al. (2019a, b) provides a comprehensive review of petroleum refinery waste management techniques.. Rezaei et al. (2017a, b) discuss various treatment techniques for petroleum wastes, including stabilization and solidification, incineration, and landfilling. Wu et al. (2019a, b) review the treatment of petroleum refinery waste sludge by stabilization and solidification. These papers and others provide valuable insights into the various techniques and their effectiveness in managing petroleum waste.

3.6 Conclusion and Future Directions

The petroleum industry generates a significant amount of waste, including solid waste, wastewater, and air emissions, which pose a potential threat to the environment and public health. There are several waste management techniques available for petroleum waste, including source reduction, reuse and recycling, composting, incineration, landfilling, stabilization and solidification, deep well injection, and reduction of hydrocarbon emissions. Each technique has its advantages and disadvantages, and the selection of a suitable technique depends on the type and characteristics of the waste and environmental and economic factors.

In India, there are various rules and national missions in place for waste management, including the Hazardous and Other Wastes (Management and Transboundary Movement) Rules, 2016, and the Swachh Bharat Abhiyan. However, there is still a need for better implementation and enforcement of these rules and regulations. In recent years, there has been a trend towards transforming petroleum waste into useful biofuels, which presents an opportunity for waste reduction and resource recovery. Several research studies and projects in India have shown promising results in this area, including the production of biodiesel, bioethanol, and biogas from petroleum waste.

Overall, effective waste management practices and innovative technologies are essential to minimize the environmental impact of petroleum waste and ensure sustainable development in the petroleum industry. There are several possible directions that India can take in managing petroleum waste and transforming it into useful biofuels. These include:

i. Implementing effective waste management strategies: India can continue to focus on implementing effective waste management strategies, including reducing waste generation, promoting reuse and recycling, and using advanced treatment technologies such as composting, incineration, landfilling, stabilization, and solidification.
ii. Promoting the use of biofuels: India can focus on promoting the use of biofuels as an alternative to traditional fossil fuels. This can be achieved by setting targets for biofuel production and consumption, providing financial incentives for biofuel production, and investing in research and development to improve the efficiency and cost-effectiveness of biofuel production technologies.
iii. Encouraging private sector investment: India can encourage private sector investment in the development of biofuel production infrastructure and technology. This can be achieved by providing tax incentives and subsidies, reducing regulatory barriers, and promoting public–private partnerships.

iv. Collaborating with international partners: India can collaborate with international partners to share knowledge and best practices in petroleum waste management and biofuel production. This can include partnerships with countries that have already established successful biofuel production industries, as well as collaboration with international organizations such as the United Nations and the World Bank.
v. Overall, it is clear that there are significant opportunities for India to transform petroleum waste into useful biofuels. By implementing effective waste management strategies and promoting the development of biofuel production infrastructure, India can reduce its dependence on traditional fossil fuels, promote sustainable development, and contribute to global efforts to address climate change.

References

Abdallah M et al (2015) Recycling of petroleum waste. In: Singh S, Singhania R, Larroche C (eds) Waste biorefinery: potential and perspectives, pp 15–43
Abdulredha MH et al (2020) A comprehensive review of the enhanced oil recovery (EOR) methods in the petroleum industry. J Pet Sci Eng 195:107605
Adeosun TA, Odeyemi OO et al (2021) Petroleum hydrocarbon pollution control and remediation measures: a review. Sustain Chem Pharm 22:100471
Almuktar SA, Dawood AS et al (2020) Impact of a refinery wastewater discharge on the water quality of the receiving bodies. Water Sci Technol 82(6):1137–1146
Ayinla WA et al (2019) Petroleum refinery waste management: a review. J Clean Prod 235:1176–1194
Ayinla WK, Ojewumi ME, Bello IA et al (2019) Petroleum refinery waste management: a review J Clean Prod 214:62–75
Chakravarty S, Yoo CK, Lee JH et al (2020) Flare gas recovery from offshore platforms: a review. J Clean Prod 257:120452
Chen J, Huang C, Liu Y et al (2021) Overview of biofuels in China: current status, policies, and prospects. Renew Energy 168:1047–1061
Chibueze P et al (2020) Heavy metal contamination of soils around an oil refinery in Nigeria. Environ Sci Pollut Res 27(17):21906–21916
Dhanapal P et al (2019) Environmental impacts of petroleum refining and petrochemical industries: a review. Int J Chem Eng 2019:1–13
Government of India (n.d.) Swachh Bharat Abhiyan
Hoda N, Saleh H et al (2017) A review of petroleum waste generation and management practices. J Environ Manag 194:208–222
Huang YJ et al (2015a) Petroleum refinery waste: a review of its characteristics and management alternatives. Process Saf Environ Prot 98:234–249
Huang YJ et al (2015b) Petroleum refinery waste: a review of its characteristics and management alternatives. Process Saf Environ Prot 98:358–368
index at www.iogp.org. (n.d.). https://www.iogp.org/
Indian Institute of Technology, Bombay (2019) IIT Bombay develops low-cost processes to convert industrial and agricultural waste into biodiesel, 14 February 2019
Indian Oil Corporation. “IndianOil Launches Indigenous Technology for Converting Plastic Waste to Diesel.” Press release, 14 June 2018

International Association of Oil and Gas Producers (IOGP), the American Petroleum Institute (API), and the U.S. Environmental Protection Agency (EPA)
International Energy Agency (2018) Bioenergy
International Energy Agency (2021) India. https://www.iea.org/countries/india
Kaur P, Sarma AK et al (2021) Refinery sludge management by pyrolysis for bio-oil production: a review. J Environ Manag 280:111681
Khan FI, Husain T et al (2004) An overview and analysis of site remediation technologies. J Environ Manag 71(2):95–122
Kumar A et al (2017) Petroleum waste management: incineration as an alternative. Int J Curr Microbiol Appl Sci 6(11):1319–1325
Kumar S, Kumar N, Tyagi VK et al (2020) Gasification of petroleum waste: a review. Energy Rep 6:182–189
Li X, Liu S, Liu Y, Zhang J et al (2020) Recovery of heavy oil from oil sands: a review. J Pet Sci Eng 187:106870
Liu J, Li Y, Zhao X et al (2021) Selective catalytic reduction of NOx over Cu-zeolite catalysts: a review. J Environ Chem Eng 9(3):105421
Lopez-Sotelo JB, Serna-Guerrero R et al (2018) Catalytic technologies for hydrocarbon emissions abatement. Catalysts 8(12):576
Lu Q, He Q, Wang Y et al (2018) Petroleum sludge composting: microbial dynamics and their effects on sludge characteristics. J Hazard Mater 348:132–139
Ministry of Environment, Forestry, and Climate Change (2016) Hazardous and Other Wastes (Management and Transboundary Movement) Rules
Ministry of Environment, Forestry, and Climate Change (n.d.) Hazardous Waste Management
Ministry of Petroleum and Natural Gas (2019) Annual report 2018–19
Mohammadi L et al (2020) Petroleum hydrocarbon removal from wastewater: a review. Processes 8(4):1–36
Mukherjee A et al (2013) Organic carbon transformations in soil composites amended with petroleum hydrocarbons. Environ Sci Technol 47(19):11005–11012
Nasrullah M, Al-Ghamdi SG, Al-Mubaddel FS et al (2020) Petroleum sludge as a potential feedstock for renewable energy: a review. J Clean Prod 273:123079
National Mission for Clean Ganga (n.d.) About NMCG
Nguyen TAH et al (2021) A review of technologies for the treatment of oily sludge from the petroleum industry. J Clean Prod 295:126346
Odeyinka SM, Adetunji OR et al (2016) Waste minimization strategies in petroleum industry: a review. J Environ Manag 183(Pt. 3):416–428
Olanipekun AA et al (2019) Petroleum refinery waste characterization and management in Nigeria: a review. J Environ Prot Ecol 20(3):1215–1230
Pandey DS, Sudhakara Reddy M et al (2018) Pyrolysis: an effective approach for conversion of waste plastics into liquid fuel. Waste Manag 79:950–967
Payne JR et al (1991) Oil-weathering behavior in arctic environments. Polar Res 10(2):631–662
Rahman MS et al (2018) Petroleum waste management: an overview. Int J Environ Sci Technol 15(2):407–420
Rezaei SM et al (2017a) Treatment of petroleum refinery waste: a review. J Hazard Mater 339:257–271
Rezaei SM, Naddafi K, Jorfi S et al (2017b) Treatment of petroleum refinery waste: a review. J Hazard Mater 340:714–727
Saleem M, Riaz U, Ali M et al (2014) Recycling waste lubricating oil into chemical feedstock or fuel oil over iron oxide catalysts. J Environ Chem Eng 2(3):1513–1519
Seshadri K et al (2020) A review of emerging technologies for wastewater treatment and reuse in the petroleum industry. J Water Process Eng 38:101553
Shafaei HZ, Karimi A et al (2021a) Petroleum sludge treatment: a review of current treatment technologies. J Environ Chem Eng 9(4):105374

Shafaei H et al (2021b) Petroleum sludge treatment: a review of current treatment technologies. J Environ Chem Eng 9(6):106271
Singh R, Arya SK et al (2019) Waste to wealth: the potential of converting petroleum sludge to value-added products. J Environ Manag 232:73–86
The Hindu Business Line "HPCL and LanzaTech will build India's first waste-gas-to-ethanol plant", 25 May 2017
U.S. Environmental Protection Agency (1995) Waste Minimization National Plan: Guidance manual
U.S. Environmental Protection Agency (2021) Resource Conservation and Recovery Act (RCRA)
U.S. Environmental Protection Agency (2022) Source reduction, 31 March 2023
Williams PT (1994) Waste minimization. Chem Eng J 54(2):91–99
Wong JWC et al (2017) A review of technology for the treatment of oily sludge from petroleum industry: towards sustainability. J Environ Manag 192:240–257
Wu Y, Wu P, Liu Y, Zhu N et al (2019) Treatment of refinery waste sludge by stabilization/solidification. J Environ Manag 241:42–49
Wu Y et al (2019b) Treatment of refinery waste sludge by stabilization/solidification. J Environ Manag 237:246–253
Yang S, Yuan X et al (2021) Recovery of heavy oil and bitumen from oil sands and oil shale: a review. J Petrol Sci Eng 202:109–178
Zhang Y, Wang X, Zhang L et al (2020) Treatment of petroleum sludge by incineration. Energy Sources Part a: Recov Utiliz Environ Effects 42(19):2386–2397

Chapter 4
Adverse Effects of Petroleum Spillage on Marine Environment During Transport

G. Koteswara Reddy, D. Harika, and V. Meghana

Abstract Transportation of oil from the petroleum reservoirs to the purchaser sites includes a high possibility of oil leakage because of damages and disasters that take place at the time of shipping, along with quarrying, eventually, clash with the ecological community. The outcome of unexpected leakage of lubricants into the water bodies gives rise to consequences of unendurable extinction of flora and fauna which devastate aquatic habitat resources. The noxious quality of fuels is based on the attentiveness and configuration of the petroleum and the liability of the marine organisms. Spilled lubricants can harm marine organisms like sea birds because their compositions of these chemicals are very toxic. Oil can have huge influence on living beings in two ways: firstly, oil enters the living beings' body through inhalation which leads to the exposure of oil inside the body of organism, secondly it leads to the external exposure like skin and eye irritation. Since most oils float on the ocean's tropopause or on coastlines, creatures that live there are most affected, like sea otters and sea birds. Biological effects of oil spillage include, phytoplankton cannot grow in the oil-spilled water, the fall of fishery resources takes place, a threat to man through eating contaminated seafood, Damage to the habitat and a reproductive rate of marine organisms can decrease in the long run. In coastal waters, the impact will be on mangroves, seaweeds, and intertidal fauna. Daily tons of oil spills are bound to happen across the globe as well as in seas and rivers. Oil slicks include unintentional discharge of fuel varying from different types of petroleum to enormous, filtered products, from massive diligent oils to lower diligent, but very dangerous oils.

Keywords Ecological community · Oil slicks · Marine organisms

---

G. Koteswara Reddy (✉) · D. Harika · V. Meghana
Department of Biotechnology, Koneru Lakshmaiah Education Foundation (Deemed to be University), Green Fields, Vaddeswaram 522502, India
e-mail: koteswarareddy@kluniversity.in

I. D. Behera and A. P. Das (eds.), *Impact of Petroleum Waste on Environmental Pollution and its Sustainable Management Through Circular Economy*, Environmental Science and Engineering, https://doi.org/10.1007/978-3-031-48220-5_4

4.1 Introduction

Oil leakages are the dangerous discharge of fuels into the marine habitat, conventionally inside the aquatic nature, periodically eliminating the regions of vegetation, plants, and animal life. Fuel happens to be the major intimate contaminant present in the oceans. Over three thousand ideal tons of fuel spoils the ocean annually (Stevens et al. 2013). Particularly, the lubricant contamination in the marine habitat comes up through the ground. Drainage and dissipation from the metropolitan areas, industries, and streams carry fuel towards the sea. Yachts produce around a tertian of fuel pollution in the aquatic ecosystems once they clear the vessels and throw away the effluents (Dodds 2008). This is the adverse derivative of depositary, retention, and delivery of lubricants and fuels that causes infrequent discharge. Lubricant spillage is extremely burdensome to neaten.

The type of fuel leak we commonly expect is the **fortuitous** and intended discharge of crude oil derivatives through the oceanic habitats. This is an outcome of anthropogenic activities (breaking, creating, conserving, shipping, garbage administration), which lies on the peak of estuaries, a distinct accumulation brought through a breeze and streams. Fuel slicks will be moderately managed by synthetic dissemination, ignition, and surface assimilation, which has catastrophic consequences on the seaside habitats (Mohamed et al. 2017) (Fig. 4.1).

Indication of the fuel spillage would be the circumstances like eruption, channel breakage, ship crashing or sinking, garbage drawing from the boats, dripping of subterranean depository vessels, and oil-adulterated water overflow from roads or streets and parking areas in time of rain storms or typhoon (Pick 2012). The ocean fuel slick causes the hazardous outcome of coastal side lubricant penetrating and its

Fig. 4.1 The impact of lubricants on the aquatic ecosystem (hotcore.info)

marine shipment. Leak handle enterprises concentrate on the avoidance, segregation, anticipation, and tidying up of a factory's fuel leakage (Hough 2004).

4.2 Characteristics of Oil Spill

Most of the spillage of crude oil extracts and their substances within the oceans takes place during the time of shipping by oil tankers, filling and emptying operations, etc. (Downs 1997; Kanungo et al. 2023). when found in the marine environment, the lubricants proceed along different types of transfigurations, presuming in physical and natural actions start to work preferably after the lubricant or petroleum has spilt on the ocean. This involves dissipation, extending, emulsification, suspension, ocean-air interchange, and precipitation (Ahluwalia 2005).

4.3 Mechanisms for Oil Spill Damage

Fuel may affect a marine habitat by the following mechanisms:

- Substantial stifling having a clash in biological functions.
- Synthetic toxicity gives rise to virulent or non-virulent effects or causes the destruction of metabolic activities (Sylves 1998).
- Environmental conversions, principally the disappearance of essential organisms from an ecosystem and the acquisition of a home by dynamic organisms.
- Unintended consequences, namely the deprivation of habitats or protection and the subsequent removal of environmentally dominant species (Świerkosz et al. 2017).

The environmental nature and most of the results of lubricant spillage depends on the inclusive constituents. These encompass the amount and kind of fuel spill, its actions in the ecosystem, the site of the spill in terms of atmospheric circumstances and substantial features, and the timing, significantly in association with the season and frequent climatic conditions (Bill Freedman 1995).

4.4 The Behaviour of Oil Spills in the Aquatic Habitat

i. **Photo-oxidation**—Spontaneous daylight with the presence of O_2 can convert various fossil fuels into hydroxyl substances like acetaldehydes, methyl ethyl ketone body, and eventually to less molar mass carbonic acid gas, and the compounds are water-insoluble, which replace the analysable nature of leaks (Bill Freedman 1995).

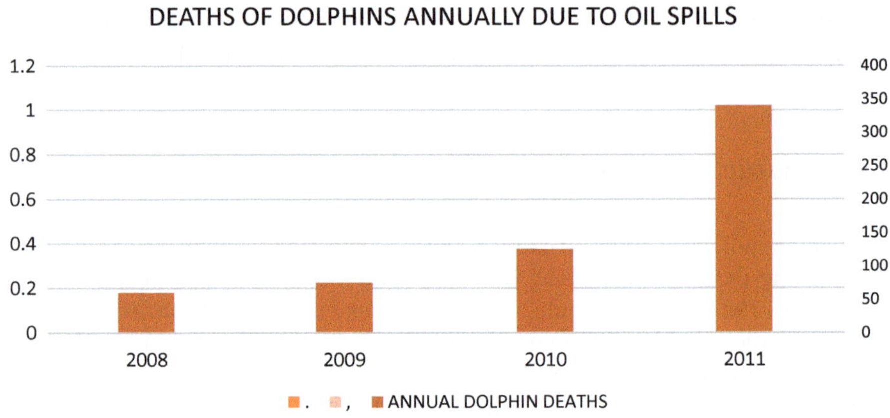

Fig. 4.2 The below bar graph shows the number of dolphins that die annually due to oil spills (Ahluwalia 2005)

ii. **Diffusion**—Diffusion is oil dispersed in water convergence that develops through the inclusion of tiny particles of lubricants into a water line. Lubricant starts to disperse straight away in exposure to H_2O, which is majorly immense within the starting of ten hours or so.

iii. **Disintegration**—Disintegration is an alternate physiological activity in which the less molar mass organic compounds, which are moderately disappear through the oil into the water lines (Bednarek 2001).

iv. **Degradation**—Utilising living microorganisms, biodegradation is an effective natural technique for removing petroleum hydrocarbon pollutants from terrestrial and aquatic environments (Behera et al. 2022). Bio-deterioration procedures determine the circumstances of petroleum in the marine ecosystem, including microbial deterioration, consumption by fauna, absorption by underwater invertebrates and vertebrates like herbivores and carnivores, etc. Along with reprecipitation, bacteria, microbes, and pathogens are efficient in disintegrating fossil fuel organic compounds and associated elements that are widely spread in the environment. The amount of microbial degeneration differs with the synthetic convolution of the crude, the bacterial community, and most habitat situations (Downs 1997). According to reports, a bioaugmentation method can increase biodegradation efficiency by performing particular microbial inoculations to remove different pollutants (Behera et al. 2020, 2021) (Fig. 4.2).

4.5 Major Oil Spills Related to Marine Environments

Facts examine that 32 lakh tonnes of petroleum per annum are being liberated from all roots into the marine surroundings (Block 1992). The bulk quantity of petroleum spillage is because of popular transportation and commercial hard

Table 4.1 The oil leakages that occurred from 1969 through 2015 (Dodds 2008)

S.no	Years	7–700 tonnes	Greater than 700 tonnes	Spilled amount*10^3 times
1	1970–1974	189	125	1114
2	1975–1979	342	117	2012
3	1980–1984	221	41	570
4	1985–1989	124	48	513
5	1990–1994	165	48	907
6	1995–1999	108	25	194
7	2000–2009	182	–	213
8	2010–2013	28	–	22

work. At the time of the war between Iran and Iraq, which happened from 1980 to 1988, around 20 tonnes of fuel were launched into the Persian Gulf Sea (Saadoun 1995). In January 1969, a petroleum platform eruption took place in Santa Barbara Channel, California; spillages occurred of approximately 151.4 lakh litres of unrefined petroleum over the period of 288 h as workers hustled to stop the flow. The U.S. Coast Defence launched records from this incident as lessons found out using the U.S. from the Torrey Canyon easy-up within the shape of a record and submitted it to the branch of Transportation. This record is taken into consideration as an essential first step when it comes to enhancing the Coast Shield's oil spill reaction skills (Pierzynski et al. 2005).

From the Table 4.1, it is observed that oil spills during marine transportation decreased over the years during the time of industrialisation, many more oil spills took place the present time. That is because of the advancement of technology, and more safety precautions are much higher than in the past (Bahrain).

Table 4.2, it is shown that Atlantic Empress 1979 spilt around 287 * 10^6 kg of unrefined lubricants in off Tobago, West Indies (SEAANDJOB 2021). This ship spilt oil because of a collision that took place between the Atlantic Express, which was about 270,000 tons of oil and the Aegean Caption, which was about 200,000 tons of crude oil (Rafferty 2009).

4.5.1 Massive Petroleum Leakages in History

- Oil spill during Operation Desert Shield in 1991
- Gulf of Mexico oil spill in 2010
- Ixtoc 1 blowout in 1979
- SS Atlantic Empress oil spill in 1979
- Fergana Valley spillage in 1992
- Russian Federation petroleum spillage in 1994 (Sea III).

Table 4.2 Significant petroleum leakage from 1965 (Slovakia)

Ship name	The year it took place	The region where the oil spillage took place	Amount of spill in (10^3) tonnes
Prestige	2002	Off Galicia, Spain	63
Khark 5	1989	120 nautical. miles off the Atlantic coast of Morocco	70
Odyssey	1988	700 nautical. miles off Nova Scotia, Canada	132
Braer	1993	Shetland Islands, UK	85
Torrey Canyon	1967	Scilly Isles, UK	119
Novo	1985	Off Kharg Island, Iran	70
Independent	1979	Bosphorus, Turkey	94
Aegean Sea	1992	La Coruna, Spain	74
Exxon Valdez	1989	Prince William Sound, Alaska, USA	37
Sea Star	1972	Gulf of Oman	115
Irenes Serenade	1980	Navarino Bay, Greece	100
Atlantic Empress	1979	Off Tobago, West Indies	287
Sea Empress	1996	Milford Haven, UK	72
Hawaiian Patriot	1977	300 nautical. miles off Honolulu	95
Katina P	1992	Off Maputo, Mozambique	67
Jakob Maersk	1975	Oporto, Portugal	88
ABT Summer	1991	700 nautical miles off Angola	260
Haven	1991	Genoa, Italy	144
Amoco Cadiz	1978	Off Brittany, France	223
Urquiola	1976	La Coruna, Spain	100
Hebel Spirit	2007	South Korea	11
Castillo de Bellver	1983	Off Saldanha Bay, South Africa	252

Below, Fig. 4.3 shows the influence of oil spills on sea birds. The main influence of oil spills on sea birds is that it is the cause of migration, decreased population size, indirect mortality, and reduced habitat occupancy (National Research Council US) (Fig. 4.4) (Gaur et al. 2018).

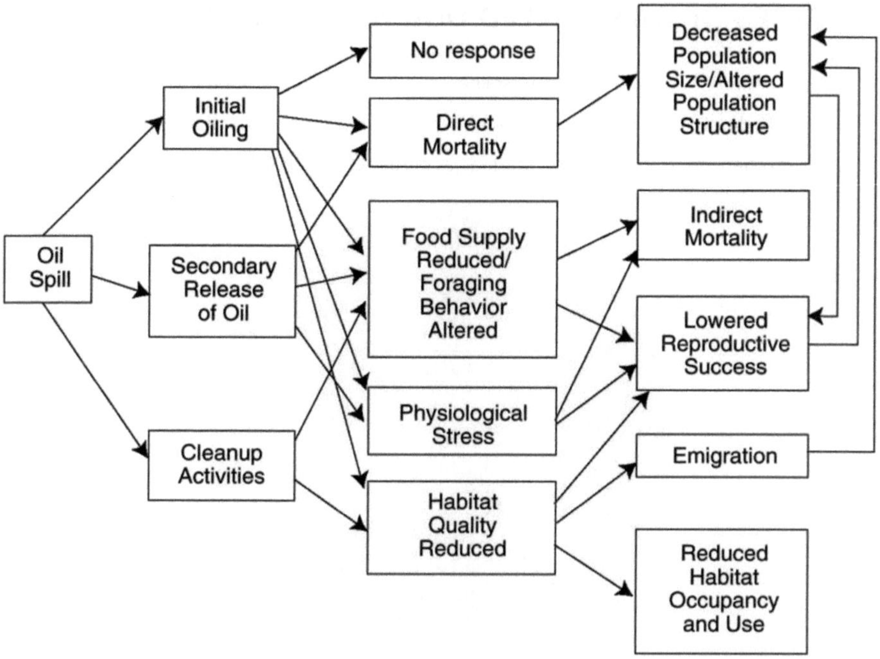

Fig. 4.3 Influence of oil spills on sea birds (Votier et al. 2008)

4.5.2 Cons of Oil Spills in the Marine Environment

- Even though petroleum makes life easier, it also causes many negative things to the environment. Marine transport plays a vital role in the global trade market (Cheng) (Little et al. 2021, Short 2003).
- Producing and transporting crude oils from one place to another has the most negative effects on the environment. Exploring and drilling for oil disrupts the land marine ecosystem (Schmidheiny 1992).
- It was reported that about 1% enlargement in spill size has been evaluated to increase the damage by some 3% (Zhang et al. 2019).
- Further, oil spills contaminate and damage aqua supplies, decreasing the purity of the aqueous habitat of the marine ecosystem (Harrison 2001).
- Oil spills will destroy fisheries, tourism industries, and other businesses that depend on a healthy marine environment.
- Economic consequences may occur to exceed the costs of cleaning the area where the oil spill took place and environmental remediation (Mohnta et al. 2023). This will negatively influence the local neighbourhood and economies for years to come.

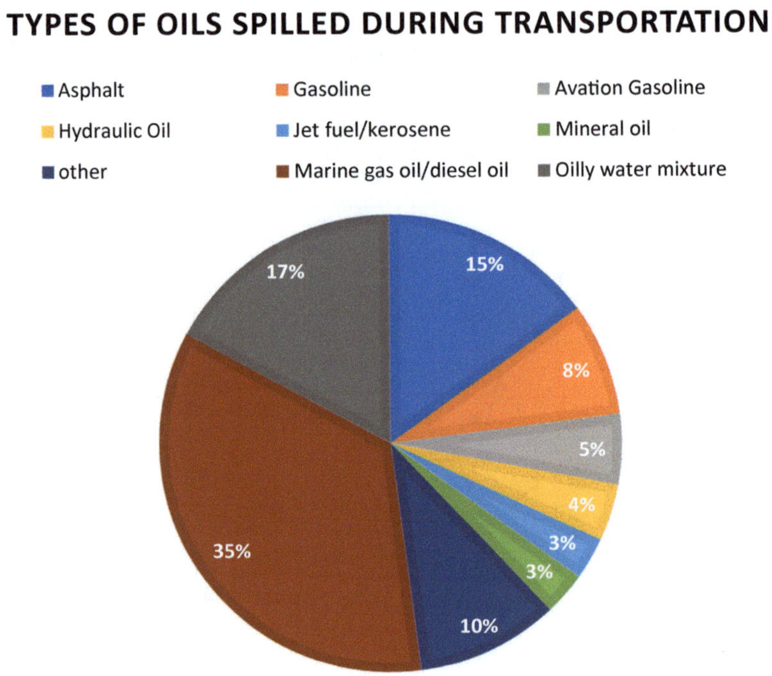

Fig. 4.4 Types of oil spills during transportation

- Petroleum is composed of various poisonous additives. Those additives can cause motive risky fitness issues like coronary heart damage, stunted growth, immune system consequences, and even loss of life in marine animals (Agarwal 2021).
- When the affected animals are eaten up by human beings, they are also affected by various health problems (Jaishankar et al. 2014).
- Fouling occurs when the lubricant bodily harms a plant or animal (Ismail et al. 2020).
- When petroleum is covered onto the bird's wings, it is very difficult or impossible for the birds to fly or to cleanse the *Enhydra lutris* skin; this also increases the vulnerability of birds to cold and frostbite.
- The extent of leaked oils can influence the survival prospects of marine organisms.
- Further, those involved in marine transport need to prioritise environmental stewardship and sustainability practices (Salafsky et al. 2008).
- These efforts are necessary to ensure a safer and more sustainable future. Effective measures and policies that can prevent or mitigate the impact of oil spills should always be implemented to address these environmental and economic challenges.

4.6 The Hazards of Oil Spillages: Present, Past, and Later

As increasing follow-up research after huge oil had been proclaimed, an impact was detected. On several rocky shorelines and in surface waters, large quantities of Chlorella were produced within a couple of months or maybe a year after the spillage (Moore et al. 2004), for example, within the after-effects of the SS. Torrey Canyon accident. The motive for the incident became misapprehended in a maximum of the early news articles, as the circumstance featured is about uncontrolled nutritive load, artificial, or as the outcome of a substantial rookery of seabirds (White and Baker 1998). Hence, it was frequently considered that enriched food had been released throughout the bio or photochemical deterioration of the fuel (Alexander 2018).

The reason for this is unique. Metaldehyde, abalone, and different types of algae grazers. Among them, the chlorella has a span of unrestricted boom. While the grazers reappeared, in a matter of abalones after an enormous lubricant leak may take time, the area that the algae have been covered has been so broad and outspread that it took a very long time for the absolute stability to renew (Zatzman 2012). For most of the animals, like mussels and lobsters that have the alga pupae, finding chlorella again has become a challenging task.

The amount of offshore drilling, transport, and storage has reached record highs due to the rising need for oil as a critical energy source (Hansen et al. 2016). Unfortunately, this has also contributed to an increase in oil spills (Economides and Wood 2009), which have negative impacts on the habitat and the financial system through a variety of measures, the inclusion of adoption of safer technologies and practices, routine equipment inspections and maintenance checks, and stronger rules for offshore drilling activities, efforts are being undertaken to prevent future oil leaks. Also, efforts are being made to actively explore the experimentation and enlargement of the latest technology to address the problem of oil spills (Burgherr 2007).

For instance, efforts are being made to increase the efficiency of lubricant spillage reaction and tidying works (OSRC), particularly by using specialised tools and methods, including skimming, sorbents, bioremediation, and dispersants. The use of autonomous and remotely piloted vehicles for oil spill monitoring and response is also gaining popularity. These initiatives are essential to reducing the potential harm that oil spills could bring to people and the environment and to lessen their financial effects on the areas they may touch (Heslin and Ochoa 2008).

4.7 Conclusion

The effect of petroleum spillage on the aquatic ecosystem and earth is very long-lasting. Therefore, it is very important to take immediate preventive measures and actions to mitigate the environmental, social, and economic damages caused by oil spills. It is also essential to continue creating and implementing new technologies and methods for preventing oil spill clean-up. Adventitious or planned releases and discharge of oil from ships, in particular tankers, offshore structures, and pipelines, are the maximum glaring and seen due to the pollutants of oil within the marine surroundings. Diverse herbal, bodily, chemical, and organic deterioration techniques are the reason for oil spills inside the marine ecosystem. Oil spill mitigation and prevention require a multifaceted strategy that includes a variety of actions. One of the most important tactics is to create more dependable and efficient technology, such as double-hull tankers and pipeline leak detection systems, for stopping and preventing oil spills. The oil and maritime firms should be subject to more vital rules, and those accountable for oil spill events should face stiffer punishments.

In conclusion, lubricant spillage causes a severe impact to the environment, its ecosystems, and its inhabitants in the maritime environment. Governments, businesses, and those involved in the transportation and distribution of oil must seriously consider and strictly implement measures to minimise and mitigate these effects. We can lessen the effects of oil spills and lower the likelihood that they will happen again by putting appropriate precautions in place.

References

Agarwal M (2021) 10 methods for oil spill clean-up at sea marine environment

Ahluwalia VK (2005) Environmental pollution, and health. The Energy and Resources Institute (TERI)

Alexander D (2018) Natural disasters. Routledge

Atlantic empress oil spill by SEAANDJOB, 31 December 2021 (2021)

Bednarek AT (2001) Undamming rivers: a review of the ecological impacts of dam removal. Environ Manag 27:803–814

Behera ID, Basak G, Kumar RR, Sen R, Meikap BC (2020) Treatment of petroleum refinery sludge by petroleum degrading bacterium Stenotrophomonas pavanii IRB19 as an efficient novel technology. J Environ Sci Health Part A 1–13. https://doi.org/10.1080/10934529.2020.1866924

Behera ID, Nayak M, Biswas S, Meikap BC, Sen R (2021) Enhanced biodegradation of total petroleum hydrocarbons by implementing a novel two-step bioaugmentation strategy using indigenous bacterial consortium. J Environ Manag 292:112746. https://doi.org/10.1016/j.jenvman.2021.112746

Behera ID, Nayak M, Mishra A, Meikap BC, Sen R (2022) Strategic implementation of integrated bioaugmentation and biostimulation for efficient mitigation of petroleum hydrocarbon pollutants from terrestrial and aquatic environments. Mar Pollut Bull 177:113492. https://doi.org/10.1016/j.marpolbul.2022.113492

Bill Freedman (1995) In: Environmental ecology, 2nd edn. Oil pollution

Block E (1992) The organosulfur chemistry of the genus allium–implications for the organic chemistry of sulphur. Angew Chem Int Ed Engl 31(9):1135–1178

Burgherr P (2007) In-depth analysis of accidental oil spills from tankers in the context of global spill trends from all sources. J Hazard Mater 140(1–2):245–256

Canyon T (2019) The world's first major oil tanker disaster by the editorial team, in accidents, maritime knowledge, pollution

Cheng J, Microplastics in the marine environment: an ecotoxicological perspective. Doctoral dissertation, Sorbonne University

Dodds WK (2008) Humanity's footprint. Momentum, impact, and our global environment. Columbia University Press

Downs GW (1997) Enforcement and the evolution of cooperation. Mich J Int 19:319
Economides MJ, Wood DA (2009) The state of natural gas. J Nat Gas Sci Eng 1(1–2):1–3
Gaur N, Narasimhulu K, PydiSetty Y (2018) Recent advances in the bio-remediation of persistent organic pollutants and its effect on the environment. J Clean Prod 10(198):1602–1631
Hansen AM, Kraus TE, Pellerin BA, Fleck JA, Downing BD, Bergamaschi BA (2016) Optical properties of dissolved organic matter (DOM): effects of biological and photolytic degradation. Limnol Oceanogr 61(3):1015–1032
Harrison RM (ed) (2001) Pollution: causes, effects, and control. Royal Society of Chemistry
Heslin PA, Ochoa JD (2008) Understanding and developing strategic corporate social responsibility. Organ Dyn 37:125–144
Hough M (2004) Cities and natural process. Routledge. https://hotcore.info/babki/oil-spill.htm
Ismail NH, Salleh WN, Ismail AF, Hasbullah H, Yusof N, Aziz F, Jaafar J (2020) Hydrophilic polymer-based membrane for oily wastewater treatment: a review. Sep Purif Technol 15(233):116007
Jaishankar M, Tseten T, Anbalagan N, Mathew BB, Beeregowda KN (2014) Toxicity, mechanism and health effects of some heavy metals. Interdiscip Toxicol 7(2):60–72. https://doi.org/10.2478/intox-2014-0009
Kanungo J, Sahoo S, Bal M, Behera ID (2023) Performance of bioremediation strategy in waste lubricating oil pollutants: a review. Geomicrobiol J. https://doi.org/10.1080/01490451.2023.2245395
Little DI, Sheppard SR, Hulme D (2021) A perspective on oil spills: what we should have learned about global warming. Ocean Coast Manag 1(202):105509
Mohamed AM, Paleologos EK (2017) Fundamentals of environmental engineering: understanding soil, water, and pollutant interaction and transport. Butterworth-Heinemann
Mohanta S, Pradhan B, Behera ID (2023) Impact and remediation of petroleum hydrocarbon pollutants on agricultural land: a review. Geomicrobiol J 1–5. https://doi.org/10.1080/01490451.2023.2243925
Moore MN, Depledge MH, Readman JW, Leonard DP (2004) An integrated biomarker-based strategy for ecotoxicological evaluation of risk in environmental management. Mutat Res/fundam Mol Mech Mutagen 552(1–2):247–268
National Oceanic and Atmospheric Administration Government agency
National Research Council (US) Committee on oil in the sea: inputs, fates, and effects
Oil in the sea III: inputs, fates, and effects
Pick D (2012) Rome or death. The obsessions of general garibaldi. Random House
Pierzynski GM, Vance GF, Sims JT (2005) Soils and environmental quality. CRC Press
Producing and transporting crude oils from one place to another has the most negative effects on the environment. Exploring and drilling for oil disrupts the land marine ecosystem
Rafferty JP (2009) 9 of the biggest oil spills in history
Rowse A, Feasibility study for commercial underwater marine viewing facilities in Western Australian marine conservation reserves
Saadoun IM (1995) Impact of oil spills on marine life. Emerging pollutants in the environment-current and further implications. In: Environmental ecology, vol 10, 2nd edn, 2 September 2015. Oil pollution. Bill Freedman, p 60455
Salafsky N, Salzer D, Stattersfield AJ, Hilton-Taylor CR, Neugarten R, Butchart SH, Collen BE, Cox N, Master LL, O'CONNOR SH, Wilkie D (2008) A standard lexicon for biodiversity conservation: unified classifications of threats and actions. Conserv Biol 22(4):897–911
Schmidheiny S (1992) Changing course: a global business perspective on development and the environment. MIT Press
Short J (2003) Long-term effects of crude oil on developing fish: lessons from the Exxon Valdez oil spill. Energy Sources 25(6):509–517
Stevens P, Kooroshy J, Lahn G, Lee B (2013) Conflict and coexistence in the extractive industries. Chatham House, London

Świerkosz K, Reczyńska K, Kuras I (2017) Increasing area of deciduous forest communities Querco-Fagetea Classas an unintended effect of regular forestry management-a study from Central Europe. Pol J Environ Stud 26(1)

Sylves RT (1998) How the Exxon Valdez disaster changed America's oil spill emergency management. Int J Mass Emerg & Disasters 16(1):13–43

The National Oceanic and Atmospheric Administration is a Washington, D.C

Understanding Oil Spills and Oil Spill Response (1999) By United States Office of Emergency EPA 540-K-99-007 Environmental Protection and Remedial OSWER 9200.5-104A Agency Response PB2000-963401, December 1999

Votier SC, Birkhead TR, Oro D, Trinder M, Grantham MJ, Clark JA, McCleery RH, Hatchwell BJ (2008) Recruitment and survival of immature seabirds in relation to oil spills and climate variability. J Anim Ecol 77(5):974–983

White IC, Baker JM (1998) The sea empress oil spill in context. In: International conference on the sea empress oil spill, pp 11–13

Wiens JA, Crist TO, Day RH, Murphy SM, Hayward GD, Effects of the Exxon Valdez oil spill on marine bird communities in Prince William Sound, Alaska

World Seaborne Trade with Oil: One of Main Causes for Oil Spills? By Department of Water Transport, University of Žilina, Univerzitná 8251/1, 010 26 Žilina, Slovakia

Zatzman GM (2012) Sustainable resource development. Wiley

Zhang B, Matchinski EJ, Chen B, Ye X, Jing L, Lee K (2019) Marine oil spills-Oil pollution, sources and effects. In World Seas: an environmental evaluation. Academic Press, pp 391–406. https://doi.org/10.1016/B978-0-12-805052-1.00024-3

Chapter 5
Emerging Petroleum Pollutants and Their Adverse Effects on the Environment

Deeptimayee Pal and Sujit Sen

Abstract The petroleum sector's contribution to the global economy is significant and remarkable, with a vast spectrum of products utilised in transportation, industry, and household consumption. However, the petroleum industry generates a vast amount of emerging petroleum pollutants. As petroleum pollutants have a detrimental effect on human health and the environment, these pollutants are rising sources of concern. One highly concerning emerging petroleum contaminant is Polycyclic aromatic hydrocarbons (PAHs). PAHs have been related to cancer, and adverse effects on development and reproduction are among other health issues. Polychlorinated biphenyls (PCBs) are another emerging pollutant extensively employed in industrial and commercial applications as coolants and lubricants in electrical equipment. But cancer and immune system malfunction are some health issues connected to PCB exposure. These pollutants are joined by perfluoroalkyl and polyfluoroalkyl substances (PFAS), utilized in various applications. They may penetrate the food chain and harm wildlife. These contaminants may harm many creatures, from tiny plankton to bigger fish and mammals, and have long-term repercussions on the environment. These toxins may accumulate in these species' tissues and damage animals who consume them. Nitrogen-containing compounds, oxygenated compounds, Endocrine disrupting compounds (EDCs), nanoparticles, and flame retardants are some of the emerging petroleum pollutants that are notoriously difficult to eliminate and pose severe human and environmental health risks. Emerging petroleum pollutants constitute a significant problem that must be addressed to safeguard human health and the natural environment. The removal of these contaminants will need substantial efforts in the areas of research, regulatory reform, and environmental cleaning.

Keywords Emerging petroleum pollutants · Polycyclic aromatic hydrocarbons · Endocrine disrupting compounds · Polychlorinated biphenyls · Perfluoroalkyl and poly-fluoroalkyl substances

D. Pal · S. Sen (✉)
Department of Chemical Engineering, National Institute of Technology Rourkela, Rourkela 769008, Odisha, India
e-mail: sensujit@nitrkl.ac.in

I. D. Behera and A. P. Das (eds.), *Impact of Petroleum Waste on Environmental Pollution and its Sustainable Management Through Circular Economy*, Environmental Science and Engineering, https://doi.org/10.1007/978-3-031-48220-5_5

5.1 Introduction

Petroleum is a complex mixture of hydrocarbons that forms naturally in geological formations underneath the surface of the Earth (Glasby 2006). Petroleum, often known as crude oil, is an organic compound mainly consisting of carbon and hydrogen, with small amounts of nitrogen, oxygen, and sulfur. Hydrocarbons are carbon- and hydrogen-based organic compounds that comprise the bulk of petroleum. Alkanes, alkenes, alkynes, and aromatics are various hydrocarbons that may be categorised depending on their molecular structure. The most prevalent hydrocarbons in petroleum are alkanes, sometimes called paraffin (Del Carmen García et al. 2000). These molecules comprise straight or branching chains of carbon atoms surrounded by hydrogen atoms. Alkenes and alkynes are two hydrocarbons containing double and triple bonds between carbon atoms. They are sometimes referred to as olefins and acetylenes, respectively. Hydrocarbons, known as aromatics, have ring-like molecular structures. In addition to carbon and hydrogen atoms, petroleum includes many other organic molecules, such as heterocyclic compounds, including nitrogen, oxygen, or sulfur atoms. Petroleum also contains sulfur-containing chemicals, including thiols and sulfides, responsible for the distinctive smell of crude oil. Lesser nitrogen-containing substances, such as pyridines and quinolines, are present. Petroleum includes trace quantities of inorganic substances like metals and salts in addition to organic substances. They may contain sodium and chloride ions and elements like iron, nickel, and vanadium (Smirnov 2014). The geological formation from which it is produced, the depth of the reservoir, and the refining procedures used to separate its components are only a few of the variables that may significantly impact the composition of petroleum. One of the problems with petroleum is that it may alter chemically over time due to reactions that occur when it is exposed to air, water, and other environmental elements. The production of new compounds and the breakdown of existing ones due to these interactions may affect petroleum's quality and physical characteristics.

Petroleum, often known as crude oil, is derived from the remnants of ancient sea creatures, including plankton and algae. The formation of petroleum is a multimillion-year process that includes the interaction of geological, chemical, and biological factors. The formation of petroleum starts with the deposition of organic-rich silt in the ocean (Kutcherov and Krayushkin 2010). Throughout time, successive sediment layers bury the sediment, which raises the warmth and pressure in the sedimentary basin. When the silt gets more deeply buried, the organic matter starts to undergo the diagenesis process. During diagenesis, organic matter is converted into kerogen, a carbon, hydrogen, nitrogen, sulfur, and oxygen-containing waxy solid (Vandenbroucke 2003).

Petroleum formation relies on several factors, including the temperature and pressure of the sedimentary basin, the kind of organic matter being buried, and the area's geological history (Staplin 1969). When the temperature and pressure continue to rise during burial, kerogen undergoes further chemical transformations, finally transforming into liquid and gaseous hydrocarbons through a process known as thermal cracking. Several kinds of hydrocarbons are formed at various temperatures throughout this process. Lower temperatures, for instance, promote the formation of heavier, longer-chain hydrocarbons like crude oil, while higher temperatures favour the formation of lighter, shorter-chain hydrocarbons like natural gas. When hydrocarbons are formed, they migrate through the surrounding rock towards locations of more significant pressure or lower density, collecting in porous reservoir rocks such as sandstone or limestone. The reservoir rocks operate as a natural barrier, inhibiting hydrocarbons from migrating and enabling them to accumulate over time. The reservoir rock is often capped by impermeable cap rock, such as shale or salt, which prevents hydrocarbons from escaping to the surface. Over time, geological pressures such as tectonic activity or erosion may cause reservoir rocks to fracture or shift, enabling hydrocarbons to migrate to the surface. If the hydrocarbons reach the surface, they appear as seepages or oil slicks; however, they are often confined under the Earth's surface. Petroleum formation is a complex process that takes millions of years (Hirsch et al. 2005).

Crude oil, a naturally occurring liquid combination of hydrocarbons, is refined to produce various valuable products, including gasoline, diesel fuel, jet fuel, lubricating oil, and numerous chemicals. As crude oil cannot be utilized unprocessed, petroleum refining is critical in producing these products. Crude oil is processed to separate its component elements using significant physical and chemical processes. Refining aims to separate the various components of crude oil into specific products with distinct qualities (Murty 2020). Crude oil is made up of various hydrocarbons of different sizes and forms.

(a) Using a process known as fractional distillation, the crude oil is first divided into various fractions. The crude oil is heated in this procedure, and as it flows down a column, the various constituents are separated according to their respective boiling points. Gasoline and other light hydrocarbons are separated at the top of the column, while heavy fuel oil and bitumen are separated at the bottom. Components with lower boiling points, such as gasoline and other light hydrocarbons, are also separated here. When the crude oil has been separated into different fractions, each component is refined using various procedures to enhance its quality and characteristics. These procedures include, among others, hydrotreating, cracking, reforming, isomerization, and alkylation.
(b) The cracking process employs pressure and heat to fragment more significant hydrocarbons into smaller ones. This method converts heavier substances like naphtha, kerosene, and gas oil into gasoline and other light hydrocarbons.
(c) Reforming alters the hydrocarbons' molecular structure to produce higher calibre and better-quality products. High-octane gasoline is made from naphtha using this method.
(d) In the process of isomerisation, the molecular structure of hydrocarbons is altered to produce molecules called isomers, which have the same chemical formula but distinct structures. High-octane fuel and other products with better qualities are produced using this process.

(e) The alkylation process involves joining smaller hydrocarbons to produce large molecules with better characteristics. High-octane gasoline and other products with better qualities are produced using this process.
(f) Hydrotreating is purifying hydrocarbons using hydrogen gas to treat contaminants, including sulfur, nitrogen, and metals. Cleaner-burning fuels and other high-quality products are produced using this process.

Petroleum refining also involves the production of numerous chemicals and additional products, such as lubricating oils, waxes, and asphalt, in addition to these processes (Murty 2020). Several techniques, such as solvent extraction, dewaxing, and polymerization, produce these products.

5.2 The Significance of the Petroleum Sector

The petroleum sector is one of the most significant sectors in the global economy (Mojarad et al. 2018). Petroleum is a vital energy source used to power vehicles, generate electricity, and manufacture various consumer products. The petroleum industry's importance is multifaceted since it significantly influences global commerce, energy security, technical developments, economic growth, and much more.

(a) One of the most significant sources of energy for nations all over the globe is petroleum. It fuels automobiles, produces energy, and warms houses and structures. Access to petroleum reserves enables nations to provide their population with a dependable supply of energy, which is crucial for the expansion and advancement of their economies. The petroleum sector is essential for maintaining national security because it helps nations meet their energy demands and lessens reliance on foreign oil.
(b) The petroleum sector contributes significantly to the world economy by generating income, employment, and investment prospects. It assists an extensive network of sectors that depend on petroleum products for fuel and raw materials, such as transportation, manufacturing, and construction. As a result of the substantial revenue generated by oil exports, nations with large petroleum reserves often have robust economies. As businesses engage in cutting-edge technology to increase productivity and sustainability, the petroleum sector also generates chances for innovation and research.
(c) The fact that crude oil and refined petroleum products are exported and imported by nations makes them globally traded. The petroleum sector has significantly fostered international commerce and economic cooperation as nations collaborate to secure a steady energy supply. As businesses traverse borders to discover and exploit petroleum reserves, the sector has also produced chances for international cooperation and investment.

(d) Technology has advanced in many fields, primarily due to the petroleum sector. For instance, the development of refineries and pipelines was sparked by the demand for gasoline generated by the early 20th-century advent of the vehicle. Recently, the industry has made significant investments in research and development to increase the productivity and sustainability of petroleum extraction and production. As a result, new technologies like hydraulic fracturing were developed, completely changing how shale gas and oil were produced.

The petroleum sector plays a significant role in the global economy by generating fuel and energy for industrial processes, including manufacturing, transportation, and heating. It has significantly shaped contemporary culture and enabled technical progress. Yet, the sector also poses tremendous risks and challenges, especially while releasing emerging pollutants, which seriously negatively affect the environment and offer serious safety issues.

5.3 Emerging Petroleum Pollutants

The usage and extraction of petroleum products, such as gasoline, diesel, and crude oil, may lead to the emission of various pollutants. Emerging petroleum pollutants are gaining increasing attention because of their potential effects on human health and the environment, while some pollutants have been extensively investigated and controlled. Emerging petroleum pollutants are not routinely monitored or controlled contaminants but can potentially harm human health and the environment. Table 5.1 lists some of the most significant emerging that has been identified.

5.4 Adverse Effects of Emerging Petroleum Pollutants

PAHs, Nitrogen-containing compounds, Oxygenated compounds, EDCs, Nanoparticles, Flame retardants, PCBs, and PFAS have various applications in the petroleum sector. However, these emerging pollutants have adverse environmental effects and are associated with various health hazards.

5.4.1 PAHs

PAHs are a class of organic chemicals formed during the incomplete combustion of fossil fuels, notably petroleum products. PAHs consist of two or more fused aromatic rings and may exhibit a spectrum of chemical characteristics and toxicity. Many PAHs, such as naphthalene, phenanthrene, anthracene, fluorene, pyrene,

Table 5.1 Most alarming emerging petroleum pollutants

Categories	Examples
PAHs	Naphthalene, Phenanthrene, Anthracene, Fluorine, Pyrene, and Benzo(a)pyrene
Nitrogen-containing compounds	Pyridine, Quinoline, and Carbazole
Oxygenated compounds	Ketones, Aldehydes, Organic acids, Esters, Alcohols, and Esters
EDCs	Bisphenol A (BPA) and Phthalates
Nanoparticles	Carbon-based nanoparticles, metal-based nanoparticles, and metal oxide nanoparticles
PCBs	Aroclor 1242, Aroclor 1254, and Aroclor 1260
Flame retardants	Brominated Flame Retardants (BFRs): Polybrominated diphenyl ethers (PBDEs), tetrabromobisphenol A (TBBPA), and Hexabromocyclododecane (HBCD)
	Chlorinated Flame Retardants (CFRs): Tris(2-chloroethyl) phosphate (TCEP), Tris(2-chloroisopropyl) phosphate (TCPP), and Hexachlorocyclopentadiene (HCCP)
	Phosphorous Flame Retardants (PFRs): Triphenyl phosphate (TPP), Tris(2-butoxyethyl) phosphate (TBEP), and Bisphenol A bis (diphenyl phosphate) (BDP)
PFAS	Perfluorooctane sulfonate (PFOS), Perfluorooctanoic acid (PFOA), Perfluorobutane sulfonic acid (PFBS), and Perfluorohexane sulfonic acid (PFHxS)

benzo(a)pyrene, etc., have been found in petroleum products, and their chemical characteristics and toxicity may vary widely. Table 5.2 lists some of the highly potent **PAHs, their physical properties, and their chemical structure**.

5.4.1.1 Naphthalene

A two-ring PAH called naphthalene is present in crude oil and processed petroleum goods like gasoline and diesel fuel. Obtainable from petroleum, naphthalene is a crystalline hydrocarbon. It is a versatile substance employed in several industrial applications, including the petroleum sector. The following are some of the applications for naphthalene in petroleum:

(a) Naphthalene is an excellent solvent for various petroleum products. It is often used as a solvent for rubber, resins, waxes, and oils and is also employed in refining petroleum to remove specific contaminants. Naphthalene is a critical ingredient that keeps insects out of clothes and other fabrics. Insects are repelled by vaporizing naphthalene, which is present in mothballs (Medha et al. 2021).
(b) Naphthalene is a fuel that the petroleum sector uses for various operations. Gas turbines and diesel engines utilize it as fuel. Naphthalene is added to lubricants

Table 5.2 List of potent PAHs and their physical properties

PAHs	Molecular structure	Physical properties
Naphthalene	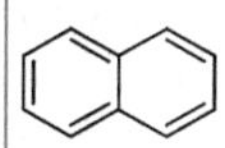	Chemical formula: $\mathbf{C_{10}H_8}$ Colour: **White** Molecular weight: **128.1705 g/mol** Density: **1.14g/cm^3**
Phenanthrene	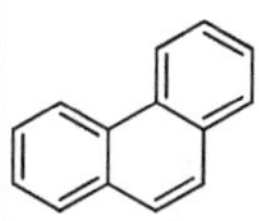	Chemical formula: $\mathbf{C_{14}H_{10}}$ Colour: **Colorless** Molecular weight: **178.23 g/mol** Density: **1.18 g/cm^3**
Anthracene	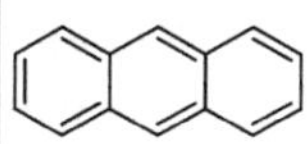	Chemical formula: $\mathbf{C_{14}H_{10}}$ Colour: **Colorless** Molecular weight: **178.23 g/mol** Density: **1.25 g/cm^3**
Fluorene	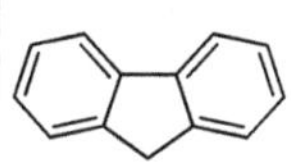	Chemical formula: $\mathbf{(C_6H_4)_2\ CH_2}$ Colour: **White** Molecular weight: **166.223 g/mol** Density: **1.2 g/cm^3**
Pyrene	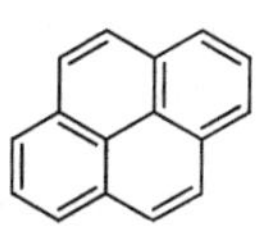	Chemical formula: $\mathbf{C_{16}H_{10}}$ Colour: **Yellow** Molecular weight: **202.25 g/mol** Density: **1.27 g/cm^3**
Benzo(a)pyrene	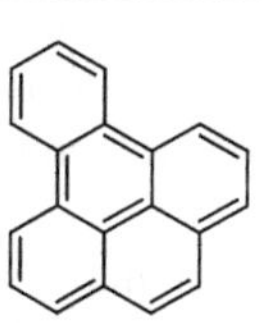	Chemical formula: $\mathbf{C_{20}H_{12}}$ Colour: **Pale yellow/orange/green** Molecular weight: **252.316 g/mol** Density: **1.24 g/cm^3**

to increase their lubricity and decrease friction. Moreover, lubricants include it as a corrosion inhibitor.

(c) Naphthalene serves as an intermediary in synthesizing other compounds, such as phthalic anhydride, which is used to create plasticizers for PVC. Also, it is used to manufacture naphthalene sulfonate formaldehyde condensates, which are dispersants utilized in manufacturing concrete, ceramics, and other products.

(d) Naphthalene is a common pesticide used to combat moths and other bugs that damage the cloth. In addition, it is used as a fumigant in agricultural goods such as grain storage. Manufacturing several dyes, including vat and dispersion dyes, uses naphthalene as an intermediary (Jia and Batterman 2010).

(e) Carbon black is a black powder used as a pigment and a strengthening agent in rubber and plastic goods. To make carbon black, naphthalene is utilized as a primary ingredient (Okoye et al. 2021).

Crystalline hydrocarbon naphthalene is made from petroleum. Although it has several industrial uses, such as in the petroleum sector, naphthalene may also harm human health and the environment (Yost et al. 2021; Preuss et al. 2003; Medha et al. 2021). The following are some of the naphthalene's adverse effects:

(a) Health effects: According to the International Agency for Research on Cancer, naphthalene may cause cancer in humans (IARC), which could result in cancer in individuals, especially if exposed to high concentrations of naphthalene over a prolonged period. Naphthalene exposure may result in additional health issues than cancer, such as respiratory issues, headaches, nausea, vomiting, and dementia.
(b) Environmental effects: Naphthalene may negatively affect the environment. Fish, amphibians, and other aquatic species may be harmed by it since it harms marine life. Animals that live on land may suffer consequences if they come into contact with polluted soil. Naphthalene may bioaccumulate in the food chain and last a very long period in the environment.
(c) Fire hazard: Naphthalene is very explosive and may provide a fire risk if improperly handled. It is hazardous to store or carry since it readily ignites and burns with an intense flame.
(d) Workplace exposure: Naphthalene exposure may occur by ingestion, skin contact, or inhalation for those who work in the petroleum industry and other sectors that employ it. Naphthalene exposure at work may have various adverse health impacts, such as headaches, skin irritation, and respiratory issues.
(e) Household exposure: Naphthalene is often used in insect repellents like mothballs and other items. But, if these goods are mishandled, naphthalene fumes may be released into the air, endangering human health. Mothballs and other products may expose people to naphthalene, and children and animals are especially vulnerable.
(f) Soil and groundwater contamination: If naphthalene is discharged into the environment, it may pollute soil and groundwater. Spills, leaks, or inappropriate product disposal containing naphthalene may cause this. Naphthalene has a prolonged half-life in the environment and may contaminate groundwater, impacting drinking water sources.

5.4.1.2 Phenanthrene

Three-ringed phenanthrene is present in crude oil and refined petroleum products. It is used to make dyes, polymers, and other compounds. It has several industrial uses, especially within the petroleum sector. These are examples of phenanthrene's usage in petroleum:

(a) Phenanthrene is used in diesel and gasoline engines as a fuel additive. It may enhance engine performance and decrease pollutants by increasing the octane rating of gasoline. Moreover, it may strengthen diesel fuel's lubricating qualities, reducing engine wear and extending engine life (Santana et al. 2006).
(b) Phenanthrene is a catalyst in manufacturing various chemicals and products, such as plastics, synthetic fibres, and medicines. It may facilitate the acceleration of chemical reactions and enhance the efficacy of chemical processes.

(c) Phenanthrene is a common solvent in the petroleum industry. It makes it easier to dissolve other substances and handle or treat them. It's beneficial in lubricants since it helps dissolve and blend different oils and additives.
(d) Phenanthrene is a carbon source for producing carbon fibres and other high-tech products. Some mechanical properties, including strength and toughness, may be improved with its help.

Phenanthrene is used as a fuel additive and a catalyst, among many other industrial purposes, but prolonged exposure to it can potentially harm human health and the environment (Olsson et al. 2010).

(a) Carcinogenic: One of the most serious risks associated with phenanthrene is the possibility that it may cause cancer in humans. The International Agency for Research on Cancer (IARC) classifies phenanthrene as a Carcinogen of Concern (Category 3) because there is insufficient evidence to determine whether or not it is carcinogenic to humans (Waigi et al. 2015). Despite its potential to cause cancer, phenanthrene exposure may also result in respiratory issues. According to studies, inhaling phenanthrene may cause respiratory system inflammation, including coughing, wheezing, and shortness of breath, resulting in more severe respiratory conditions, such as asthma or chronic obstructive pulmonary disease (COPD).
(b) Health effects: Phenanthrene exposure may also result in irritation, erythema, and itching because phenanthrene may be absorbed via the skin, which can induce cell damage and inflammation. Moreover, prolonged or repeated contact with phenanthrene may lead to dry, cracked skin, increasing the risk of infection.
(c) Environmental effects: Besides its impact on human health, phenanthrene may have negative ecological repercussions. When phenanthrene is discharged into the environment, it may contaminate soil and water, where it can remain for prolonged durations, harming plants and animals and affecting whole ecologies (Gao et al. 2006).
(d) Marine habitat: Phenanthrene is hazardous to fish and other aquatic species in aquatic habitats, causing stunted development, poor reproduction, and even fatal. It may also result in bioaccumulation, in which phenanthrene accumulates in the tissues of species further up the food chain, resulting in even higher toxicity levels (Mu et al. 2014).

5.4.1.3 Anthracene

Anthracene has the chemical formula $C_{14}H_{10}$, a PAH. It is colourless and odorless and is composed of three fused benzene rings. Anthracene is an important organic molecule found in coal tar, crude oil, and other fossil fuels. Its distinctive molecular structure makes it useful for various uses, especially in the petroleum sector (Van Damme and Du Prez 2018). The following are some of the benefits of anthracene:

- Anthracene is primarily employed in petroleum as a feedstock for synthesizing chemicals, which is one of its significant uses. Many substances, including anthraquinone, employed to make dyes and other organic compounds, may be synthesized from anthracene. Naphthalene is a crucial chemical produced from anthracene and employed in manufacturing several industrial goods.

- In oil reservoirs, anthracene may also be utilized as a tracer. An oil reservoir may monitor the flow of oil and water by injecting anthracene into it. Using this expertise will increase oil production efficiency and maximize oil recovery. Anthracene is also used as a luminous tracer in petroleum. In oil and gas exploration, anthracene may be employed as a luminous tracer to find the existence of hydrocarbons. Anthracene exhibits a distinctive fluorescence signal that may be identified and quantified when it is activated by UV light. This method may determine probable oil and gas deposits and track hydrocarbon transport underground.
- Coke may be made with the help of anthracene. In the absence of air, coke is produced by heating coal to high temperatures. Anthracene, a byproduct of this process, is a fuel that may be used to produce coke. In making coke, anthracene may also be used as a binding agent. The coal particles become more tightly bound, producing a solid, cohesive mass that may be burned to produce coke.

Although anthracene has several applications in various sectors, it poses various risks to human and environmental health. Anthracene's adverse effects include its toxicity, possible carcinogenicity, and ecological impact (Armstrong et al. 2004).

(a) Toxicity: Anthracene is a toxin that may affect people by ingestion, inhalation, and skin contact. Anthracene may induce stomach discomfort, vomiting, and diarrhoea when ingested. Anthracene may irritate the respiratory system when inhaled, resulting in coughing, wheezing, and shortness of breath. Anthracene contact may result in irritation and dermatitis. Anthracene's toxicity emanates from its potential to impair cellular activity. Anthracene may enter cells and interact with biological components such as lipids and proteins, causing damage and malfunction. This disturbance may eventually result in oxidative damage, inflammation, and cell death.

(b) Possible Carcinogenicity: The International Agency for Research on Cancer has categorized anthracene as a potential carcinogen (IARC). According to studies, anthracene exposure increases the risk of cancer, notably lung and skin cancer. Anthracene may induce DNA damage and mutations that can contribute to cancer development (Sun et al. 2020; Gunter et al. 2007).

(c) Environmental Impact: Anthracene may enter the environment through various sources, including industrial operations, such as manufacturing coal tar and coke and using fossil fuels. Burning fossil fuels releases anthracene and other PAHs into the atmosphere. They may be transported long distances and deposited in soils and aquatic bodies.

(d) Anthracene is a persistent organic pollutant with long-lasting environmental impacts. It is very toxic to aquatic species like fish and algae. Exposure to anthracene may induce physiological and behavioural changes in these species, resulting in diminished development and reproduction. Moreover, anthracene may collect in soil and sediment, which might remain for years. This deposit may contaminate groundwater and surface water, harming the health of plants and animals that rely on these water sources.

5.4.1.4 Fluorene

A three-ringed PAH called fluorene is present in crude oil and refined petroleum products. Pharmaceuticals, dyes, and other chemicals are made with it. In the petroleum sector, it is used for a variety of significant purposes, some of which are listed below, such as a fuel additive, a solvent, and a precursor to other chemicals:

(a) Fluorene has been used as a fuel additive in the petroleum industry to improve the quality of gasoline and diesel fuels. When fluorene is added to gasoline, the fuel's octane rating may be raised, enhancing engine performance and lowering engine knock. By reducing the growth of gum and other deposits that might block fuel lines and injectors, fluorene can help improve the stability of gasoline. Adding fluorene to diesel fuel can raise the fuel's cetane rating, leading to better engine noise reduction and ignition. Moreover, fluorene may reduce the number of pollutants and particle matter that diesel engines emit.
(b) Fluorene is a versatile solvent with many uses in the petroleum sector. Due to its capacity to dissolve petroleum products like lubricants and waxes, it is frequently utilized as a solvent for these substances. Additionally, other chemicals like dyes and resins can be dissolved in fluorene.
(c) Moreover, fluorene may be used in petroleum products as a dispersion. The dispersion of solid particles in liquids, such as carbon black and other fillers, may enhance the performance of these goods. Fluorene can also be a precursor to other chemicals in the petroleum industry. It may be converted into several other PAHs, including pyrene and anthracene, with important industrial uses.
(d) Fluorene serves as a precursor in the synthesis of fluorenone. Pharmaceuticals, including anti-inflammatories and antidepressants, are synthesized using fluorenone as a critical intermediate. There are several ways to convert fluorene to fluorenone, including hydrogen peroxide or potassium permanganate. Furthermore, fluorescent pigments and dyes may be produced using fluorene as a precursor. These colours and pigments are used in various processes, including producing textiles, paintings, and polymers.

Fluorene has several essential applications in the petroleum industry but also negatively affects human health and the environment (Lawal 2017; Bach et al. 2003). The significant impacts of fluorene are as follows:

(a) Carcinogenic: Long-term fluorene exposure has also been linked to increased cancer risk. The International Agency for Research on Cancer (IARC) has classified fluorene as a possible human carcinogen based on studies showing an increased incidence of lung cancer in workers exposed to high levels of fluorene.
(b) Neurotoxicity: Fluorene may affect the central nervous system neurotically. Research has revealed that exposure to fluorene may induce damage to nerve cells, resulting in alterations in behaviour and cognitive function. In animal

experiments, exposure to high doses of fluorene has been proven to produce seizures, convulsions, and other neurological symptoms.

(c) Health effects: Fluorene exposure can negatively affect human health, particularly the respiratory and nervous systems. Inhalation of fluorene vapours or dust can cause respiratory irritation, coughing, and shortness of breath. Prolonged or repeated exposure to high quantities of fluorene may also cause damage to the lungs and lead to chronic respiratory disorders, such as bronchitis and emphysema.

(d) Environmental Effects: Fluorene, notably aquatic creatures and ecosystems, may harm the environment. Fluorene is highly persistent in the environment and can bioaccumulate in the tissues of marine organisms, leading to potential toxicity and long-term impacts on the ecosystem's health.

(e) Marine habitat: Exposure to fluorene can also adversely affect fish and other aquatic organisms. Studies have shown that exposure to fluorene can cause changes in fish's behaviour and reproductive function and damage their organs and tissues. In addition, fluorene may alter the structure and function of aquatic ecosystems by interfering with the growth and development of aquatic plants.

5.4.1.5 Pyrene

Crude oil and other petroleum products include PAH pyrene. The solid's molecular composition is $C_{16}H_{10}$, which is yellowish-white. Pyrene is a significant chemical compound with several applications in the petroleum sector, including its usage as a marker for detecting petroleum products in the environment and as an energy source (Overton et al. 1981).

(a) Detecting petroleum products in the environment is one of the main applications of pyrene in petroleum. When pyrene is subjected to ultraviolet (UV) light, a highly fluorescent chemical, it emits light. Because of this characteristic, petroleum compounds in soil, sediment, and water samples may be found using pyrene as a marker. Pyrene is chemical researchers and environmentalists use to identify and determine the origin of oil spills, leaks from underground storage tanks, and other environmental sources of petroleum pollution.

(b) Energy production is another significant usage of pyrene in petroleum. Pyrene is a chemical that may be used as a fuel for electricity production. It is highly flammable. It is often used as a component of coal and other solid fuels but may also be burnt directly as a liquid fuel. The energy released when pyrene burns may be converted into electricity and used for industrial activities. This energy is also released as heat and light.

Although it has various benefits, pyrene is also a toxic compound that may harm the environment and human health. Moreover, pyrene may pollute soil and water, harming aquatic and terrestrial ecosystems. It is a widely used compound linked to adverse environmental and human health impacts (Diggs et al. 2011).

(a) Carcinogenic: Pyrene can potentially cause cancer in people since it is a recognized human carcinogen. Pyrene has been categorized as a Category 2A carcinogen by the International Agency for Research on Cancer (IARC), which indicates that it is probably harmful to humans. Pyrene's potential to damage DNA is regarded to be predominantly responsible for its carcinogenic effects (Garcia-Suastegui et al. 2011). Pyrene can bind to DNA and trigger mutations that may result in cancer growth. Humans exposed to pyrene have an increased chance of developing skin, bladder, and lung cancer.
(b) Human health: Pyrene may have numerous harmful impacts on human health and be carcinogenic. Shortness of breath, coughing, and respiratory irritation may all be brought on by prolonged exposure to high amounts of pyrene. Chronic bronchitis and other respiratory issues may result from pyrene exposure over an extended period.
(c) Environmental effects: Pyrene may harm the environment, especially aquatic habitats. Pyrene is a very persistent chemical in the environment and may build up in the tissues of fish, crustaceans, and other marine creatures. Aquatic organisms exposed to pyrene may have many adverse consequences, such as decreased growth and reproduction, changed behaviour, and changes in metabolism. Pyrene may also lead to improper fish and other aquatic creature growth. Pyrene may have indirect impacts on the environment in addition to its direct effects on marine creatures. By migrating over great distances in air and water currents, pyrene and other PAH chemicals may contaminate isolated ecosystems far from the source of the pollution.

5.4.1.6 Benzo(a)pyrene

Crude oil and refined petroleum products include the five-ringed PAH benzo(a)pyrene. The various uses of benzo(a)pyrene are listed below (Kinik et al. 2021):

(a) The primary usage of benzo(a)pyrene in the petroleum industry is as a crude oil quality indicator. The presence of additional, more dangerous PAH chemicals in crude oil is often indicated by benzo(a)pyrene. Refineries can assess the degree of contamination and take measures to eliminate harmful compounds from the oil by evaluating the quantities of benzo(a)pyrene in crude oil.
(b) As a reference material for environmental testing, benzo(a)pyrene is another significant petroleum usage. The effects of benzo(a)pyrene on the environment have been thoroughly investigated and tested. Scientists and regulators may use benzo(a)pyrene as a reference material to compare the impacts of other PAH chemicals to those of benzo(a)pyrene and to make better-informed decisions about the environmental effects of petroleum products.
(c) Benzo(a)pyrene has industrial applications in synthesizing several compounds and is used as a marker and reference material. Several PAH chemicals used to make plastics, dyes, and other industrial products may be synthesized using benzo(a)pyrene as a starting material.

Although benzo(a)pyrene has several applications; it is associated with several environmental issues and health hazards (Collins et al. 1991; Unwin et al. 2006) such as:

(a) Carcinogenic: Carcinogenicity is one of the main side effects of benzo(a)pyrene. Research has shown that benzo(a)pyrene exposure may lead to cancer in people, notably skin and lung cancer. Long-term exposure to benzo(a)pyrene may increase cancer risk, especially when inhaled.
(b) Health effects: Human and animal developmental abnormalities have been linked to benzo(a)pyrene. In animal studies, Benzo(a)pyrene exposure during pregnancy has been linked to lower birth weight, skeletal deformities, and other developmental issues. Benzo(a)pyrene exposure during pregnancy has been linked to an increased risk of low birth weight and premature delivery in humans.
(c) Marine habitat: Fish and other vertebrates are particularly hazardous to benzo(a)pyrene-containing aquatic species. Benzo(a)pyrene exposure may harm the reproductive system, stunt growth, and induce developmental abnormalities in fish and other marine species; significantly impacting the condition of aquatic habitats and the health of the fish and other animals that rely on them.
(d) Environmental effects: Besides, benzo(a)pyrene may generate other forms of air pollution, such as photochemical smog. When released into the environment, benzo(a)pyrene may combine with primary pollutants to develop secondary pollutants, such as ozone and particulate matter.

5.4.2 Nitrogen-Containing Compounds

Both crude oil and refined petroleum products have been shown to include nitrogen-containing molecules. Some examples of these chemicals are pyridine, quinoline, and carbazole. These nitrogen-containing compounds have widespread use in the petroleum industry, serving many functions, including solvents, corrosion inhibitors, and lubricant additives.

5.4.2.1 Pyridine

A heterocyclic aromatic compound called pyridine is often used in the petroleum sector. It is highly soluble in organic solvents like ethanol and diethyl ether and has a six-membered ring structure with five carbon atoms and one nitrogen atom (Scriven and Murugan 2005). Pyridine is used as a solvent, catalyst, and corrosion inhibitor in the petroleum industry in many processes.

(a) Pyridine is excellent for extracting and purifying different components in crude oil since it is particularly efficient at dissolving various organic molecules.
(b) Alkylation of hydrocarbons, a vital step in manufacturing high-octane gasoline, is one of the many chemical processes pyridines may catalyse.

(c) By developing a shielding coating on metal surfaces that inhibits the interaction of the metal with corrosive chemicals, pyridine may be added to petroleum products to stop corrosion.

If mishandled, pyridine is a highly hazardous substance that may adversely affect human health and the environment (Lin et al. 2020). The following are examples of pyridine's adverse effects:

(a) Toxicity: Pyridine is highly toxic and may have various negative health consequences, including irritation, burns, organ damage, and even death. Pyridine vapour inhalation may lead to chest discomfort, shortness of breath, coughing, and other respiratory issues. Pyridine exposure may lead to burns, rashes, and skin irritation. Gastrointestinal distress, nausea, vomiting, and diarrhoea may result from pyridine ingestion.
(b) Carcinogenicity: According to the International Agency for Research on Cancer, pyridine may cause human cancer (IARC). Long-term pyridine exposure has been associated with a higher risk of developing various cancers, including lung and bladder cancer.
(c) Pyridine has a long-lasting influence on the environment since it is a very persistent chemical that may develop up there. Exposure to pyridine may harm ecosystems and biodiversity and is hazardous to aquatic life.
(d) Fire and explosion risk: Pyridine is volatile and may cause fires and explosions if improperly handled. It burns toxically and may catch fire at shallow temperatures.

5.4.2.2 Quinoline

Quinolines are heterocyclic compounds with a fused benzene ring and pyridine ring. They serve several different functions in the petroleum sector. They act as corrosion inhibitors, antifouling agents, solvent additives, catalysts, fuel additives, fluorescent tracers, and surfactants. The following are some of the quinoline's various applications:

(a) Quinolines are employed in the petroleum sector as corrosion inhibitors to resist corrosion on metal surfaces. They form a shielding coating on the metal's surface to prevent corrosive substances from coming into contact with it. Quinolines inhibit corrosion in acidizing treatments because they function incredibly well in acidic conditions (Verma et al. 2020).
(b) Quinolines are used as antifouling agents in the petroleum sector to avoid fouling. They prevent the development of microbes, the primary source of fouling in petroleum systems. Quinolines are used as solvent additives in the petroleum sector to increase the solubility of hydrocarbons in solvents. They work exceptionally well for extracting aromatics and olefins from hydrocarbon mixtures.

(c) In several processes within the petroleum industry, quinolines serve as catalysts. They are very effective in the alkylation of aromatics, a critical factor in producing high-octane gasoline.
(d) Quinolines are employed as fuel additives in the petroleum sector to enhance the performance of fuels. They may raise the cetane number of diesel fuels, which increases combustion efficiency and lowers emissions. Quinolines may also be added to gasoline as octane boosters, which boosts the fuel's octane rating and enhances its performance.
(e) Quinolines are used as fluorescent tracers in the petroleum sector to monitor reservoir fluid flow. They are injected into the reservoir, and the fluid flow is monitored by monitoring their fluorescence. Quinolines effectively identify water movement in oil reservoirs, vital for enhanced oil recovery.
(f) Quinolines are employed as surfactants in the petroleum industry to enhance the emulsification of water and oil. They perform exceptionally well when producing crude oil emulsions since such are more accessible to carry than pure crude oil. Quinolines may also be employed as surfactants to create foam, which is crucial for well drilling and completion.

Quinolines are adaptable substances with several uses in the petroleum industry. Quinolines may negatively affect human health and the environment (Bachmann et al. 2014).

(a) Neurotoxicity: Quinolines can cause neurotoxicity or harm the neurological system. Tremors, convulsions, and disorientation are a few symptoms that may result from this. Quinolines' capacity to bind metal ions, which might interfere with typical cellular processes, is suggested to be a contributing factor to their neurotoxicity. Both in humans and in animals, quinolines have been linked to neurotoxicity.
(b) Phototoxicity: Quinolines can potentially harm skin and eyes when exposed to sunlight, known as phototoxicity; this occurs due to quinolines' capacity to absorb UV light and produce reactive oxygen species, which may harm cellular structures. Blisters, swelling, and redness of the skin are only a few signs of phototoxicity.
(c) Cardiotoxicity: Cardiotoxicity, or damage to the heart, may be caused by quinolones. Heart failure, arrhythmias, and cardiomyopathy are just a few symptoms that might result from this. The potential of quinolines to produce reactive oxygen species, which may harm the cellular structures in the heart, is considered a contributing factor to their cardiotoxicity.
(d) Hepatotoxicity: Hepatotoxicity, or liver damage, may be brought on by quinolines, resulting in symptoms including hepatitis, liver failure, and jaundice. The capacity of quinolines to produce reactive oxygen species, which may harm cellular structures in the liver, contributes to their hepatotoxicity.
(e) Genotoxicity: Damage to a cell's genetic structure, or genotoxicity, may be brought on by quinolines resulting in cancer, chromosomal abnormalities, and mutations. Quinolines have the potential to intercalate into DNA and alter its structure, which is linked to their genotoxicity.

(f) Ecotoxicity: Quinolines may harm the environment, especially aquatic organisms. Quinolines may harm fish by impairing their ability to grow normally, preventing them from reproducing, and even killing them. Quinolines may also negatively impact aquatic microbial populations, disturbing ecological processes.

5.4.2.3 Carbazole

A heterocyclic aromatic chemical called carbazole is often used in the petroleum sector. It is highly soluble in organic solvents like ethanol and diethyl ether and has a seven-membered ring structure with two nitrogen and five carbon atoms. The petroleum industry uses carbazole for various uses such as lubricant additives, corrosion inhibitors, and solvents (Han et al. 2020).

(a) Carbazole is an excellent solvent for various organic molecules and may be used to extract and purify different components present in crude oil.
(b) In the petroleum industry, carbazole is also a corrosion inhibitor. By developing a shielding coating on metal surfaces that blocks the interaction of the metal with corrosive chemicals, carbazole may be added to petroleum products to inhibit corrosion.
(c) In the petroleum industry, carbazole is also a lubricant additive. Lubricants that have included it perform better and last longer under difficult operating circumstances. Carbazole has been shown to reduce friction and wear in engine parts, resulting in extended equipment life and increased effectiveness.

Although carbazole has many advantageous uses in the petroleum industry, it is also linked to a variety of negative impacts on both human health and the environment (Mumbo et al. 2015). The following are some of the primary drawbacks of carbazole:

(a) Toxicology: Carbazole is regarded as hazardous and linked to several adverse health consequences in people and animals. The effects of carbazole exposure might include respiratory issues, gastrointestinal discomfort, and skin and eye irritation. More severe health consequences from carbazole use include cancer and liver damage.
(b) Environmental effects: Carbazole is a persistent organic contaminant that may build up in soil, water, and the atmosphere. It may have long-lasting consequences on ecosystems and biodiversity and is hazardous to aquatic life. Carbazole may affect other marine species, including crustaceans and molluscs, and has been shown to have a detrimental effect on fish growth and reproduction.
(c) Combustibility: If carbazole is improperly handled, it might ignite and cause a fire. It burns toxically and may catch fire at shallow temperatures.
(d) Chemical reactivity: In the presence of other chemicals, carbazole is a reactive compound that may undergo various chemical reactions, making it challenging to govern and cause unexpected responses and results.

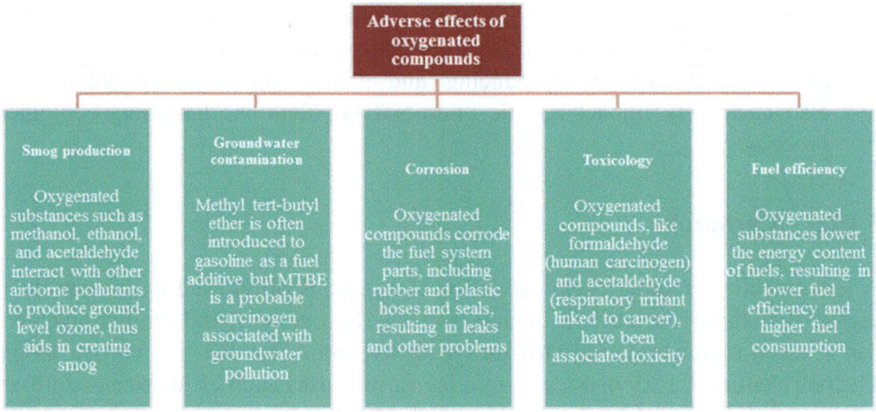

Fig. 5.1 Environmental and health impacts of oxygenated compounds

5.4.3 *Oxygenated Compounds*

Organic molecules that include oxygen atoms in their molecular structure are called oxygenated compounds. Some widely used oxygenated compounds are Ketones, Aldehydes, Organic acids, Esters, Alcohols, and Esters. These substances may be found in petroleum and products made from petroleum. They are often added to gasoline and other fuels to enhance their qualities, such as lowering emissions and improving combustion efficiency (Talmadge et al. 2014). Yet, some of these compounds have been associated with adverse environmental and human health impacts, as illustrated in Fig. 5.1.

5.4.4 *EDCs*

Chemicals that can potentially disrupt the endocrine systems of living organisms are known as EDCs. It has been determined that certain products derived from petroleum, such as BPA and phthalates, are known to be EDCs and can cause harm to both human health and the environment.

5.4.4.1 BPA

BPA is a commonly used synthetic organic compound in manufacturing polycarbonate plastics, epoxy resins, and other products (Kumar et al. 2018). Moreover, BPA is used in various industrial applications, including the petroleum sector. Following are some of the petroleum applications for BPA:

(a) BPA is an essential component in manufacturing epoxy resins, which are used in various petroleum sector applications. These materials resist chemicals, abrasion, and high temperatures, making them suitable for extreme use.
(b) Drilling fluids, used to lubricate drill bits and minimize friction during drilling operations, contain BPA. It has been shown that BPA-based drilling fluids offer superior lubrication and reduce equipment wear, which may lead to increased drilling efficiency and less downtime.
(c) BPA is used to manufacture thermoplastic elastomers (TPEs) in various petroleum-related applications. TPEs are flexible, long-lasting polymers that may manufacture seals, gaskets, and other components for pipelines, tanks, and other equipment.
(d) Certain petroleum goods, such as cable insulation and electrical connections, employ BPA as a flame retardant. It has been shown that flame retardants containing BPA have high fire resistance, which may aid in preventing fires and limiting their spread in the case of an accident.
(e) Certain petroleum goods, such as PVC pipes and electrical wires, employ BPA as a plasticizer. BPA-based plasticizers have been found to provide superior flexibility and durability, making them suited for various applications.

Despite its extensive usage in the petroleum sector, concerns exist about the potential health and environmental effects of BPA. BPA exposure has been related to several health issues, including cancer, reproductive and developmental, and metabolic abnormalities (Rochester 2013; Vandenberg et al. 2007). In addition, BPA is a persistent contaminant that accumulates in the environment and has long-term effects on ecosystems (Chen et al. 2016).

(a) Water contamination: BPA may seep into freshwater and marine habitats through landfills, sewage treatment facilities, and industrial effluents, causing pollution. Moreover, it may pollute groundwater and soil, which poses a threat to the ecology.
(b) Air pollution: Incineration of BPA-containing products, such as plastics, may release toxic chemicals and particulate matter into the air, contributing to air pollution and respiratory ailments.
(c) Soil contamination: BPA-containing products that wind up in landfills may degrade over time and release toxic chemicals into the soil, possibly impacting plant development and animal health.
(d) Bioaccumulation: Over time, BPA may accumulate in the tissues of animals, including humans, potentially causing health consequences. It may also accumulate in the food chain, with the highest concentrations reported in apex predators.
(e) Endocrine disruption: BPA has been found to disrupt the endocrine system, which regulates the body's synthesis and distribution of hormones. Particularly, BPA may mimic estrogen's actions and disrupt the body's average hormonal balance, resulting in various health issues, such as reproductive difficulties, developmental abnormalities, and cancer.

(f) Reproductive disorders: Several reproductive disorders, including infertility, miscarriage, and decreased sperm count in men, have been linked to BPA exposure. Polycystic ovary syndrome (PCOS) is an ovarian condition linked to BPA and may contribute to infertility and other health problems.
(g) Developmental abnormalities: Prenatal exposure to BPA has been linked to various neurological and behavioral problems in children, including ADHD and autism. Furthermore, cleft lip and palate, among other congenital disabilities, have been linked to prenatal exposure to BPA.
(h) Health effects:

- Breast, prostate, and other cancers have all been linked to BPA. BPA's potential to imitate estrogen's effects raises concerns that it might promote the growth of cancer cells.
- Obesity, diabetes, and metabolic syndrome are only some metabolic issues linked to BPA. BPA exposure may cause insulin resistance and other metabolic problems by interfering with insulin signaling pathways.
- The risk of hypertension and coronary artery disease, among other cardiovascular ailments, has been linked to BPA exposure. The hormone-like effects of BPA have been linked to an increase in the risk of developing atherosclerosis, which may lead to cardiovascular problems, including heart attacks and strokes.
- Asthma and other respiratory issues may be more likely in those exposed to BPA. BPA has been linked to immune system dysfunction, which in turn may lead to airway inflammation, respiratory symptoms, and an exacerbation of asthma.
- Depression, anxiety, and cognitive impairment are just some of the neurological issues that have been linked to BPA exposure. BPA may cause irreversible harm to the nervous system by interfering with average brain growth and function.
- Allergies and autoimmune illnesses are immune system problems linked to BPA exposure. The body's natural immunological response may be disrupted by BPA, leading to a hyperactive immune system, which may lead to chronic inflammation and other health problems.

5.4.4.2 Phthalates

Plasticizers, such as phthalates, are often added to plastics to make them more malleable and long-lasting. They are used as solvents in many industries, including petroleum. DEHP, or di(2-ethylhexyl) phthalate, is a common plasticizer used in the petroleum sector. PVC pipes are often used for transporting and storing oil and gas (Thomas et al. 1984). Among the many uses for phthalates in petroleum, operations are as solvents and transporters for chemicals and additives like colors, perfumes, and lubricants:

(a) One of the most common uses of phthalates in the petroleum industry is in producing polyvinyl chloride (PVC) pipes and other PVC products. Electrical insulation, packaging, and building materials are just a few of the various applications of PVC plastic. Phthalates are added to PVC to make it more malleable, stable, and easy to work with.
(b) Phthalates have various uses, but the most common is as a solvent for other compounds like resins, rubber, and adhesives. These compounds are dissolved and diluted with phthalates, which makes them simpler to handle and manage. Moreover, they are used in crude oil as flow enhancers, which lessen the oil's viscosity and facilitate pipeline transportation.
(c) Other polymers in the petroleum industry, such as polyethylene and polypropylene, include phthalates as plasticizers. Many products, such as packaging, gasoline tanks, and automobile components, employ these polymers.

Although phthalates are often utilized in the petroleum sector, growing concerns about their possible health implications have been raised. Phthalates have been associated with many health issues, notably effects on reproduction and development. Due to their well-known endocrine-disrupting characteristics, they have the potential to affect the body's hormonal equilibrium. Owing to the various impacts such as (Hauser and Calafat 2005; Lottrup et al. 2006):

(a) Reproductive issues: Researchers discovered that phthalates negatively affect fertility in both sexes. Males exposed to phthalates have a higher risk of developing erectile dysfunction, having less sperm, and having fewer active sperm. Studies indicate a link between phthalate exposure to accelerated puberty, menstrual problems, and decreased fertility in females.
(b) Developmental abnormalities. Phthalate exposure has been linked to early puberty, menstrual issues, and reduced fertility in women. Phthalates have also been connected to many developmental problems, notably in youngsters. Low birth weight, neurological and behavioral impairments, ADHD and autism spectrum disorders, and improper development of the male reproductive system have all been linked to phthalate exposure during pregnancy.
(c) Disruption of the endocrine system: Phthalates may affect the body's hormonal balance, especially estrogen synthesis. Many health issues, such as breast cancer, prostate cancer, and obesity, may result from this.
(d) Asthma and allergies: According to some research, exposure to phthalates may make it more likely for people to develop asthma and allergies, especially in youngsters. Phthalate exposure has been linked to liver and kidney damage, especially in laboratory animals. There is some evidence that phthalates may cause cancer, especially in certain cancers like liver cancer.
(e) Soil pollution: Phthalates may enter the soil by dumping garbage, including items like plastic toys or packaging containing phthalates. They may linger in the soil and seep into the groundwater, endangering plants, and soil-dwelling microbes.

(f) Water contamination: Phthalates may enter the water system via industrial and domestic wastewater, household garbage made of plastic, and runoff from landfills. Once entering the water, they may hurt fish and other aquatic life.
(g) Air pollution: While manufacturing and disposing of phthalate-containing items, phthalates may be discharged into the air, which adds to air pollution.
(h) Bioaccumulation: Phthalates can accumulate over time in the tissues of animals, including humans, which might pose health risks. Moreover, they may build up in the food chain, with predators at the top exhibiting more significant amounts.

5.4.5 Nanoparticles

Particles that are less than 100 nm are referred to as nanoparticles. Increased surface area and altered chemical reactivity are two examples of nanoparticles exhibiting features distinct from their bulk-form counterparts. The presence of these particles in petroleum and products produced from petroleum may be attributed to the natural processes in these environments or to the industrial activities in those environments. Nanoparticles based on carbon, nanoparticles based on metal, and nanoparticles based on metal oxide are some of the most often employed nanoparticles in petroleum (Zhou et al. 2020). While nanoparticles have unique properties that make them desirable for various applications, they may also negatively impact the environment and human health (Nowack 2009). The adverse effects of nanoparticles are illustrated in Fig. 5.2.

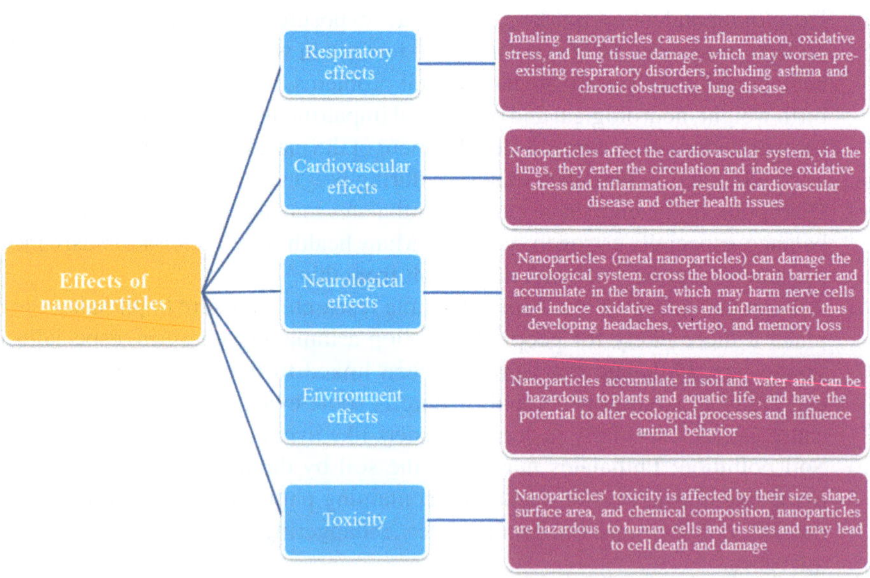

Fig. 5.2 Adverse effects of nanoparticles on humans and the ecosystem

5.4.6 PCBs

PCBs are a class of chemicals formerly used in the electrical industry, serving as lubricants and coolants for various machinery components. These chemicals have a long half-life and are widespread throughout the environment, including in products derived from petroleum (Erickson and Kaley 2011). PCBs such as Aroclor 1242, Aroclor 1254, and Aroclor 1260 are the most common types used in the petroleum industry. The significant applications of Aroclor 1242, Aroclor 1254, and Aroclor 1260 in the petroleum sector are listed below in Table 5.3 (Shanahan et al. 2015):

The distinctive chemical and physical characteristics of Aroclor 1242, 1254, and 1260 made them a popular compound in the petroleum industry. Yet, because of their toxicity and environmental durability, PCBs have been outlawed in several nations; the various adverse effects of PCBs are (Ross 2004; Carpenter 2006):

(a) Neurological problems: It is well known that PCBs negatively affect the neurological system. PCB exposure has been related to cognitive and memory deficiencies in adults and developmental delays in toddlers and babies. PCBs may also impair the immune system's ability to fight off infections, which raises the danger of neurological damage.
(b) Effects on the reproductive and endocrine systems: PCBs may harm the reproductive system and interfere with the endocrine system. Exposure to PCBs has been associated with a higher incidence of miscarriage and stillbirth and decreased fertility in both men and women. Various health issues may result from PCBs impacting the body's hormone synthesis and function. PCBs may disrupt embryonic development by entering the placenta, resulting in low birth weight, cognitive impairments, and behavioral issues.
(c) Cancer: According to the International Agency for Research on Cancer, PCBs are likely human carcinogens (IARC). Many cancers, including kidney, liver, and non-lymphoma, Hodgkin's have been associated in studies with PCB exposure.

Table 5.3 Applications of PCBs

PCBs	Applications
Aroclor 1242	• One of the principal uses of Aroclor 1242 in the petroleum industry was as a dielectric fluid in electrical equipment • The petroleum industry often used Aroclor 1242 as a hydraulic fluid • Aroclor 1242 was used in the petroleum industry as a heat transfer fluid
Aroclor 1254	• As a dielectric fluid in electrical machinery like transformers and capacitors, Aroclor 1254 is utilized in the petroleum sector • Moreover, Aroclor 1254 serves as a heat transfer medium in industrial operations • In certain circumstances, the lubricant additive Aroclor 1254 has been used
Aroclor 1260	• In the petroleum industry, Aroclor 1260 is used as a constituent of transformer oil • As a component of lubricants, hydraulic fluids, and heat transfer fluids, Aroclor 1260 has also been utilized

(d) PCBs negatively impact the environment because they are persistent and may build up in soil, sediment, and streams. PCBs may weaken the immune system, cause developmental delays, and harm the aquatic environment and species' health. Moreover, PCBs may go up the food chain, which might impact human health if people eat contaminated fish and other seafood.
(e) Other health effects: Several additional health concerns, such as skin diseases, respiratory difficulties, and cardiovascular illness, have also been connected to PCB exposure. PCBs may impair the immune system, rendering people more prone to infections and disorders. PCBs may accumulate in the liver and cause damage, leading to liver disease and other health issues. PCB exposure has been linked to an increased risk of cardiovascular diseases like hypertension and coronary artery disease.

5.4.7 *Flame Retardants*

Chemicals known as flame retardants are mixed into various materials to impede or prevent the occurrence of a fire. Products derived from petroleum may include a variety of flame retardants, including brominated flame retardants, chlorinated flame retardants, flame retardants based on phosphorus or nitrogen, and inorganic flame retardants.

(a) Explosions and fires may occur in the petroleum sector when drilling, producing, transporting, and storing petroleum products. Flame retardants aid in avoiding flames and explosions, which may be disastrous in various circumstances.
(b) Flame retardants may aid in lowering the intensity of flames and minimizing the damage they cause.
(c) Flame retardants may be used to comply with safety regulations and industry standards.

Although the implications of flame retardants in the petroleum sector are quite high, it is associated with adverse environmental effects (Darnerud 2003). The negative impact of various flame retardants is depicted in Fig. 5.3.

5.4.8 *PFAS*

PFAS refers to manufactured compounds that do not occur naturally in the environment. Several consumer goods, such as non-stick cookware, stain-resistant textiles, and firefighting foam, have all been known to use PFAS (Glüge et al. 2020). Nevertheless, PFAS include perfluorooctane sulfonate, perfluorooctanoic acid, perfluorobutane sulfonic acid, and perfluorohexane sulfonic acid, are also found in some petroleum products, such as gasoline and jet fuel, since they are used as surfactants and other additives in these goods (Li et al. 2019). The significant applications of PFAS are listed in Table 5.4.

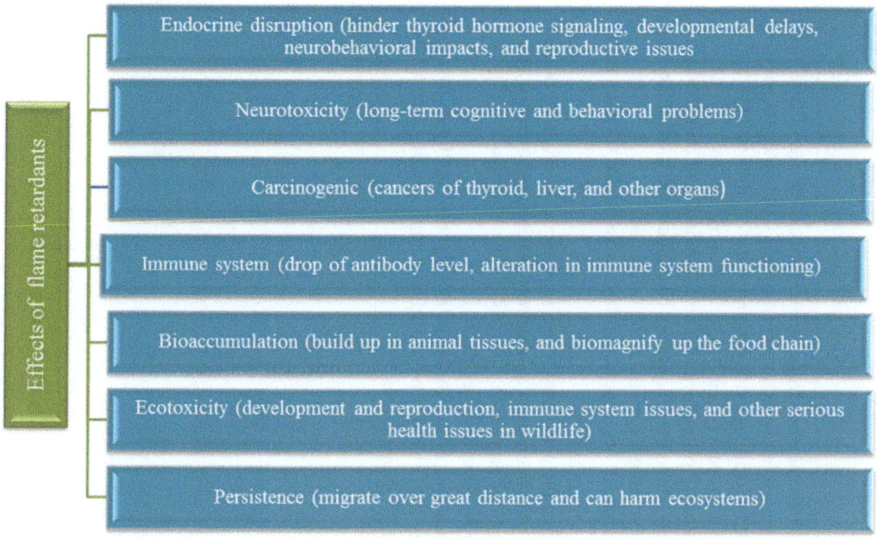

Fig. 5.3 Health complications and environmental consequences of flame retardants

5.4.8.1 Perfluorooctane Sulfonate (PFOS) and Perfluorooctanoic Acid (PFOA)

There are several different kinds of PFAS, but one of the most common is PFOS, which has been used in firefighting foam and stain-resistant coatings. It has also been detected in petroleum products such as gasoline and jet fuel. Similarly, one kind of PFAS, known as PFOA, has widespread use in consumer goods, including non-stick cookware, waterproof clothes, and other similar items. Moreover, it has been identified in a few goods derived from petroleum, such as hydraulic fluid and lubricating oil. Still, PFOS and PFOA are associated with various adverse effects on human health and the environment, such as (Wang et al. 2022; Steenland et al. 2010):

(a) Since it is so resistant to degradation, PFOS may remain in the environment for a long time, indicating that once released into the environment, it can accumulate in the soil, the water, and many living species, including people. Consequently, a person's contact with PFOS may be ongoing and persistent.
(b) Exposure to PFOS has been linked to prenatal and reproductive problems and human liver and kidney damage. An increased liver, pancreatic, and testicular cancer risk have been linked to PFOS exposure.
(c) According to studies, fish and other aquatic animals may have impaired immune function and slower development and reproduction if exposed to PFOS.

Table 5.4 Applications of PFAS

PFAS	Applications
Perfluorooctane sulfonate (PFOS)	• The surfactant qualities of PFOS allow it to aid in lowering the surface tension between two substances, such as oil and water. This characteristic of PFOS has been used in the petroleum sector to assist in oil extraction from wells and other sources • Drilling through rigid rock formations is simpler when PFOS is added to drilling fluids because it lowers friction between the drill bit and the surrounding rock, accelerating and increasing the effectiveness of the drilling operation • PFOS may be utilized in oil recovery procedures to aid in the separation of oil and water. The combination of oil and water that results when water is injected into an oil well to build pressure and drive the oil out may be challenging to separate. The oil and water combination may be stabilised using PFOS, making separating the oil from the water easier • PFOS can be employed in firefighting foams as it can effectively extinguish fires (fuels and flammable liquids) (Zhang et al. 2012)
Perfluorooctanoic acid (PFOA)	• PFOA has been chiefly used as a surfactant in drilling fluids and during the petroleum sector's enhanced oil recovery process (EOR). PFOA may be utilized in the EOR technique to increase the effectiveness of oil recovery. To enhance pressure and drive oil out of wells, EOR entails pumping fluids into the well. The EOR process may be made more effective by using PFOA to lessen the quantity of fluid that gets trapped in the rock formations (Enick et al. 2012) • The drilling process may be accelerated and made more effective by adding PFOA to the drilling fluids to minimize friction between the drill bit and the surrounding rock. Moreover, PFOA may aid in stabilizing the drilling fluid, which might lessen the possibility of gas hydrate formation, a possible issue in certain drilling operations

(continued)

Table 5.4 (continued)

PFAS	Applications
Perfluorobutane sulfonic acid (PFBS)	• Drilling fluids are used throughout the drilling process to lubricate and cool the drill bit and to remove drill cuttings from the borehole. Since it can form stable emulsions even at high temperatures and pressures, which are typical during drilling operations, PFBS is the perfect emulsifier for drilling fluids • PFBS is an excellent surfactant for EOR fluids because of its ability to lower the surface tension between the oil and the rock matrix, hence facilitating the displacement of the oil • By developing a protective layer on the surface of the metal, PFBS may limit corrosion by preventing corrosive chemicals from coming into contact with the metal • PFBS is used to aid in stabilizing emulsions generated during the production and processing of crude oil and natural gas. Foam formation may be a concern during the production and processing of natural gas; PFBS can assist in avoiding this • Since it may lessen the surface tension of the lubricant or grease, PFBS is an excellent wetting agent allowing the lubricant or grease to spread more freely • PFBS is an effective flame retardant because it reduces the combustibility of petroleum products, making them safer to carry and handle (Glüge et al. 2020)
Perfluorohexane sulfonic acid (PFHxS)	• PFHxS is an effective surfactant and emulsifier for drilling fluids due to its ability to generate stable emulsions even at high temperatures and frequent pressures during drilling operations • Since it may lower the surface tension between the oil and the rock matrix, PFHxS is an excellent surfactant and emulsifier for EOR fluids, allowing the oil to be more readily displaced • Foam formation may be a concern during the production and processing of natural gas; PFHxS can assist in avoiding this • Since it may lessen the surface tension of the lubricant or grease, PFHxS is a suitable wetting agent that allows the lubricant or grease to spread more freely • PFHxS is an effective flame retardant because it reduces the combustibility of petroleum products, making them safer to carry and handle (Glüge et al. 2020)

5.4.8.2 Perfluorobutane Sulfonic Acid (PFBS)

Surfactants like the perfluoroalkyl sulfonate (PFAS) form known as PFBS may be found in various products, such as detergents and fire extinguishing foams. Some petroleum-based goods, including petrol and diesel fuel, have also been shown to contain it. The various associated health complications are listed below (Haverinen et al. 2021):

(a) According to studies, PFBS exposure may impair an animal's ability to grow and reproduce. For instance, PFBS lowered fetal weight, led to skeletal deformities, and reduced litter number in research on mice exposed to the chemical. Another study on rats exposed to PFBS revealed that the substance delayed sexual development and lowered fertility and sperm count.
(b) Evidence suggests that exposure to PFBS may cause immunotoxicity or immune system impairment. For instance, when PFBS was administered to rats, researchers discovered that the chemical decreased immune cell activity and increased vulnerability to infection.
(c) According to studies, exposure to PFBS may cause animals' livers to become damaged. For instance, research on rats given PFBS exposure discovered that the substance increased liver weight and impaired liver function.
(d) Since PFBS is very bioaccumulative, it may gradually accumulate in the tissues of living things affecting the ecosystem and the health and reproductive success of these creatures.
(e) By a multitude of distinct channels, such as the release of industrial effluent or the breakdown of other PFAS compounds in the environment, PFBS may pollute water resources. This poisoning may affect the safety of the water used for drinking, the well-being of aquatic life, and the ecosystem.
(f) By a range of distinct channels, such as the application of PFAS-containing goods or the deposition of PFAS-containing dust, PFBS may also contaminate the soil. Plants and other species that depend on soil for existence may suffer from this contamination, which may also affect their productivity.

5.4.8.3 Perfluorohexane Sulfonic Acid (PFHxS)

PFHxS is a form of the PFAS chemical family used in various consumer goods, such as coatings designed to prevent stains and food packaging. Moreover, it has been discovered in a few goods derived from petroleum, such as diesel and aviation gasoline. PFHxS has been associated with various environmental and human health hazards (Schulz et al. 2020). Some of the adverse effects of PFHxS:

(a) Research conducted on animals that were administered PFHxS showed that the compound harmed the growth and development of the fetus, including a drop in fetal body weight, a smaller skeletal size, and delayed ossification. In addition, research has demonstrated that exposure to PFHxS may result in reproductive toxicity, manifesting in various ways, including decreased fertility, and altered hormone levels.

(b) According to the findings of many studies, exposure to PFHxS is related to immune system suppression, which manifests itself as lower levels of antibody production and reduced immune cell activity. This consequence may result in a greater vulnerability to illnesses caused by infectious organisms.
(c) Studies have revealed that exposure to PFHxS may induce liver damage and malfunction, including alterations in liver histology and elevated levels of liver enzymes. This injury carries with it the possibility of causing liver disease.
(d) It has been shown that PFAS, especially PFHxS, may affect thyroid hormone levels which, in turn, can lead to thyroid hormone-mediated developmental impacts and metabolic and neurobehavioral effects.
(e) PFHxS may remain in the environment for long periods and accumulate in living things. According to several studies, PFHxS may accumulate in fish and other animals, harming their ability to reproduce and survive.
(f) Groundwater may be contaminated with PFHxS from various sources, including waste from industrial processes and foams used in firefighting. This pollution can negatively impact human health, including the possibility of exposure to drinking water.
(g) Using items that include PFAS, such as firefighting foams and fertilisers, may also cause PFHxS to pollute the soil. This pollution may have a negative impact not only on the development and health of plants but also on the species that are dependent on the plants for their existence.
(h) PFAS, particularly PFHxS, can be discharged into the air during the compounds' manufacture, use, and disposal, resulting in air pollution and exposure to the chemical by breathing.

5.5 Regulations and Mitigation Strategies

Since petroleum pollutants may harm ecosystems and human health, they are a serious environmental threat. Consequently, there are stringent laws in place to control how petroleum products are handled, stored, and disposed of to reduce the danger of contamination.

- Depending on the nation, regulations for petroleum pollution are often enforced at the national or regional level. For instance, the Environmental Protection Agency (EPA) in the United States enforces laws about petroleum pollution. These rules refer to various operations, including drilling, production, transportation, and storage.
- The United States federal statute, known as the Clean Water Act, passed in 1972, is one of the most significant controls for petroleum pollution. The discharge of contaminants into surface waterways, such as lakes, rivers, and streams, is governed by this regulation. The Clean Water Act restricts the types and quantities of pollutants that may be emitted while requiring industries to get licenses before doing so.

- The Oil Pollution Act, passed in 1990 as a reaction to the Exxon Valdez oil disaster, is another important legislation for petroleum pollutants. This legislation creates accountability for spills and other oil discharges into the environment and mandates the creation of spill response plans by businesses. The Oil Spill Liability Trust Fund, which provides funds for spill response and cleaning activities, was also created under the Oil Pollution Act.
- Regulations cover the transportation of petroleum products for petroleum pollution. In the US, the transportation of hazardous materials, such as petroleum products, is governed by the Pipeline and Hazardous Materials Safety Administration (PHMSA). The PHMSA establishes guidelines for the safe handling and transportation of hazardous products and standards for the design, building, and operation of pipelines and other transportation infrastructure.
- Managing waste products is a crucial component of rules for petroleum pollutants. Petroleum products must be appropriately disposed of when they are no longer useable to protect the environment, including recycling, treatment, or landfill disposal. Federal legislation known as the Resource Conservation and Recovery Act (RCRA) governs the disposal of hazardous waste, which includes petroleum products. The RCRA imposes rules for treating, disposing, and appropriately labelling and storing hazardous waste in businesses.
- Rules also cover the cleaning of spills and other discharges for petroleum pollution. The EPA and other organisations have created guidelines for reacting to spills and releases. These standards include containment, cleaning, and restoration procedures of impacted regions. These recommendations seek to reduce the adverse effects of spills and other discharges on the environment and human health.

In addition to laws, mitigation methods are crucial for reducing the adverse effects of spills and other petroleum releases (Odeyemi and Ogunseitan 1985).

- Prevention: The best strategy to lessen the effects of petroleum pollutants is to prevent them from being released or spilt in the first place. Leaks and spills may be avoided with regular inspections and maintenance that identifies problems early on. Furthermore, companies may prepare for a leak by setting up spill prevention procedures.
- Containment: When a spill or discharge does occur, the next step is to prevent the pollution from spreading by containing it. Booms, barriers, and absorbent materials may confine the spills; strategies for controlling pollution aim to halt it before spreading to vulnerable regions like rivers or neighbourhoods. Timely responses are essential for maintaining containment.
- Cleanup: When the pollution has been confined, the area must be cleaned up. Depending on the nature and location of the spill, many cleaning methods may be used. Cleanup methods such as in situ burning, mechanical skimming, and bioremediation are only a few examples.
- Restoration: When the pollutant has been removed, the region must be restored to its pre-pollution condition. Vegetation planting, wetland restoration, and other

habitat revitalisation are all examples of restoration activities. These measures depend on helping the ecology recover from the spill's effects.

- Monitoring: Continuous monitoring is required to ensure the ecosystem completely recovers when the cleaning and restoration operations are done. Testing the water and soil, in addition to biological monitoring, may help determine how effectively an ecosystem is recovering. Further preventative steps may be required if problems are uncovered during monitoring.
- Research and development: Efforts are constantly being made to find new ways to reduce petroleum pollution. The objective of this research is to find novel approaches to the problems of oil spill prevention, pollution containment, and cleanup.
- Communication and public outreach: Success in mitigating an issue depends on your ability to get the word out and engage the public. The public has a right to know about the preventative measures and emergency procedures of companies that deal with petroleum products. More people will be aware of the dangers posed by petroleum pollution, and more people will be motivated to reduce pollution in their areas of successful public education initiatives.

5.6 Conclusions

The emergence of new petroleum pollutants is a growing challenge that demands rapid action from policymakers, industries, and individuals. The production and consumption of petroleum-based products are growing at the same rate as the demand for energy, which is continuing to rise. These items, if not managed and disposed of appropriately, have the potential to have significant effects on both human and environmental health. The list of environmental contaminants already in existence has been expanded due to the discovery of new pollutants derived from petroleum, such as PAHs, PCBs, PFAS, EDCs, and flame retardants. Since these contaminants are pervasive, toxic, and have the potential to accumulate in the food chain, they provide a substantial risk to both the health of humans and marine life. The growing use of hydraulic fracturing (fracking) and deep-sea drilling further adds to the situation's complexity by raising the likelihood of leaks and spills.

Emerging pollutants derived from petroleum have a negative impact not only on the economy but also on human health and the environment. The ever-increasing need for energy and the continued use of goods derived from petroleum are the underlying causes of this pollution. It is critically essential to comprehend the detrimental effects of petroleum pollutants and take action to avoid future damage as new types of these pollutants continue to develop. Adopting a holistic strategy to reduce the negative consequences caused by newly discovered petroleum contaminants is necessary. This strategy should involve the implementation of more stringent regulations and penalties for industries that engage in practices that are not responsible, investments in research and development of alternative, sustainable products, and the promotion of environmentally conscious behaviour by individuals. In addition, it is of the utmost importance to acknowledge that the consequences of emerging petroleum pollutants are not confined to specific geographical areas or populations. These contaminants have the potential to disperse throughout the globe via air and water currents, having a negative impact not just on people's health but also on their livelihoods and the ecosystems on which they depend. As a result, it is essential to adopt a global approach to tackling these concerns and to collaborate to stop any more damage.

To effectively control rising petroleum pollutants, a collaborative effort on the part of the industry, governments, and individual citizens is required. Industry executives are responsible for implementing environmentally friendly practices and making financial investments in research and development to discover viable alternatives to petroleum-based products. Governments must impose stringent rules and sanctions to prevent businesses from participating in environmentally harmful behaviours. In addition, it is essential to understand that growing petroleum contaminants are not solitary problems. They are a component of a more widespread environmental catastrophe that demands an all-encompassing solution. There is a connection between climate change, the loss of biodiversity, and pollution, and immediate action is required. If these challenges are not addressed, there will be significant repercussions for both the Earth and the generations to come in the future.

To summarise, emerging petroleum pollutants significantly harm the environment, human health, and the economy. The damaging effects of these pollutants on the environment are widespread and can have inevitable consequence that persists for a long time. It is indispensable for all parties involved to accept responsibility and collaborate toward the goal of minimising additional damage. We can build a more sustainable future and guarantee that developing petroleum pollutants does not harm our world more if we implement sustainable behaviours, promote environmental awareness, and enforce stricter rules. These are the three pillars of the sustainability movement.

References

Armstrong B, Hutchinson E, Unwin J, Fletcher T (2004) Lung cancer risk after exposure to polycyclic aromatic hydrocarbons: a review and meta-analysis. Environ Health Perspect 112(9):970–978. https://doi.org/10.1289/ehp.6895

Bach PB, Kelley MJ, Tate RC, McCrory DC (2003) Screening for lung cancer: a review of the current literature. Chest 123:72S-82S. https://doi.org/10.1378/chest.123.1

Bachmann RT, Johnson AC, Edyvean RGJ (2014) Biotechnology in the petroleum industry: an overview. Int Biodeterior Biodegrad 86:225–237. https://doi.org/10.1016/j.ibiod.2013.09.011

Carpenter DO (2006) Polychlorinated biphenyls (PCBs): routes of exposure and effects on human health. Rev Environ Health 21(1):1–23. https://doi.org/10.1515/reveh.2006.21.1.1

Chen D, Kannan K, Tan H, Zheng Z, Feng YL, Wu Y, Widelka M (2016) Bisphenol analogues other than BPA: environmental occurrence, human exposure, and toxicity - a review. Environ Sci Technol 50(11):5438–5453. https://doi.org/10.1021/acs.est.5b05387

Collins JF, Brown JP, Dawson SV, Marty MA (1991) Risk assessment for benzo[a]pyrene. Regul Toxicol Pharmacol 13(2):170–184. https://doi.org/10.1016/0273-2300(91)90020-V

Darnerud PO (2003) Toxic effects of brominated flame retardants in man and in wildlife. Environ Int 29(6):841–853. https://doi.org/10.1016/S0160-4120(03)00107-7

Del Carmen García M, Carbognani L, Orea M, Urbina A (2000) The influence of alkane class-types on crude oil wax crystallization and inhibitors efficiency. J Petrol Sci Eng 25(3–4):99–105. https://doi.org/10.1016/S0920-4105(99)00057-1

Diggs DL, Huderson AC, Harris KL, Myers JN, Banks LD, Rekhadevi PV, Niaz MS, Ramesh A (2011) Polycyclic aromatic hydrocarbons and digestive tract cancers: a perspective. J Environ Sci Health C 29(4):324–357. https://doi.org/10.1080/10590501.2011.629974

Enick RM, Olsen D, Ammer J, Schuller W (2012) Mobility and conformance control for CO2 EOR via thickeners, foams, and gels - a literature review of 40 years of research and pilot tests. SPE - DOE Improv Oil Recov Symp Proc 2:910–921. https://doi.org/10.2118/154122-ms

Erickson MD, Kaley RG (2011) Applications of polychlorinated biphenyls. Environ Sci Pollut Res 18(2):135–151. https://doi.org/10.1007/s11356-010-0392-1

Gao Y, Xiong W, Ling W, Xu J (2006) Sorption of phenanthrene by soils contaminated with heavy metals. Chemosphere 65(8):1355–1361. https://doi.org/10.1016/j.chemosphere.2006.04.030

Garcia-Suastegui WA, Huerta-Chagoya A, Carrasco-Colin KL, Pratt MM, John K, Petrosyan P, Rubio J, Poirier MC, Gonsebatt ME (2011) Seasonal variations in the levels of PAH-DNA adducts in young adults living in Mexico City. Mutagenesis 26(3):385–391. https://doi.org/10.1093/mutage/geq104

Glasby GP (2006) Abiogenic origin of hydrocarbons: an historical overview. Resour Geol 56(1):83–96. https://doi.org/10.1111/j.1751-3928.2006.tb00271.x

Glüge J, Scheringer M, Cousins IT, Dewitt JC, Goldenman G, Herzke D, Lohmann R, Ng CA, Trier X, Wang Z (2020) An overview of the uses of per- and polyfluoroalkyl substances (PFAS). Environ Sci Process Impacts 22(12):2345–2373. https://doi.org/10.1039/d0em00291g

Gunter MJ, Divi RL, Kulldorff M, Vermeulen R, Haverkos KJ, Kuo MM, Strickland P, Poirier MC, Rothman N, Sinha R (2007) Leukocyte polycyclic aromatic hydrocarbon–DNA adduct formation and colorectal adenoma. Carcinogenesis 28(7):1426–1429. https://doi.org/10.1093/carcin/bgm022

Han Q, Li M, Liu X, Jiang W, Shi S, Tang Y, He D (2020) Fractionation of alkylated carbazoles in petroleum during subsurface migration: evidence from molecular simulation and application in sandstone reservoirs. J Petrol Sci Eng 191:107308. https://doi.org/10.1016/j.petrol.2020.107308

Hauser R, Calafat AM (2005) Phthalates and human health. Occup Environ Med 62(11):806–818. https://doi.org/10.1136/oem.2004.017590

Haverinen E, Fernandez MF, Mustieles V, Tolonen H (2021) Metabolic syndrome and endocrine disrupting chemicals: an overview of exposure and health effects. Int J Environ Res Public Health 18(24). https://doi.org/10.3390/ijerph182413047

Hirsch RL, Bezdek R, Wendling R (2005) Peaking of world oil production: Impacts, mitigation, & risk management. Report for the US Department of Energy. https://doi.org/10.2172/939271

Jia C, Batterman S (2010) A critical review of naphthalene sources and exposures relevant to indoor and outdoor air. Int J Environ Res Public Health 7(7):2903–2939. https://doi.org/10.3390/ijerph7072903

Kinik FP, Ortega-Guerrero A, Ongari D, Ireland CP, Smit B (2021) Pyrene-based metal organic frameworks: from synthesis to applications. Chem Soc Rev 50(5):3143–3177. https://doi.org/10.1039/d0cs00424c

Kumar S, Krishnan S, Samal SK, Mohanty S, Nayak SK (2018) Toughening of petroleum based (DGEBA) epoxy resins with various renewable resources based flexible chains for high performance applications: a review. Ind Eng Chem Res 57(8):2711–2726. https://doi.org/10.1021/acs.iecr.7b04495

Kutcherov VG, Krayushkin VA (2010) Deep-seated abiogenic origin of petroleum: from geological assessment to physical theory. Rev Geophys 48(1):1–30. https://doi.org/10.1029/2008RG000270

Lawal AT (2017) Polycyclic aromatic hydrocarbons. A review. Cogent Environ Sci 3(1):1339841. https://doi.org/10.1080/23311843.2017.1339841

Li R, Munoz G, Liu Y, Sauvé S, Ghoshal S, Liu J (2019) Transformation of novel polyfluoroalkyl substances (PFASs) as co-contaminants during biopile remediation of petroleum hydrocarbons. J Hazard Mater 362:140–147. https://doi.org/10.1016/j.jhazmat.2018.09.021

Lin SX, Curtis MA, Sperry J (2020) Pyridine alkaloids with activity in the central nervous system. Bioorg Med Chem 28(24):115820. https://doi.org/10.1016/j.bmc.2020.115820

Lottrup G, Andersson AM, Leffers H, Mortensen GK, Toppari J, Skakkebæk NE, Main KM, De Muinck Keizer-Schrama S, Main KM, Swan S, Main KM, Foster P, Astrup Jensen A (2006) Possible impact of phthalates on infant reproductive health. Int J Androl 29(1):172–180. https://doi.org/10.1111/j.1365-2605.2005.00642.x

Medha K, Rajna S, Devi LJ, Samant L, Jose S (2021) A comprehensive review on moth repellent finishing of woolen textiles. J Cult Herit 49:260–271. https://doi.org/10.1016/j.culher.2021.02.006

Mojarad AAS, Atashbari V, Tantau A (2018) Challenges for sustainable development strategies in oil and gas industries. Proc Int Conf Bus Excell 12(1):626–638. https://doi.org/10.2478/picbe-2018-0056

Mu J, Wang J, Jin F, Wang X, Hong H (2014) Comparative embryotoxicity of phenanthrene and alkyl-phenanthrene to marine medaka (Oryzias melastigma). Mar Pollut Bull 85(2):505–515. https://doi.org/10.1016/j.marpolbul.2014.01.040

Mumbo J, Henkelmann B, Abdelaziz A, Pfister G, Nguyen N, Schroll R, Munch JC, Schramm K-W (2015) Persistence and dioxin-like toxicity of carbazole and chlorocarbazoles in soil. Environ Sci Pollut Res 22(2):1344–1356. https://doi.org/10.1007/s11356-014-3386-6

Murty KG (ed) (2020) History of crude oil refining. Springer, pp 1–8. https://doi.org/10.1007/978-3-030-40212-9_1

Nowack B (2009) The behavior and effects of nanoparticles in the environment. Environ Pollut 157(4):1063–1064. https://doi.org/10.1016/j.envpol.2008.12.019

Odeyemi O, Ogunseitan OA (1985) Petroleum industry and its pollution potential in Nigeria. Oil Petrochem Pollut 2(3):223–229. https://doi.org/10.1016/S0143-7127(85)90218-2

Okoye CO, Jones I, Zhu M, Zhang Z, Zhang D (2021) Manufacturing of carbon black from spent tyre pyrolysis oil – a literature review. J Clean Prod 279:123336. https://doi.org/10.1016/j.jclepro.2020.123336

Olsson AC, Fevotte J, Fletcher T, Cassidy A, 't Mannetje A, Zaridze D, Szeszenia-Dabrowska N, Rudnai P, Lissowska J, Fabianova E, Mates D, Bencko V, Foretova L, Janout V, Brennan P, Boffetta P (2010) Occupational exposure to polycyclic aromatic hydrocarbons and lung cancer risk: a multicenter study in Europe. Occup Environ Med 67(2):98–103. https://doi.org/10.1136/oem.2009.046680

Overton EB, McFall JA, Mascarella SW, Steele CF, Antoine SA, Politzer IR, Laseter JL (1981) Identification of petroleum residue sources after a fire and oil spill. In: International oil spill conference proceedings, pp 541–546. https://doi.org/10.7901/2169-3358-1981-1-541

Preuss R, Angerer J, Drexler H (2003) Naphthalene - an environmental and occupational toxicant. Int Arch Occup Environ Health 76(8):556–576. https://doi.org/10.1007/s00420-003-0458-1

Rochester JR (2013) Bisphenol A and human health: a review of the literature. Reprod Toxicol 42:132–155. https://doi.org/10.1016/j.reprotox.2013.08.008

Ross G (2004) The public health implications of polychlorinated biphenyls (PCBs) in the environment. Ecotoxicol Environ Saf 59(3):275–291. https://doi.org/10.1016/j.ecoenv.2004.06.003

Santana RC, Do PT, Santikunaporn M, Alvarez WE, Taylor JD, Sughrue EL, Resasco DE (2006) Evaluation of different reaction strategies for the improvement of cetane number in diesel fuels. Fuel 85(5–6):643–656. https://doi.org/10.1016/j.fuel.2005.08.028

Schulz K, Silva MR, Klaper R (2020) Distribution and effects of branched versus linear isomers of PFOA, PFOS, and PFHxS: a review of recent literature. Sci Total Environ 733:139186. https://doi.org/10.1016/j.scitotenv.2020.139186

Scriven EFV, Murugan R (2005) Pyridine and pyridine derivatives. In: Kirk-Othmer encyclopedia of chemical technology. John Wiley & Sons, Inc., pp 1–53. https://doi.org/10.1002/0471238961.1625180919031809.a01.pub2

Shanahan CE, Spak SN, Martinez A, Hornbuckle KC (2015) Inventory of PCBs in Chicago and opportunities for reduction in airborne emissions and human exposure. Environ Sci Technol 49(23):13878–13888. https://doi.org/10.1021/acs.est.5b00906

Smirnov NN (2014) Chemical composition. In: Physiology of the cladocera, 2nd edn. Elsevier, pp 13–32. https://doi.org/10.1016/B978-0-12-396953-8.00003-3

Staplin F (1969) Sedimentary organic matter, organic metamorphism, and oil and gas occurrence. Bull Can Pet Geol 17(1):47–66. https://doi.org/10.35767/gscpgbull.17.1.047

Steenland K, Fletcher T, Savitz DA (2010) Epidemiologic evidence on the health effects of perfluorooctanoic acid (PFOA). Environ Health Perspect 118(8):1100–1108. https://doi.org/10.1289/ehp.0901827

Sun K, Song Y, Zong W, Tang J, Liu R (2020) Anthracene-induced DNA damage and oxidative stress: a combined study at molecular and cellular levels. Environ Sci Pollut Res 27(33):41458–41474. https://doi.org/10.1007/s11356-020-10049-y

Talmadge MS, Baldwin RM, Biddy MJ, McCormick RL, Beckham GT, Ferguson GA, Czernik S, Magrini-Bair KA, Foust TD, Metelski PD, Hetrick C, Nimlos MR (2014) A perspective on oxygenated species in the refinery integration of pyrolysis oil. Green Chem 16(2):407–453. https://doi.org/10.1039/c3gc41951g

Thomas JA, Thomas MJ, Gangolli SD (1984) Biological effects of DI-(2-Ethylhexyl) phthalate and other phthalic acid esters. CRC Crit Rev Toxicol 13(4):283–317. https://doi.org/10.3109/10408448409023761

Unwin J, Cocker J, Scobbie E, Chambers H (2006) An assessment of occupational exposure to polycyclic aromatic hydrocarbons in the UK. Ann Occup Hyg 50(4):395–403. https://doi.org/10.1093/annhyg/mel010

Van Damme J, Du Prez F (2018) Anthracene-containing polymers toward high-end applications. Prog Polym Sci 82:92–119. https://doi.org/10.1016/j.progpolymsci.2018.02.002

Vandenberg LN, Hauser R, Marcus M, Olea N, Welshons WV (2007) Human exposure to bisphenol A (BPA). Reprod Toxicol 24(2):139–177. https://doi.org/10.1016/j.reprotox.2007.07.010

Vandenbroucke M (2003) Kerogan: from types to models of chemical structure. Oil Gas Sci Technol 58(2):243–269. https://doi.org/10.2516/ogst:2003016

Verma C, Quraishi MA, Ebenso EE (2020) Quinoline and its derivatives as corrosion inhibitors: a review. Surf Interfaces 21:100634. https://doi.org/10.1016/j.surfin.2020.100634

Waigi MG, Kang F, Goikavi C, Ling W, Gao Y (2015) Phenanthrene biodegradation by sphingomonads and its application in the contaminated soils and sediments: a review. Int Biodeterior Biodegrad 104:333–349. https://doi.org/10.1016/j.ibiod.2015.06.008

Wang P, Liu D, Yan S, Cui J, Liang Y, Ren S (2022) Adverse effects of perfluorooctane sulfonate on the liver and relevant mechanisms. Toxics 10(5):265. https://doi.org/10.3390/toxics10050265

Yost EE, Galizia A, Kapraun DF, Persad AS, Vulimiri SV, Angrish M, Lee JS, Druwe IL (2021) Health effects of naphthalene exposure: a systematic evidence map and analysis of potential considerations for dose–response evaluation. Environ Health Perspect 129(7). https://doi.org/10.1289/EHP7381

Zhang L, Liu J, Hu J, Liu C, Guo W, Wang Q, Wang H (2012) The inventory of sources, environmental releases and risk assessment for perfluorooctane sulfonate in China. Environ Pollut 165:193–198. https://doi.org/10.1016/j.envpol.2011.09.001

Zhou K, Zhou X, Liu J, Huang Z (2020) Application of magnetic nanoparticles in petroleum industry: a review. J Petrol Sci Eng 188:106943. https://doi.org/10.1016/j.petrol.2020.106943

Chapter 6
Environmental Fate and Microbial Reactions to Petroleum Hydrocarbon Contamination in Terrestrial Ecosystems

Pankaj Parmar, Rashmi Dhurandhar, and Sriya Naik

Abstract Petroleum hydrocarbons are the most prevalent contamination in the environment. The ecosystem's functionality is diminished when petroleum hydrocarbons are introduced into a healthy environment because they swiftly change the ecology's properties. Natural attenuation is the only and the most significant biological activity that cleans up the environment by removing petroleum hydrocarbons. The microorganisms already existing without the assistance of exogenous bioremediation enhancers like electron suppliers, electron acceptors, other organisms, or nutrients break down the organic pollutants at the site. Because of how well this innate attenuation mechanism works, environmental biotechnology has progressively developed bioremediation in the previous 50 years. Bioremediation is based on the naturally occurring biodegradation. Petroleum hydrocarbon pollution is the most prevalent type. Petroleum hydrocarbons rapidly modify the ecology's properties, lowering the ecosystem's functionality when introduced into a healthy setting. Natural attenuation is the only biological process that is most significant for removing petroleum hydrocarbons from the environment. The on-site microorganisms can break down the organic pollutants without using outside agents that improve bioremediation, like electron donors, electron acceptors, extra bacteria, or nutrients. Environmental biotechnology has been progressively expanding bioremediation based on this organic biodegradation process for the past 50 years due to the efficacy of this attenuation process in nature. Petroleum hydrocarbons begin to interact with their surroundings when they pollute the land. These interactions can be physical (dispersion), biological (plant and microbial catabolism of hydrocarbons), chemical (photo-oxidation, auto-oxidation), or physiochemical (evaporation, dissolution, sorption) processes. Investigations on the microbial communities inside polluted soils are crucial for any bioremediation project because microorganisms (including bacteria

P. Parmar (✉) · S. Naik
Chemical Engineering Department, Indian Institute of Technology Kharagpur, Kharagpur 721302, West Bengal, India
e-mail: pankajuecu23@gmail.com

R. Dhurandhar
Chemical Engineering Department, National Institute of Technology Durgapur, Durgapur 713209, West Bengal, India

I. D. Behera and A. P. Das (eds.), *Impact of Petroleum Waste on Environmental Pollution and its Sustainable Management Through Circular Economy*, Environmental Science and Engineering, https://doi.org/10.1007/978-3-031-48220-5_6

and fungus) play a significant role in the breakdown of petroleum hydrocarbons. This article emphasizes the fate of petroleum hydrocarbons in tertial habitats and the contributions of various microbial consortia for the best potential for petroleum hydrocarbon bioremediation. It also highlights how high-throughput metagenomic sequencing affects the identification of the underlying mechanisms of deterioration. This information will support the creation of commercial bioremediation systems that are more effective and affordable.

Keywords Microbiological consortia · Natural attenuation · Bioremediation · Petroleum hydrocarbon

6.1 Introduction

The addition of petroleum hydrocarbons instantly alters a once-clean environment. The added hydrocarbons kill or inhibit numerous microbial species, which changes how well the microbial community and, consequently, the ecosystem function. The direct toxicity of hydrocarbons to plants, the denial of access to light, the difficulty obtaining nutrients and water because oil restricts their flow through the soil matrix and all these effects considerably reduce plant productivity (Nie et al. 2011). The contaminated habitat has a restricted capacity to host higher-order life forms because it lacks primary producers and functional microbial biogeochemical networks.

The weathering process starts when hydrocarbons from petroleum are released into the environment. Petroleum hydrocarbons are weathered in the environment by a variety of processes: biological (microbial and plant catabolism of hydrocarbons), chemical (photo-oxidation, auto-oxidation), or physiochemical (evaporation, dissolution, sorption) processes. This article's main topic, biodegradation, is mainly handled by the natural microbial community of the soil. Natural attenuation is the only and most effective biodegradative method for eliminating petroleum-related environmental hydrocarbons. Without the assistance of exogenous bioremediation enhancers (for instance, electron donors, electron acceptors, additional microorganisms, or nutrients), the site's existing microorganisms break down the organic contaminants (Ellis 2011). An ecologically balanced environment is maintained in large part by microorganisms. They are in charge of controlling several soil ecosystem activities, such as the reuse of nutrients, the decomposition of organic matter, and the development of symbiotic partnerships with plants (Van Der Heijdan et al. 2008). Microorganisms are exposed to situations that have characteristics they would not often encounter in natural settings because of petroleum hydrocarbon contamination. It is crucial to comprehend how anthropogenic chemical pollutants interact with microbial populations in soil and how they interact with one another.

Different microorganisms have different ways of adjusting to and degrading petroleum hydrocarbons. The degradation of both inorganic and organic contaminants by enzymes is a mechanism for hydrocarbon degradation that is possessed by individual microorganisms; however, other species may contribute to this process through symbiotic relationships (For instance, the release of glucose to encourage the growth of species that break down hydrocarbons or the secretion of surfactants to increase the bioavailability of the oil (Bak et al. 2015). This natural attenuation process is so effective that it has served as the foundation for environmental biotechnology, or bioremediation, continuously developing in the previous 50 years. This review will explain how petroleum hydrocarbons behave in the environment and how they interact with aerobically and anaerobically biotic and abiotic elements. The examination of natural defences against petroleum hydrocarbon contamination will also take place. A thorough analysis of these processes will lead to a better possibility of natural attenuation is understood to eliminate all petroleum hydrocarbon fractions. This is then utilized for updating environmentally friendly and sustainable bioremediation techniques currently utilized in industrial bioremediation. Creating custom microbial consortia for various hydrocarbon classes and site-specific circumstances is suggested. Additionally, a closely watched site clean-up strategy introducing groups of microorganisms at various clean-up stages may encourage total microbial hydrocarbon catabolism. Another essential requirement is to combine combination of microbial functional analysis with next-generation sequencing to understand how microbial consortia detoxify petrogenic hydrocarbons at various locales.

6.2 Upon Entering the Environment, Oil Composition Changes

Petrogenic hydrocarbons are a diverse class of contaminants, including polyaromatic hydrocarbons (PAHs), cyclo-alkanes, n-alkanes, and branched alkanes, commonly released into the environment. These numerous hydrocarbon compounds, which range in size and structure from C1 (methane) to n-C40+, interact with their surroundings in a variety of ways that are typically based on each substance's molecular weight and properties. Petroleum hydrocarbons are classed based on their chemical composition into saturates, aromatics, resins, and asphaltenes. As a result of (Dispersion) physical), physiochemical (dissolution, sorption and evaporation), chemical (auto-oxidation, photo-oxidation), and biological (microbial and plant breakdown of hydrocarbons) influences, petroleum hydrocarbons weather in the environment (Boehm et al. 1982). In (Fig. 6.1). Each site will experience a different level and type of weathering, and it could significantly influence the potential for oil to degrade.

6.2.1 Volatilization

Its lighter components volatilize into the atmosphere as the oil moves across the ground. Petroleum hydrocarbons can behave differently from one another. If other environmental interactions do not prevent light crudes and petrol from evaporating

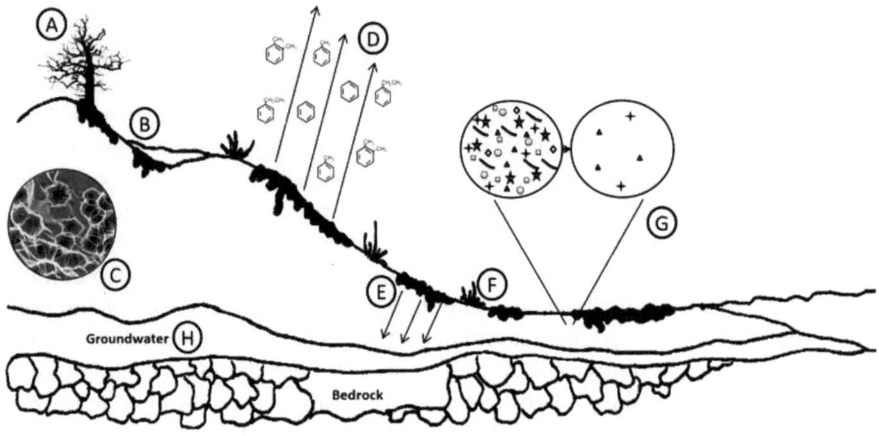

Fig. 6.1 Changes in site characteristics and oil's fate after a terrestrial oil spill: Plant demise, anaerobic zones, altered soil structure, volatilization, hydrocarbon percolation, aerobic zones, early decline in microbial populations and diversity, and hydrocarbon-contaminated groundwater are only a few examples of the effects that can occur (Truskewycz et al. 2019)

in warm temperatures. However, the volatilization of bunker oil may only reach a small percent of the total. Lighter aromatics, such as BTEX and various straightforward ringed structures, are liberated from more intricate oil mixes by volatilization (Fig. 6.1). It is thought that 1–5 ring saturated hydrocarbons configurations are flammable or unstable. As volatilization rate and capability increase with temperature, chain length decreases volatility (Truskewycz et al. 2019). Volatilization can happen in the subsurface in addition to oil that is surface-dwelling.

Essaid et al. (2011) observed and calculated the 25-year evolution from the hydrocarbons a ruptured oil pipeline for crude Minnesota, Bemidji, which polluted the environment with 1.7 million liters of oil. The oil that had permeated the soil and was still evaporating created at the saturated zone, a plume of soil vapour (Van Metre et al. 2012).

6.2.2 Dissolution

A hydrocarbon's solubility in water diminishes with increasing hydrocarbon chain length or aromatic ring count. The composition of the oil determines the number of petroleum hydrocarbons that are dissolved in water. Less dissolvable fractions may exist in large quantities of viscous crude oil than in smaller amounts of lighter-class oil. However, the C34 + n-alkane hydrocarbons' solubility in oils can be increased by polar nonhydrocarbon compounds that are common in the environment. Hydrocarbon chains below C_8 are frequently present in the dissolvable fractions at 20°C. Additionally, BTEX, toluene, benzene, xylene, ethyl benzene, and naphthalene are among the numerous light aromatic and polycyclic aromatic chemicals frequently discovered in contaminated groundwater (Essaid et al. 2011). The solubility of hydrocarbons can be significantly impacted by salinity.

The displacement of up to 95% of all hydrocarbons from the solution is possible at salinities of 350,000 ppm. The ability of hydrocarbons to bind to soils and form a solution in water can be influenced by temperature. Research shows that hydrocarbon solubility decreases logarithmically when a hydrocarbon's boiling point rises (De Hemptinne et al. 1998).

6.2.3 Desorption and Sorption

Numerous hydrophobic hydrocarbon fractions have a connection to bind to soil constituents through several methods. These comprise the diffusion of hydrocarbons into nano pores, which are inaccessible to microorganisms, part of the division of hydrocarbons adding organic materials to soil, and the attachment and establishment of powerful connections with sites on soil organic material. The sorption and desorption kinetics are probably significantly influenced by the structure and content of the soil. Increases in clay content, organic matter, and hydrophobicity of the hydrocarbons have all been demonstrated to increase hydrocarbon sorption in the soil. The lower water solubility and inaccessibility of older PAH fractions attached to soil components increase their recalcitrance (Landmeyer and Effinger 2016). It has been possible to desorb firmly bound hydrocarbons from soil matrices by interacting oil with plant-secreted bioactive chemicals.

Additionally, aggregates can be damaged by plant roots, which can release hydrocarbons that are bonded by nano pores. A surfactant is frequently used to lower oil viscosity during bioremediation of contaminated areas, facilitating oil release from sorption in soil. Hydrocarbon desorption from soil matrices is facilitated by increased content of sand, increased soil moisture and temperature (Khan et al. 2013).

6.3 Hydrocarbon Toxicology of Microbes and Communities of Microbes

Specific petroleum hydrocarbon fractions released into the environment have been related to causing major health issues in humans, including cancer, as well as high toxicity to soil biota. The variety and evenness of a microbial community are decreased by petroleum hydrocarbon pollution of a place (Yadav et al. 2016) for a number of factors, including:

- Regarding the chemical compounds' direct poisoning,
- the sequestering of nutrients and water, denying microbes access to crucial components for proliferation,

- Trouble adapting to extremely nonpolar circumstances, which might rupture microbial cells by dissolving the lipids in the cytoplasm membrane (Yadav et al. 2016).

According to Mukherjee et al. (2014), less microbial variety and evenness was seen in creosote-contaminated soils, but there was also an increase in overall microbial activity. This was probably brought about by the micro flora of the soil adaptation to and enrichment with hydrocarbon-degrading abilities (Yadav et al. 2016). There will be microorganism species that can tolerate and degrade hydrocarbons that will live and multiply at a given place even though most of them will not be able to (Nie et al. 2011). Numerous pathways exist to exchange nutrition, genetic material, or metabolites that might cause gene activation between microorganisms.

These broadly including straight transfer (like a transfer mediated by contact), direct exclusive exchange (like a phagosome-transfer through mediation), chemical diffusion-based indirect transfer, and electron transfer. This exchange of metabolites, nutrients, and other cell–cell interactions may improve the fitness of organisms that break down hydrocarbons or may favour parasites, decreasing microbial life diversity. Khudur et al. (2015) observed modifications in colonies of bacteria in soil following 14 days of exposure 40 mL/kg of gasoline added to diesel. Microbial evenness and diversity were discovered to have drastically declined. Alterations in biological balances are changed due to hydrocarbon contamination at a site, and a small number of specialized species develop to dominate it. However, by 4 weeks of contamination, swiftly increasing number of dominant species, most likely because of the production of numerous intermediates in degradation and generally decreased toxicity (Khudur et al. 2015).

When a colony of microbes adapts to incident of contamination, many changes in the community's features take place. This could involve adjustments to a cell's physiology and morphology and to the community's whole ecological dynamics. Natural stress tolerance and physiological characteristics like growth rates can have an impact on adaption. The diversity of microbial populations may produce positive genetic variants, some of which may be advantageous to the entire community (Yan et al. 2016). To enhance their capacity to move through the soil matrix and reach contaminants, microbial cells' morphological traits, sometimes referred to as structural features, may change in size and shape.

Any biological species can develop adaption mechanisms to employ a particular pollutant as a substrate based on the concentration and physio-chemical characteristics of the pollutant (Yan et al. 2016). Several factors can a hydrocarbon's level of danger contamination is to the biota, including:

i. The actual hydrocarbon fraction: The length of the chain is closely connected to saturated hydrocarbons' inherent toxicity. Lighter molecules structural fractions (C_6 to C_{20}) have been determined to be more dangerous because they are highly bioavailable, yet more significant chain-saturated hydrocarbons have a higher potential for mutation (Boehm et al. 1982). Unsaturated hydrocarbon structures may have more predictable toxicity. It may depend on active reactive groups, the solubility of the substance viscosity, interactions with the membrane, and water and membranes and its constituents, and other elements.

ii. The nutrient concentration: It has been demonstrated that the variety of the soil's microbiological populations with greater concentrations of N and P increases the size and composition of the microbial community activity that degrades hydrocarbons (Johnsen et al. 2005).

iii. Contamination by additional substances (such as heavy metals): When heavy metals are present, petroleum's toxicity to hydrocarbons as a co-contaminant to soil microorganisms is higher than when the sole contamination is a petroleum hydrocarbon. According to Khudur et al. (2019), lead (Pb) contamination made more resistant to microbial bioremediation are hydrocarbon-contaminated soils. Lead stresses species that break down hydrocarbons and inhibit several metabolic pathways, including many bacteria's enzymatic and respiratory processes. Reducing variety and overall biomass can have substantial, immediate negative effects on the native microorganisms in the soil (Sutton et al. 2013).

However, there are a number of adaptable methods used by the soil microbial population. Short-lived bacteria can quickly adjust to environmental changes by sporulating, creating secondary metabolites, forming mutualistic connections with other microorganism species, etc. Soil-based microbiological biofilms are naturally resistant to disturbances because they can conserve their contents through along with organic contaminants and slower diffusion rates (Burmølle et al. 2014).

6.4 Physiochemical Elements Affecting Petrogenic Hydrocarbons' Natural Attenuation

In addition to biotic elements like salinity, the availability of moisture, nutrients, the chemistry of the soil, and factor abiotic also impact the possibility of naturally reducing soil pollution by petroleum hydrocarbons. The potential for microbial pollution catabolism is substantially affected by these abiotic variables, which are altered by hydrocarbon contamination.

6.4.1 Nutrients and Additives

Access to nutrients is necessary for the microbial utilization of oil. Oils can lock up nutrients, making them unavailable to plants and microbes. The production of new plant and microbial cells' structural and metabolic building blocks and the support of all these processes depend on nutrients. Additionally, anaerobic hydrocarbon degradation (common in some nutrients may be used as the final microbial catabolism pathways' electron acceptors (for hydrocarbon breakdown) (Bento et al.

2005). Nutrient concentration changes also affect the ecological interactions and relationships among microbial populations, such as parasitism or communalism.

It has been demonstrated that bio stimulation, or the introduction of nutrients, causes increased hydrocarbon degradation. Still, it has also been shown that this reduces the fitness of other hydrocarbon-degrading species and benefits pathogenic bacterial species. Complex organic substrates with high humid content, such as mature compost, might encourage oil to desorb from the soil, boosting the greasy substance's bioavailability components and the rate of bioremediation (Plaza et al. 2009).

6.4.2 *Salinity*

Due to variations nutrients are locked away and osmotic pressure changes occur, high salt concentrations provide a selection pressure that renders the environment unfavourable for a variety of microbial species. When salts make hydrophobic organic molecules less soluble and available, this is known as "salting out". Oil fields frequently become salinized; this can cause microbial respiration to drop, which can slow the bioremediation rate for a site. Certain microbial consortiums are resistant to diesel breakdown from saline soils and exhibit functional redundancy. If the salinity was less than 15% (w/w), according to Riis et al., diesel might be biodegraded utilizing consortia from the species Halo Monas, Bacillus, Cellulomonas, and Diet Zia in soil (Riis et al. 2003).

With increases in salinity up to 20% (w/w), Kleinsteuber et al. revealed functional redundancy for the breakdown of diesel in colonies of naturally occurring microbes in salinised soils (6.4% w/w). Bacteria frequently "salt-in cytoplasm," increasing the amount of chloride and potassium ions in their cell to cope with halo stress. In intertidal sediments, biostimulation with K^+ and Ca^{2+} ions can increase the hydrocarbon elastic activity of microorganisms and help them buffer against osmotic stress. It has been demonstrated that soil polluted with motor oil has a lower capacity for microbial oil utilisation by 44% in soils with clay loam and 20% in Sand and clay soils under extreme salinity conditions (200 dS/m) generated by oil field brine (Rhykerd et al. 1995).

6.4.3 *Drought/Moisture/Rainfall*

Water is involved in almost all cellular activities and is essential for the intake of nutrient uptake and waste product synthesis in microbial cells. Water molecules stabilise proteins, DNA, and lipids, preserving microbial cells' structural integrity. Water is also necessary for the extracellular transport of macromolecules generated by microbes that interact with the environment. One of these instances is transferring the genetic code using plasmids from one type of microbe to another. By impeding their ability to interact with one another and the environment, soil microbial communities are significantly impacted by a water shortage, which may decrease abundance, variety, and structure (Sheik et al. 2011).

This effect was seen by those who used southern Spain's semi-arid agricultural soil to replicate drought conditions for just 180 days before noticing a significant decline in the microbial community in the soil's activities and a reduction in the mineralisation of C and N. By incorporating 100 t ha1 of compost into the soil, the harmful effects of the drought were reduced. For the clean-up of oil contamination during drought circumstances, combining microbial-driven bioremediation with phytoremediation might be beneficial. According to Phillips et al.'s research, wild rye plants in the Altai Mountains had Endophytic hexadecane degraders that were 100 times more numerous in their rhizospheres, unlike other plants when it was dry (Phillips et al. 2009). Based on responses of microbial communities to repeated severe rewetting and desiccation cycles.

Wet/dry cycles can cause small, yet potentially significant, changes in community structure differences in the proportional large amounts of some significant microbial species in neighbourhoods of soil. Although the possible impacts are little understood, oil bioremediation is affected by drought, droughts are probably to slow down the rate of oil bioremediation, according to extrapolations from studies on general microbial responses (Taccari et al. 2012). By sopping up the soil and lowering the amount of O_2 that may dissolve into the water, too much moisture can also hinder bioremediation. As a result, aerobic soils turn anaerobic. In a series of microcosm studies, Lahel et al. showed a >20% drop in the remediation efficacy when the humidity was raised from 10 to 30% of diesel (Lahel et al. 2016).

6.5 Relationships Between Microbes

Microorganisms are incredibly adept at modifying their surroundings to create favourable conditions for growth. Numerous microbial interactions allow them to achieve a competitive edge in specific ecosystem niches and support the microbial community's survival in challenging situations. Among these interactions are mutualism, amensalism, competition, parasitism, and predation. An extraordinary variety of oxidative and hydrolytic enzymes, as well as primary and secondary bioactive compounds (metabolites), are produced by microbial communities, which include bacteria and fungi (of various species), in natural habitats. These organic compounds have been linked to the mineralisation of multiple hydrocarbon fractions and are necessary for cell growth (Sabra et al. 2010).

The variety of hydrocarbon degradation processes that may be discovered inside a consortium is absent from single-strained microbial cultures. It has been established that interactions between mixed Microbiological communities can foster favourable survival conditions. One illustration is the relationship between the commensal local cyanobacterial species and different types of bacteria. Through photosynthesis, Blue-green bacterial mats have been found to raise oxygen that is dissolved level in the vicinity of them, forming breathing pockets used by local bacteria that can break down hydrocarbons (Abed and Köster 2005). The genes from various bacteria linked to hydrocarbon breakdown are listed in Table 6.1.

It is challenging to map microorganisms' interactions across time in a complex ecosystem. However, it is crucial for creating effective for the consortia' particular site and contaminant remediation. Deng et al. used various network techniques to determine the mutual exclusion (wrong) or co-occurrence (good) relationships within

Table 6.1 Hydrocarbon degradation-related microbial genes, their functions and pathways, and typical hosts (Truskewycz et al. 2019)

Genes (Location)	Function	Example organisms of origin
HTP (Chromosome)	Lignin breakdown by fungi Heme-thiolate peroxidase (encoding for heme-peroxidase)	Placenta rhodonia
Dyp (Chromosome)	Lignin breakdown by fungi Heme-peroxidase, a dye-decolorizing peroxidase	*Thamatephorus cucumeris*
POD Genes (Chromosome)	Degradation of lignin by fungi (heme-peroxidase encoding)	*Vjerkamdera adjusta, Gamoderma* sp.
LiP Genes (A, D, Lip) (Chromosome)	Lignin breakdown by fungi	*Phamerochaete*
MnP (Chromosome)	Peroxidase of manganese	*Phamerochaete chrysosporium*
Tmo (Chromosome)	Aerobic, benzene, toluene, ethylbenzene, and xylem (VTEX) multicomponent monooxygenase enzyme complexes	*Ralstonia picketti* PKO
Xyle € (Chromosome)	The ring cleavage reaction in PAH degradation is catechol 2, 3-dioxygenase	*Sphingomonas yamoikuae*
Tbu Genes (Chromosome)		
Pah (Plasmid)	Genes for naphthalene-phenanthrene dioxygenase	*Pseudomonas. Pseudomonas aeruginosa Pak1* and *Putida* OUS82
Nag gene (Plasmid)	Genes for naphthalene-dioxygenase	Strain U2 of *Ralstonia* sp.
Phn genes (C, I, H) (Plasmid)	Genes for phenanthrene dioxygenase	AFK2 strain of *Alcaligenes faecalis*
Dox genes (bphA bphE bphF bphG) (Plasmid)	Oxidation of dibenzothiophene (meta cleavage route)	*Falciparum* Sp.C18

groundwater microbial communities. Negative interactions can be linked to predation, competition, and amensalism, but positive relationships are more likely caused by Commensalism or mutualism (Deng et al. 2016).

6.6 Utilisation of Microorganisms for Hydrocarbons

6.6.1 Catabolism of Aerobic Hydrocarbons

Microbial hydrocarbon breakdown under aerobic circumstances is more rapid than in anaerobic environments. There are no thermodynamically undesirable requirements to hydrate the hydrocarbon to supply oxygen, influencing enhanced respiration in aerobic conditions. Aliphatic hydrocarbons undergo microbial aerobic catabolism when the contaminant's carbon backbone is broken or when a functional substitution occurs due to the oxygen molecule losing an electron. Bacterial mono-oxygenase enzymes carry out these electron-carrier-dependent processes, transforming n-alkane hydrocarbon to its corresponding alcohol. (Rojo 2009). Following additional oxidation, which gradually breaks away C–C bonds, these hydroxylated products reach the bacterial cell's peripheral metabolic pathways, resulting in more minor components that enter the cell's major metabolic pathway through -oxidation.

It is well known that bacteria can break down aliphatic chemicals. However, they must use separate metabolic pathways to break down aromatic molecules. Increased complexity is accompanied by increased resistance to the deterioration. Low-molecular-weight PAHs (LMW PAHs) have a higher potential for bioremediation since they are often more soluble. Alternative degradation pathways might be required because high-molecular-weight PAHs (HMW PAHs) may be too big to enter into many enzymes' active sites (Moreno and Rojo 2017). Additionally, the decline in their solubility in water and the rise in their possible carcinogenicity are detrimental to microbial breakdown. The last hydroxylation's electron acceptor and substrate and oxygenolytic ring breakage events in the bacterial aerobic breakdown of aromatic compounds is oxygen (Carmona et al. 2009). Essentially, oxygenase enzymes—most commonly monooxygenases or dioxygenases—are used in the bacterial aerobic PAH breakdown.

Through the action of O_2 oxidising enzymes, the aromatic ring is hydroxylated to create cis-dihydro diol, which is subsequently converted by a dehydrogenase into a diol intermediate. The aromatic ring is then destroyed by O_2 oxidising enzymes in routes for ortho- or meta-cleavage, resulting in daughter products (catechols, which eventually change into citric acid cycle intermediates). The P450-mediated process is an alternate bacterial pathway for PAH oxidation that non-ligninolytic fungi can also use. Several other routes have been described in the past (Fuchs et al. 2011), all of which entail hydrolysis-induced saturated ring cleavage.

6.6.2 Anaerobic Hydrocarbon Catabolism

In anaerobic environments, bacteria degrade hydrocarbons using metabolic pathways that use electron acceptors besides oxygen (O_2), such as metal ions, nitrate, and sulphate, to mention a few. The redox zones, microorganisms, and environmental site features can all affect the type of electron acceptor. Nitrate is the initial alternative electron acceptor that bacteria often utilise for hydrocarbon catabolism after O_2 in the atmosphere levels are reduced because microorganisms that reduce nitrate are frequently able-bodied anaerobes. Because the redox potentials of manganese, iron, and sulphate are all lower than those of nitrate, similar species of bacteria that reduce inorganic ions can use them.

For the purpose of producing methane, acetogenic and fermentative bacterial species must create by-products such as format/acetate and H_2/CO_2 because it has a lower redox potential compared to sulphur and metal ions. However, minor modifications of these may also occur. Five major routes for anaerobic bacterial hydrocarbon breakdown have been discovered. The addition of fumarate is one of these anaerobic processes to the hydrocarbon chain. Additionally, aromatics can be carboxylate, alkenes, and alkynes can be hydrated, and reverse methanogenesis is another anaerobic pathway (Abbasian et al. 2016).

6.6.2.1 Fumarate Increase in the Hydrocarbon Chain

Anaerobic bacterial species that add fumarate to a hydrocarbon chain cause a series of hydrocarbon chain rearrangements that are ultimately the oxidation pathway used to process, culminating in the creation of alkyl succinates. The ligation to coenzyme A, the reorganisation of the carbon skeleton, and beta-oxidation can all be used to break down these alkyl succinates. This process can degrade both aromatic and aliphatic hydrocarbons (Musat 2015). Alkanes in anaerobic oil reservoirs were degraded by the addition of fumarate by Bian et al. (2015). Alkyl succinate synthetase facilitates the joining of an n-alkane to fumarate's either terminal or sub-terminal carbon (using propane) in this hydrocarbon-degrading process.

2-(1-methylalkyl) succinates, also known as 2-alkylsuccinates, are the end products. Later, 2-(1-Methylalkyl) succinate is rearranged to form (2-methyl alkyl)-malonyl-CoA, which is subsequently decarboxylated to form a derivative of 4-methyl alkyl-CoA that proceeds through -oxidation. Thauera aromatic and Desulfobacula toluolica are bacteria that nitrate and sulphate reduce, respectively, that can use toluene as their only breakdown and the carbon source by adding fumarate to the carbon chain. Toluene is bound to the fumarate in this process with the help of benzyl succinate synthase (BSS), generating (R)-benzyl succinate, which is then metabolised through the -oxidation route (Qiao and Marsh 2005) and by using a sulphate-reducing enrichment culture, It was demonstrated by Safinowski, Meckenstock, and others that the PAH 2-methyl-naphthalene was broken down anaerobically. In this study, 2-methylnaphthalene's methyl group was fused with fumarate to produce 2-napthylmethylsuccinic acid was then transformed into 2-naphthoic acid via the -oxidation pathway (Safinowski and Meckenstock 2006).

6.6.2.2 The Hydrocarbon Chain is Hydroxylated Without Oxygen

Numerous aromatics and heteroaromatic hydrocarbons replaced with ethyl and propyl undergo hydroxylation of the carbon chain not requiring oxygen during degradation. Alcohols are produced by the enzyme ethyl benzene dehydrogenase (EBDH) from aromatic mono- and bicyclic ring-substituted compounds molecules. (e.g., converting ethyl benzene to (S)-1-phenyl ethanol) by catalysing using water (opposed to oxygen with aerobic settings) to hydroxylate the chains of hydrocarbons adjacent toward the ring). The structure is subsequently broken by thiolysis into acetyl-CoA and benzoyl-CoA after a series of enzymatic dehydrogenase, carboxylase, and ligase processes (Rabus et al. 2016).

6.6.2.3 Carboxylation of Aromatics

The discovery of carboxylic acids with aromatic hydrocarbons carboxyl groups derived from CO_2 demonstrates that hydrocarbons are activated by carboxylation, even though there needs to be more evidence in the literature to confirm the certitude that carboxylation is a hydrocarbon-activating process (Widdel and Musat 2010).

6.6.2.4 Hydration of Alkynes/Alkenes

In anaerobic environments, unsaturated double or triple bonds in alkene and alkyne hydrocarbons can be broken down by adding a water molecule, resulting in the parent hydrocarbons being changed into ketones, alcohols, or aldehydes. Squalene was broken down by microorganisms from marine sediments, according to Rontani et al. They proposed a microbial degradation process in which the hydrated double bonds, form ketone bodies and tertiary alcohols (Rontani et al. 2002).

6.6.2.5 Reverse Methanogenesis

Although there are some exceptions, reverse methanogenesis by archaea primarily facilitates the microbial breakdown of methane. When there is none of Archaea, Ettwig et al. demonstrated methane degradation under anaerobic circumstances and claimed that the group of bacteria connected the methane's anaerobic oxidation leads to denitrification. The terminal electron acceptors anaerobic methane oxidizers use to break down methane include iron, manganese, sulphate, and nitrate/nitrite. Since the process is anticipated to vary depending on the electron acceptors used, the precise mechanisms for reverse methanogenesis have yet to be completely defined. As a rule, the reaction moves slowly because it depends on the relationships between microbial groups that are symbiotic and syntrophic, which are challenging or impossible to cultivate in a lab setting (Cui et al. 2015).

It is believed that methyl-coenzyme M reductase, a crucial methanogenesis enzyme, is used in reverse by archaea's domain anaerobic methanotrophs to activate methane. Methane monooxygenases can attack methane because the bacteria Methylomirabilis oxyfera can convert NO into N_2 and O_2 from reduced nitrite (Fan et al. 2012).

6.6.3 Syntrophy

Syntrophy is the mechanism by which another uses the end products or intermediates from the metabolism of one species to promote proliferation (Fig. 6.2). Bacteria that produce methane, acetogenic products, syntrophic fermentation products, and syntrophic interactions frequently do so for microbial anaerobic hydrocarbon breakdown. These interactions aid in developing processes that would otherwise be energy-unfavourable, such as the breakdown of hydrocarbons through several metabolic pathways. Hydrolytic bacteria may break down hydrocarbons, proteins, and polysaccharides in an anoxic hydrocarbon-contaminated environment, producing amino acids, fatty acids, sugars, and starches. The species of bacteria that ferment food that creates carbon dioxide, hydrogen gas, butyrate, formate, propionate, butyrate, and lactate (Hanson and Hanson 1996) can then use these breakdown products (Fig. 6.2).

Methanogens use these substrates to produce methane because the conditions are perfect for them. Methane can be transformed to release electrons and carbon dioxide by reverse methanogens, such as methanotrophic bacteria (type II), which bacteria can reduce sulphur use to make hydrogen gas (in the absence of sulphate) (Musat et al. 2010). A feedback loop is created by the evolution of hydrogen gas, which helps methanogenic species increase (Plugge et al. 2011). A particular type of microorganism's by-products helps another type thrive and proliferate (derived from Wintermute and Silver 2010, with permission from CSHL Press). Polyaromatic hydrocarbons, or PAHs.

Alkanes have been observed to degrade in anaerobic environments using various electron acceptors and methanogenesis. Information about procedures and consortiums taking part in the breakdown which these hydrocarbons was gathered in a review by Mbadinga et al. The underlying principles governing the early Alkane activation in anaerobic settings were carboxylation and fumarate addition, and the former was verified as the only process under conditions of sulphate reduction. When grown utilising only cyclohexane as the carbon source, Geobacter spp.-dominated enrichments with anammox bacteria (Plantomycetales) have shown alkyl benzene nitrate reduction and anaerobic degradation with symbiotic interactions (Musat et al. 2010). Two primary degradation mechanisms have been documented for alkanes in methanogenic settings.

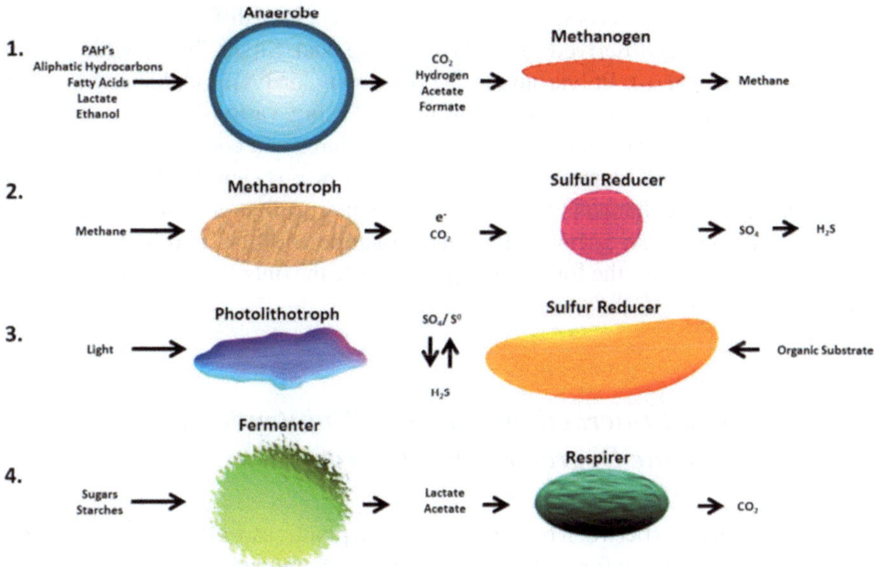

Fig. 6.2 Synoptic interactions: by-products from one species of microorganism help a different one grow and proliferate (derived from Wintermute and Silver 2010, with permission from CSHL Press)—polyaromatic hydrocarbons (Truskewycz et al. 2019)

Syntrophic bacteria first break down the hydrocarbon to produce acetate and H_2, after which acetoclastic methanogens like Methanosatea concilii can produce methane. The second pathway uses methanogens that produce hydrogen, such as methanospirillum and methanoculleus, which break down CO_2 and H_2 to produce methane. Numerous aquifers worldwide have been utilized to research the anaerobic breakdown of BTEX and various benzene degradation processes have also been postulated (Ulrich et al. 2005). Methanogenesis frequently occurs under conditions that reduce nitrate, sulphate, and iron. Thus, it has been demonstrated that various phylogenetically distinct microbes may utilize hydrocarbons under anoxic circumstances. Syntrophic interactions predominate amongst these species, and during the degradation process, carbon and electrons are shared.

The taxa engaged in hydrocarbon degradation have already been listed, and Gieg and colleagues have collated data on Methanogenic cultures that break down hydrocarbons and associated syntrophic processes. Deltaproteobacteria, Pelotomaculum, and Clostridium (all belonging to the Peptococcaceae genus) are a few of the bacteria that participated in the first attack by hydrocarbons. Numerous contaminated locations have yielded Synthropus, including Syntrophoceae and Smithella. Like those shown in Fig. 6.2, Acetoclastic methanogens frequently carry out the final phases of hydrocarbon breakdown in anaerobic environments (Coates et al. 2002).

A particular type of microorganism's by-products helps a different microorganism thrive and proliferate (derived from Wintermute and Silver 2010, with permission from CSHL Press). Polyaromatic hydrocarbons, or PAHs. Alkanes have been observed to degrade in anaerobic environments using various electron acceptors and methanogenesis. Information about the procedures and consortiums implicated in these hydrocarbons' degradation was gathered in a review by Mbadinga et al. The fundamental mechanisms that cause the alkanes to first get activated in anaerobic environments were carboxylation and fumarate addition, and under sulphate-reducing circumstances, the former was proven to be the only mechanism (Gray et al. 2010).

6.6.4 Microbial Interactions that May Unintentionally Promote the Degradation of Hydrocarbons

In nature, interactions between bacteria, fungi, and plants, as well as between bacteria and fungi, are the driving forces behind various environmental defense mechanisms. Bacterial species can communicate both with and without the need for physical contact. Grassroots gram-negative cells release N-acyl homoserine lactones (AHLs). They can control the generation of antibiotics, protein secretion, genetic transfer, cell aggregation, cell differentiation, and biofilm, and other processes. Small oligopeptides and proteins are frequently used by gram-positive bacteria to communicate with one another. Normally, oligopeptides interact with two-component extracellular adaptive response proteins after passing through the cell membrane. These proteins serve as receptors and relay information for sporulation, pathogenicity, genetic control, and other processes (Aburto et al. 2009).

Several microbial species, like Vibrio harveyi, have also adopted both methods for evolutionary proliferation. Microbial species that may produce biofilms are frequently discovered in extracellular polymeric materials (EPS). Furthermore, it has been demonstrated that EPS can concentrate BTEX contaminants and remove matrices made of soil or water (Sabra et al. 2010). This obstruction can defend against various extreme environmental conditions, shear forces, antibiotics, acid and UV damage, predators, dehydration, and high levels of dangerous contaminants and chemicals. It can harbour a single species or a group of species that interact. In addition to Mechanisms of cellular communication and chemical signals sent between cells (quorum sensing), biofilms have an excellent capacity for information transfer. Air-containing spaces are present in certain bacterial biofilms (such as Alcaligenes sp. strain d2), which facilitates the growth of micro colonies.

This event makes it possible to move oxygen and biomolecules to various community members. Several indirect processes can help oil deterioration caused by microbial interactions. One typical method bacteria use to dissolve viscous oils is the formation of bio surfactants. This changes them into a form more absorbed by the body (Sabra et al. 2010). Bacteria can also produce glucose to promote growth in the population among degraders of hydrocarbons, heavy metal chelators to lessen metal toxicity, extracellular enzymes that can degrade refractory hydrocarbons, and other biology compounds that promote expression of a functioning gene for reducing predator tolerance to and decomposition of hydrocarbons by producing antibiotics. Moreover, several microorganisms can release chemicals that communicate with plants and fungus. One such instance is the rhizobia bacterium synthesizing 1-aminocyclopropane-1-carboxylate (ACC) deaminase. As a result of this molecule's alteration in plant ethylene production, plants become more resilient to environmental challenges (Aburto-Medina and Ball 2015).

6.7 Conclusions and Plans for the Future

Petroleum hydrocarbons harm human health, economic well-being, and ecosystem health. Therefore, remediation is required to make these places functional again, whether for urban development or environmental preservation. The existing methods for cleaning up settings damaged by petroleum hydrocarbons are expensive and frequently carried out off-site, necessitating the first physical removal of polluted soil. All in-situ technologies have drawbacks, including high costs (flushing of the soil), the inability to clean up an entire contaminated site because solvent extraction from the soil is only suitable for certain types of soil, the potential for the site to not be a development tool for stabilization and solidification after remedial (Electro kinetic remediation), the creation of unwanted by-products, and the likelihood of process-induced adverse effects (ch).

Natural attenuation cannot effectively clean up locations where petrochemicals have been heavily contaminated. Therefore, complementary technologies have been created, like bioremediation, which is less expensive and more environmentally benign than traditional treatment methods. The understanding of environmental and microbiological science has historically restricted the use of bioremediation approaches. Chemical and physical characteristics have traditionally served as the foundation for models and assessments to apply bioremediation specifically since they are the aspects that are well understood. The utilization of accessible and Practical high-throughput techniques, such as next-generation sequencing (NGS) technology and microarrays, has only recently made it possible to comprehend the intricate dynamics of microbial consortiums in the breakdown of resistant hydrocarbon combinations.

Although research in Community interaction analysis is still in its infancy for complicated environmental systems like polluted soil, remediation techniques can only be developed with an understanding of the structural and functional roles played by microorganisms throughout entire communities.

References

Abbasian F, Lockington R, Megharaj M, Naidu R (2016) A review on the genetics of aliphatic and aromatic hydrocarbon degradation. Appl Biochem Biotechnol 178:224–250

Abed RM, Köster J (2005) The direct role of aerobic heterotrophic bacteria associated with cyanobacteria in the degradation of oil compounds. Int Biodeterior Biodegrad 55(1):29–37. https://doi.org/10.1016/j.ibiod.2004.07.001

Aburto A, Fahy A, Coulon F, Lethbridge G, Timmis KN, Ball AS, McGenity TJ (2009) Mixed aerobic and anaerobic microbial communities in benzene-contaminated groundwater. J Appl Microbiol 106(1):317–328. https://doi.org/10.1111/j.1365-2672.2008.04005.x

Aburto-Medina A, Ball AS (2015) Microorganisms involved in anaerobic benzene degradation. Ann Microbiol 65(3):1201–1213

Bak F, Bonnichsen L, Jørgensen NO, Nicolaisen MH, Nybroe O (2015) The bio surfactant viscos in transiently stimulates n-hexadecane mineralization by a bacterial consortium. Appl Microbiol Biotechnol 99:1475–1483

Bento FM, Camargo FA, Okeke BC, Frankenberger WT (2005) Comparative bioremediation of soils contaminated with diesel oil by natural attenuation, biostimulation and bioaugmentation. Biores Technol 96(9):1049–1055. https://doi.org/10.1016/j.biortech.2004.09.008

Bian XY, Maurice Mbadinga S, Liu YF et al (2015) Insights into the anaerobic biodegradation pathway of n-Alkanes in oil reservoirs by detection of signature metabolites. Sci Rep 5:9801. https://doi.org/10.1038/srep09801

Boehm PD, Fiest DL, Mackay D, Paterson S (1982) Physical-chemical weathering of petroleum hydrocarbons from the IXTOC I blowout: chemical measurements and a weathering model. Environ Sci Technol 16(8):498–505. https://doi.org/10.1021/es00102a014

Burmølle M, Ren D, Bjarnsholt T, Sørensen SJ (2014) Interactions in multispecies biofilms: do they actually matter? Trends Microbiol 22(2):84–91. https://doi.org/10.1016/j.tim.2013.12.004

Carmona M, Zamarro MT, Blázquez B, Durante-Rodríguez G, Juárez JF, Valderrama JA, Díaz E (2009) Anaerobic catabolism of aromatic compounds: a genetic and genomic view. Microbiol Mol Biol Rev 73(1):71–133. https://doi.org/10.1128/mmbr.00021-08

Coates JD, Chakraborty R, McInerney MJ (2002) Anaerobic benzene biodegradation—a new era. Res Microbiol 153(10):621–628. https://doi.org/10.1016/S0923-2508(02)01378-5

Cui M, Ma A, Qi H, Zhuang X, Zhuang G (2015) Anaerobic oxidation of methane: an "active" microbial process. Microbiol Open 4(1):1–11. https://doi.org/10.1002/mbo3.232

De Hemptinne JC, Delepine H, Jose C, Jose J (1998) Aqueous solubility of hydrocarbon mixtures. Revue De L'institut Français Du Pétrole 53(4):409–419

Deng Y, Zhang P, Qin Y, Tu Q, Yang Y, He Z, Zhou J (2016) Network succession reveals the importance of competition in response to emulsified vegetable oil amendment for uranium bioremediation. Environ Microbiol 18(1):205–218. https://doi.org/10.1111/1462-2920.12981

Ellis EC (2011) Anthropogenic transformation of the terrestrial biosphere. Philos Trans R Soc a: Math Phys Eng Sci 369(1938):1010–1035. https://doi.org/10.1098/rsta.2010.0331

Essaid HI, Bekins BA, Herkelrath WN, Delin GN (2011) Crude oil at the Bemidji site: 25 years of monitoring, modeling, and understanding. Groundwater 49(5):706–726

Fan L, Reynolds D, Liu M, Stark M, Kjelleberg S, Webster NS, Thomas T (2012) Functional equivalence and evolutionary convergence in complex communities of microbial sponge symbionts. Proc Natl Acad Sci 109(27):E1878–E1887. https://doi.org/10.1073/pnas.1203287109

Fuchs G, Boll M, Heider J (2011) Microbial degradation of aromatic compounds—from one strategy to four. Nat Rev Microbiol 9(11):803–816

Gray ND, Sherry A, Hubert C, Dolfing J, Head IM (2010) Methanogenic degradation of petroleum hydrocarbons in subsurface environments: remediation, heavy oil formation, and energy recovery. Adv Appl Microbiol 72:137–161. https://doi.org/10.1016/S0065-2164(10)72005-0

Hanson RS, Hanson TE (1996) Methanotrophic bacteria. Microbiol Rev 60(2):439–471. https://doi.org/10.1128/mr.60.2.439-471.1996

Johnsen AR, Wick LY, Harms H (2005) Principles of microbial PAH-degradation in soil. Environ Pollut 133(1):71–84. https://doi.org/10.1016/j.envpol.2004.04.015

Khan S, Afzal M, Iqbal S, Khan QM (2013) Plant–bacteria partnerships for the remediation of hydrocarbon contaminated soils. Chemosphere 90(4):1317–1332. https://doi.org/10.1016/j.chemosphere.2012.09.045

Khudur LS, Shahsavari E, Miranda AF, Morrison PD, Nugegoda D, Ball AS (2015) Evaluating the efficacy of bioremediating a diesel-contaminated soil using ecotoxicological and bacterial community indices. Environ Sci Pollut Res 22:14809–14819

Khudur LS et al (2019) The impact of lead co-contamination on ecotoxicity and the bacterial community during the bioremediation of total petroleum hydrocarbon-contaminated soils. Environ Pollut 253:939–948. https://doi.org/10.1016/j.envpol.2019.07.107

Lahel A, Fanta AB, Sergienko N, Shakya M, López ME, Behera SK, Park HS (2016) Effect of process parameters on the bioremediation of diesel contaminated soil by mixed microbial consortia. Int Biodeterior Biodegrad 113:375–385. https://doi.org/10.1016/j.ibiod.2016.05.005

Landmeyer JE, Effinger TN (2016) Effect of phytoremediation on concentrations of benzene, toluene, naphthalene, and dissolved oxygen in groundwater at a former manufactured gas plant site, Charleston, South Carolina, USA, 1998–2014. Environ Earth Sci 75:1–13

Moreno R, Rojo F (2017) Enzymes for aerobic degradation of alkanes in bacteria. In: Aerobic utilization of hydrocarbons, oils and lipids. Handbook of hydrocarbon and lipid microbiology, pp 1–25

Mukherjee S, Juottonen H, Siivonen P et al (2014) Spatial patterns of microbial diversity and activity in an aged creosote-contaminated site. ISME J 8:2131–2142. https://doi.org/10.1038/ismej.2014.151

Musat F (2015) The anaerobic degradation of gaseous, non-methane alkanes—from in situ processes to microorganisms. Comput Struct Biotechnnol J 13:222–228. https://doi.org/10.1016/j.csbj.2015.03.002

Musat F, Wilkes H, Behrends A, Woebken D, Widdel F (2010) Microbial nitrate-dependent cyclohexane degradation coupled with anaerobic ammonium oxidation. ISME J 4(10):1290–1301

Nie M, Wang Y, Yu J, Xiao M, Jiang L, Yang J, Li B (2011) Understanding plant-microbe interactions for phytoremediation of petroleum-polluted soil. PLoS ONE 6(3):e17961. https://doi.org/10.1371/journal.pone.0017961

Phillips LA, Greer CW, Farrell RE, Germida JJ (2009) Field-scale assessment of weathered hydrocarbon degradation by mixed and single plant treatments. Appl Soil Ecol 42(1):9–17. https://doi.org/10.1016/j.apsoil.2009.01.002

Plaza C, Xing B, Fernandez JM, Senesi N, Polo A (2009) Binding of polycyclic aromatic hydrocarbons by humic acids formed during composting. Environ Pollut 157(1):257–263. https://doi.org/10.1016/j.envpol.2008.07.016

Plugge CM, Zhang W, Scholten JC, Stams AJ (2011) Metabolic flexibility of sulfate-reducing bacteria. Front Microbiol 2:81. https://doi.org/10.3389/fmicb.2011.00081

Qiao C, Marsh ENG (2005) Mechanism of benzylsuccinate synthase: stereochemistry of toluene addition to fumarate and maleate. J Am Chem Soc 127(24):8608–8609. https://doi.org/10.1021/ja051972f

Rabus R, Boll M, Heider J, Meckenstock RU, Buckel W, Einsle O, Wilkes H (2016) Anaerobic microbial degradation of hydrocarbons: from enzymatic reactions to the environment. Microbial Physiol 26(1–3):5–28. https://doi.org/10.1159/000443997

Rhykerd RL, Weaver RW, McInnes KJ (1995) Influence of salinity on bioremediation of oil in soil. Environ Pollut 90(1):127–130. https://doi.org/10.1016/0269-7491(94)00087-T

Riis V, Kleinsteuber S, Babel W (2003) Influence of high salinities on the degradation of diesel fuel by bacterial consortia. Can J Microbiol 49(11):713–721. https://doi.org/10.1139/w03-083

Rojo F (2009) Degradation of alkanes by bacteria. Environ Microbiol 11(10):2477–2490. https://doi.org/10.1111/j.1462-2920.2009.01948.x

Rontani JF, Mouzdahir A, Michotey V, Bonin P (2002) Aerobic and anaerobic metabolism of squalene by a denitrifying bacterium isolated from marine sediment. Arch Microbiol 178:279–287

Sabra W, Dietz D, Tjahjasari D, Zeng AP (2010) Biosystems analysis and engineering of microbial consortia for industrial biotechnology. Eng Life Sci 10(5):407–421. https://doi.org/10.1002/elsc.201000111

Safinowski M, Meckenstock RU (2006) Methylation is the initial reaction in anaerobic naphthalene degradation by a sulfate-reducing enrichment culture. Environ Microbiol 8(2):347–352. https://doi.org/10.1111/j.1462-2920.2005.00900.x

Sheik CS, Beasley WH, Elshahed MS, Zhou X, Luo Y, Krumholz LR (2011) Effect of warming and drought on grassland microbial communities. ISME J 5(10):1692–1700

Sutton NB, Maphosa F, Morillo JA, Abu Al-Soud W, Langenhoff AA, Grotenhuis T, Smidt H (2013) Impact of long-term diesel contamination on soil microbial community structure. Appl Environ Microbiol 79(2):619–630. https://doi.org/10.1128/AEM.02747-12

Taccari M, Milanovic V, Comitini F, Casucci C, Ciani M (2012) Effects of biostimulation and bioaugmentation on diesel removal and bacterial community. Int Biodeterior Biodegrad 66(1):39–46. https://doi.org/10.1016/j.ibiod.2011.09.012

Truskewycz A, Gundry TD, Khudur LS, Kolobaric A, Taha M, Aburto-Medina A, Shahsavari E (2019) Petroleum hydrocarbon contamination in terrestrial ecosystems—fate and microbial responses. Molecules 24(18):3400. https://doi.org/10.3390/molecules24183400

Ulrich AC, Beller HR, Edwards EA (2005) Metabolites detected during biodegradation of 13C6-benzene in nitrate-reducing and methanogenic enrichment cultures. Environ Sci Technol 39(17):6681–6691. https://doi.org/10.1021/es050294u

Van Der Heijden MG, Bardgett RD, Van Straalen NM (2008) The unseen majority: soil microbes as drivers of plant diversity and productivity in terrestrial ecosystems. Ecol Lett 11(3):296–310

Van Metre PC, Majewski MS, Mahler BJ, Foreman WT, Braun CL, Wilson JT, Burbank TL (2012) Volatilization of polycyclic aromatic hydrocarbons from coal-tar-sealed pavement. Chemosphere 88(1):1–7. https://doi.org/10.1016/j.chemosphere.2011.12.072

Widdel F, Musat F (2010) Diversity and common principles in enzymatic activation of hydrocarbons. In: Handbook of hydrocarbon and lipid microbiology. https://doi.org/10.1007/978-3-540-77587-4_70

Wintermute EH, Silver PA (2010) Dynamics in the mixed microbial concourse. Genes Dev 24(23):2603–2614. http://genesdev.cshlp.org/lookup/doi/10.1101/gad.1985210

Yadav AK, Manna S, Pandiyan K, Singh A, Kumar M, Chakdar H, Srivastava AK (2016) Isolation and characterization of biosurfactant producing Bacillus sp. from diesel fuel-contaminated site. Microbiology 85:56–62

Yan L, Sinkko H, Penttinen P, Lindström K (2016) Characterization of successional changes in bacterial community composition during bioremediation of used motor oil-contaminated soil in a boreal climate. Sci Total Environ 542:817–825. https://doi.org/10.1016/j.scitotenv.2015.10.144

Chapter 7
Environmental Petroleum Waste: Pollution, Toxicity, Sustainable Remediation

Sudeshna Dey, Akankshya Das, Krishnamayee Mallick, Aishwarya Sahu, and Alok Prasad Das

Abstract The rapid expansion of the petroleum industry has resulted in large amounts of waste that require appropriate disposal and valuation. The potential use of petroleum industry waste in the construction industry and the complexity, sluggish progress, and high expense of various remediation techniques have gathered a great deal of interest. One of the most valuable and adaptable natural resources, petroleum, has a variety of components and a wide range of industrial uses. They are a global and persistent soil and water contaminant causing a threat to people, animals, and marine life due to their uninvited and extensive use. A large number of hydrocarbons are present in petroleum waste, and it is challenging for the microorganisms or microbial communities to break down all of the hydrocarbons, aromatics, and fused ring structures found in petroleum or crude oil. Numerous conventional remediation methods have been used, but are costly, time-consuming, and ineffective. The most promising technique for effectively utilising beneficial bacteria to break down hazardous compounds is known as bioremediation. It proposes an environmentally friendly method by utilising the microbial system to expel environmental contaminants. To reduce the harmful effect of petroleum waste on the environment and ultimately on human beings, various cost-effective and sustainable treatment strategies need to be adopted to eliminate the effluent from the petroleum sector. Therefore, this chapter provides an overview of the existing approaches to treating wastewater from the petroleum sector, emphasising their viability and efficacy. Consequently, this chapter will also explore the bioremediation of harmful contaminants using the power of various microorganisms.

Keywords Petroleum waste · Toxicity · Contaminants · Bioremediation

S. Dey · A. Das · K. Mallick · A. Sahu · A. P. Das (✉)
Department of Life Sciences, Rama Devi Women's University, Bhubaneswar 751022, Odisha, India
e-mail: alok1503@gmail.com

I. D. Behera and A. P. Das (eds.), *Impact of Petroleum Waste on Environmental Pollution and its Sustainable Management Through Circular Economy*, Environmental Science and Engineering, https://doi.org/10.1007/978-3-031-48220-5_7

7.1 Introduction

With the fast rise of petroleum prospecting, crude oil exploration, and downstream petroleum sectors such as petrochemical refinery firms, soil contamination has been identified as one of the most serious environmental issues. It is mostly brought on by the components of oil from the collecting, transportation, and storage processes, particularly from leaks and spills (Xu et al. 2018). The contamination of the environment has grown over the past few decades as a result of increasing human activity, including population expansion, dangerous farming practices, unplanned urbanisation, deforestation, fast manufacturing, and reckless use of energy resources. Crude oil, emulsion, solids, and slurry are just a few of the many components that make up hydrocarbons and their byproducts. Hydrocarbon and petroleum wastes may harm the environment, aquatic and terrestrial life, and human health when they are extracted, processed, and transported (Mishra et al. 2021a, b, c). Chemical fertilizers, heavy metals, radioactive waste, pesticides, herbicides, insecticides, greenhouse gases, and hydrocarbons are some of the pollutants whose toxicity raises issues for the environment and public health. It is expected that more hazardous waste sites will be discovered in the ensuing decades. Thousands of hazardous waste sites have already been located. Pollutants are illegally dumped into the environment by chemical corporations and other sectors. Recent research has shown that microfibers are ubiquitous in the atmosphere and can be found in sludge, topsoil, rivers, oceans, and numerous water sources (Mishra et al. 2019).

Hydrocarbon and petroleum wastes are occasionally dumped and discharged, which causes soil contamination and due to this, the petroleum wastes that contaminate the soil do not support flora and remain infertile for decades before being remedied. According to reports, long-term exposure to this waste can lead to malignancies, heart disease, brain damage, stomach, liver, and kidney damage, as well as reproductive dysfunction and inflammatory lung disease in the human body (Singh et al. 2020a). The amount of hazard varies depending on the specific waste and its solubility under biological conditions, as well as on prolonged exposure to this waste can cause a serious threat to living beings (Das and Singh 2011; Ghosh et al. 2018). Therefore, to evaluate the impact of waste we need various sophisticated microscopic and spectroscopic techniques to identify, characterize, and quantify the synthetic fibers in order to assess the true environmental impact of this pollution (Mishra et al. 2022; Tripathy et al. 2022). The overuse of petroleum resources also pollutes water supplies, particularly small agricultural canals, rivers, and streams close to oil rigs and processing facilities. Many petroleum hydrocarbons and their derivatives can be found in soils which makes the soil more contaminated and because of their various structural configurations, limited biodegradability, hydrophobicity, potent sorption phenomena, and high persistence, polycyclic aromatic hydrocarbons (PAHs) are found more among them which causes a serious threat to the ecosystem.

Halogenated hydrocarbons, which are generated from petroleum hydrocarbons, are a different group of contaminants that cause obnoxious odors and are poisonous. Petroleum hydrocarbons and their derivatives may alter the quality of the soil in addition to polluting it. Pollutant releases such as synthetic dyes, and metal pollution can induce toxicity, mutagenicity, and carcinogenicity in people as well as a range of environmental health risks. Synthetic dyes are frequently utilized because of their significantly reduced cost despite these substantial health issues (Bhattacharjee et al. 2021; Das and Mishra 2009; Das and Swain 2013; Mishra and Das 2008). The majority of these are harmful and have a detrimental influence on human health as well as plant growth. Increased exposure to these unbiodegradable wastes, particularly petroleum waste hydrocarbons (PWHCs), can impair human health and ecological diversity (Vandana et al. 2022). During the exploration and processing of crude oil, a significant amount of petroleum discharge and persistent organic pollutants are produced and when petroleum waste hydrocarbons are thrown in an open pit, the soil becomes poisoned, posing major threats to human health and the agro-geo-environmental ecosystem. The petroleum industry is wide-ranging has so many applications and poses a significant threat to the environment. During the extraction, refinement, and transportation of oil and gas, substantial amounts of toxic and non-toxic waste are produced. Several industrial wastes, including volatile organic chemicals, nitrogen and sulfur compounds, and spilled oil, can pollute the air, water, and soil to harmful amounts when handled improperly. The emissions of greenhouse gases such as carbon dioxide (CO_2) and methane, as well as micro-particulate aerosols such as black carbon, from the industries, are the main source of global effects such as ocean acidification, sea level rise, and climate warming. The toxicity of various oils and petroleum-related products varies. Weathering, solubility, and chemical composition are just a few of the many variables that might affect the levels of toxicity. It is well known that a significant amount of petroleum waste is produced during the exploration of crude oil, and this trash reveals the properties and make-up of the source rocks. Due to high energy and capital demands, the recovery of these metal and petroleum products from soil using various metallurgical techniques is excessively expensive. A huge amount of solid mine wastes and petroleum effluents were generated as a result of substantial mining, metallurgical, extraction, and other human operations (Das and Ghosh 2022; Das et al. 2011, 2015b).

The ecosystem could be seriously endangered if this hazardous garbage is not disposed of properly. There a numerous techniques used for processing and getting rid of petroleum sludge, including thermal processing, mechanical, biological, and chemical ones, however, these methods are often expensive. Numerous methods, such as solvent extraction, freezing and thawing, electrokinetic method, microwave, and ultrasonic irradiation, have been used to extract oil from petroleum sludge. Following oil recovery, there are several ways to dispose of it, including burning, oxidation, solidification/stabilization, and biodegradation (Mishra et al. 2023a, b). Reusing oily sludge as fuel or material resources without treatment in the building sector is a cost-effective and environmentally friendly strategy. Since certain petroleum waste is volatile, affects the air quality, and contributes to other forms of air pollution, such as smog, it is vital to establish anti-air pollution techniques. In addition to nose and throat allergies, nausea, and serious lung damage in both people and other living creatures, the presence of hydrocarbon and petroleum wastes in the air can also cause. Lung problems, both malignant and benign, may develop as a result of heavy hydrocarbon exposure. Chemical oxidation, leaching, incineration, landfilling, microbiological treatment, and leaching are now employed to clean up soil contaminated with petroleum. In a contaminated environment, these technologies can extract, remove, alter, or mineralize petroleum contaminants, turning them into a less harmful, innocuous, and stable form.

7.2 Petroleum Waste Pollution

Pollution is the addition of pollutants, causing harmful adverse effects on a group of different species as well as their surroundings (Pradhan et al. 2023). Petroleum, which is also known as crude oil and ranges in colour from black to yellow, is mostly made up of aromatic hydrocarbons. The leftovers of an ancient sea creature, including plants, bacteria, and algae, which were buried beneath the sedimentary rock and subjected to massive pressure and temperature, are what eventually became crude oil. The importance of the petroleum sector increases daily as more and more uses for petroleum are discovered. It is necessary to reduce the health and environmental risks associated with this release of environmental pollutants since it has a major adverse impact on our environment (Mishra et al. 2019; Mishra and Das 2022). Due to the increase in population, the requirement for petroleum and its products has increased rapidly over 150 years. Oil contamination in coastal seas has negative effects on the environment, the economy, and other factors. The purpose of environmental sensitivity and risk assessment mapping for oil spills is to determine how vulnerable coastal resources are and to support efficient decision-making for reducing the long-term impact. The major sources of petroleum pollution are mainly produced from automobiles, industries, surveys, exploration, drilling, extraction and evolution, manufacturing of petroleum products, and maintenance as illustrated in Fig. 7.1. Several airborne pollutants are released during the exploration, development, and production of petroleum, including carbon dioxide (CO_2), hydrogen sulfide (H_2S), hydrocarbons (like CH_4), nitrogen oxides (NOx), partially combusted hydrocarbons (like carbon monoxide and particulates), polyaromatic hydrocarbons (PAHs), sulfur oxides (SOx), volatile organic compounds (VOCs), and other pollutants of a similar nature. Through combustion, activities, inadvertent emissions, and site cleanup, the exploration, development, and manufacturing of petroleum can produce air pollutants. Water, cuttings, chemicals used for well treatment, processing and washing water, drainage water, unintentional spills, leaks, and sewage are just a few of the water-based pollutants produced throughout this process.

When oil is extracted from an oil well, water, gas, and sometimes oil are also generated. After the oil is removed, much of the wastewater, also known as formation water or produced water, is purposefully released into the local environment untreated. Produced water may also contain various chemical additions such as organic solvents, lead, chromium, nickel, zinc, cadmium, mercury, arsenic, cyanide, and barium in addition to 8–10% oil. Several techniques can be used to treat surface

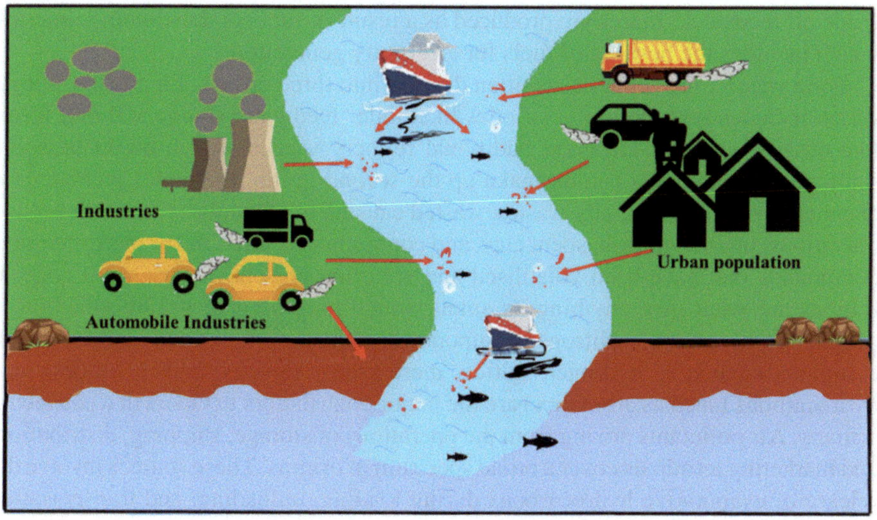

Fig. 7.1 Describes the sources and impacts of petroleum pollution on the environment

deposits of oil, but remediating subsurface oil is more difficult. Methods include low-pressure water flushing, although care has to be taken not to cause erosion. Rapid deployment of sorbents could reduce penetration into the substrate, but collection and disposal of oiled sorbent material may be a problem. Burning of oiled vegetation may be an option in some circumstances, but burning may increase the hydrocarbon content in the underlying sediment (Al-Rubaye et al. 2023). Cutting off oiled vegetation, particularly if leaving oil might be a threat to wildlife, is an option. However, the logistics of operating in soft sediment and removing and disposing of oiled material would need to be considered. Refineries may produce a mixture of effluent streams that include both soluble and insoluble contaminants. Various micropollutants that can be found include ammoniacal nitrogen, benzene, biochemical oxygen demand (BOD), chemical oxygen demand (COD), cyanides, ethylbenzene, fluorides, lead, mercury, and vanadium (special metals), pH (acids, alkalis), phenols, phosphates, sulfides, heat, taste, and odor producers, Toluene, total hydrocarbon content (THC), total metals, total nitrogen, total organic carbon (TOC), total petroleum hydrocarbon index (TPH-index), total suspended solids (TSS), and xylene (BTEX). Refineries and petrochemical facilities now have a significant challenge when it comes to environmental management. The petroleum business is an established one, and various pollution abatement procedures have been implemented at most refineries for a considerable amount of time. As a result, the emissions produced by refineries per ton of oil processed have decreased and are still decreasing. (Nelson 2013) investigated that air pollutants and hazardous air pollutants from oil refineries decreased between 1990 and 2010, there was a simultaneous increase in the crude oil distillation capacity of the industry, alongside a notable trend of higher density and sulfur content in the crude oil feedstock. Flares are produced as a result of fuel combustion in refineries, which includes the burning of fuels for electricity generation.

Solid waste from refineries is often divided into three groups. First is the sludge, which includes both non-oily varieties like boiler feedwater sludge and oily forms like desalter sludges. Different additional refinery wastes, which might be solid, semi-solid, or liquid in nature, make up the second group. Examples include used chemicals, polluted soil, oily wastes, wasted caustic, used catalysts from conversion operations, incinerator ash, spent clay, and contaminated soil. One or more chemical compounds that have been volatilised and are often present in very small concentrations are what give anything its smell, which may be detected by olfaction in people or other animals. Nitrogen compounds (such as ammonia and amines), sulfur compounds (such as hydrogen sulfide, mercaptans, sulfides, and disulfides), and hydrocarbons (such as aromatics) are the major contributors to Odors in a petroleum refinery. Air pollutants arising from the operations of storage, shipping, distribution, and marketing terminals cover a broad spectrum of origins. These sources involve the release of evaporative hydrocarbons during loading, unloading, and transportation activities, which are often referred to as "breathing losses." These emissions result from storage tank activities and the movement of vehicles like rail tank cars, tank trucks, marine vessels, and motor vehicle tanks. Evaporation losses may manifest at various stages along the entire journey of petroleum oil, starting from its extraction processes to the transportation to a refinery. Furthermore, refined products face evaporation losses during their transit from the refinery to fuel marketing terminals and petrochemical plants. Similarly, fuels encounter evaporation losses during their journey from fuel-marketing ports to service stations, commercial accounts, and local bulk storage plants. Contaminants whether unintentional or intentional, oil spills involve the release of oil onto waterways, generating a unique mass that is moved by the wind, currents, and tides. These spills are reduced using several techniques, including chemical dispersion, burning, mechanical containment, and adsorption, yet they still threaten coastal environments. An oil spill, which mostly affects maritime environments owing to human activity, is the inadvertent discharge of a liquid petroleum hydrocarbon into the environment. This kind of pollution usually relates to marine oil spills, in which oil is released into the sea and coastal waters. Among the causes are the spills of refined petroleum products (such as petrol), the release of crude oil from ships, offshore platforms, drilling rigs, and wells, and as well as spills of refined petroleum products (e.g., gasoline, diesel) and their byproducts, heavier fuels like bunker fuel used by large ships, or the leakage of any oily refuse or waste oil.

Crude oil is made up of a wide range of hydrocarbon molecules, the majority of which include carbon and hydrogen in various proportions. The components known as saturates, aromatics, resin, and asphaltenes, commonly referred to as SARA, represent the four constituents found in crude oil, which can be categorized as either heavy or light, like naphthene and paraffin, for instance. Additionally, it contains trace amounts of organic molecules and various metals such as copper, iron, nickel, and vanadium. Moreover, sulfur, oxygen, and nitrogen are also present in the crude oil composition. The elemental crude oil composition is approximately 83–87% carbon,

10–14% hydrogen, 0.05–6% sulfur, 0.1–2% nitrogen, and 0.05–1.5% oxygen, Nickel < 120ppm, Vanadium < 1200ppm. Crude oils exhibit a wide range of consistencies, from highly fluid and volatile liquids to thick, semisolid substances. Their color can vary, usually appearing as black or black with a greenish tint, though reddish, greenish-yellow, light yellow and translucent hues are also found. The smell of crude oil can differ, ranging from gasoline-like (sweet crude) to unpleasant (sour crude) or even fruity (when rich in aromatic compounds). Natural gas, on the other hand, is a colourless and odourless gas that can be used as fuel. However, an artificial odorant is added before it is sold for safety purposes. Sometimes, certain crude oil exists in a gaseous state, mixed with natural gas beneath the surface due to high temperature and pressure, and when brought to the surface during production, the temperature and pressure drop, causing the gas to condense into a liquid known as condensate (Jafarinejad 2017).

7.3 Toxicity of Petroleum Waste

All of these contaminants have hazardous impacts on various types of life, which can be chronic or acute. It is crucial that waste petroleum is not uniform and can significantly vary depending on the geographic location and the underlying geological conditions. Various forms of industrial effluents, petroleum waste, agricultural operations, and municipal trash raise the amount of hydrocarbon pollutants in the environment. Mammals, amphibians, reptiles, birds, pisces, and microbes are all highly mutagenic, toxic, and or carcinogenic to certain PAHs and related epoxy compounds. Petroleum waste affects various aspects as follows.

7.3.1 Effect on the Soil

The chemical, physical, and compositional characteristics of petroleum hydrocarbons play a crucial role in determining their effects on the environment and soil. Adsorption, biodegradation, and leaching are the main mechanisms that alter the soil's properties. A study by Jingchun Tang et al. found that the aging process of petroleum-contaminated soil is characterized by evaluating pH, total petroleum hydrocarbon concentration, and microbial population, the pH increases at the start and lowers as it ages, and total hydrocarbon is also degraded day by day but a negative correlation is seen in the population of microbe in soil. They could make their way into the food chain and harm people. Agriculture soil is essential for maintaining ecological equilibrium and assuring that there is enough food for everyone, but pollution of the soil hurts its fertility as well as its physical and chemical characteristics. There have been observations indicating that the introduction of petroleum hydrocarbons into topsoil and subsoil can lead to detrimental effects on soil texture and structure, causing a reduction in pore spaces and saturated hydraulic conductivity. This, in turn, can indirectly affect the nutrient availability to plants, consequently influencing the biological aspects of the soil, such as the microbial population and enzymatic activity (Tang et al. 2012).

7.3.2 Effect on the Plants

Crop plants and a wide variety of other vegetation have been documented to be hazardous at greater petroleum hydrocarbon concentrations. Sunflower crops root length and leaf area are largely affected by soil aeration and water penetration that is hindered by oil. After 14 days of exposure, no barley seeds were observed to sprout, even though the contaminated soil was treated with leaching and salt-relieving methods. Oil contamination results in oxidative stress, H_2O_2-accumulation, deformed leaf colors, and even plant cell death. Maize's root-shoot development, leaf lengths, and cell expansion were all inhibited (Ambaye et al. 2022).

7.3.3 Effect on Marine Animals

The effect of petroleum pollution on marine animals is severe. The involvement of petroleum in the marine environment is mainly through oil spills. It causes major behavioral changes as well as affects the development of marine species. However, these effects are not easy to quantify as several species show several different kinds of effects where multiple toxins are present. The major behavioral changes because of the pollutant are loss of motility, avoidance, changes in burrowing and feeding, and changes in reproductive behavior. In the case of phytoplankton loss of productivity was noticed. Invertebrates show changes in deterioration of growth within young ones and low reproductive capability within adults. (Loya and Rinkevich 1980) pointed out that oil spills cause various damages to coral reefs limiting their growth rate and rates of requirement. The adults of fish are more sensitive toward these effluents than the juveniles. It causes the fish to spawn fewer eggs, spawns less frequently, and delays spawning it also causes respiratory distress.

Further, the effect of crude oil on marine birds is through direct contact. When the crude oil coats their feather it causes them to lose water-repellence capability and thermal insulation. This causes the bird to sink drown and die. During oil spills, birds usually ingest some amount of oil which causes gastrointestinal damage. There is a variable effect of crude oil on marine mammals. Mammals having fur lose their ability to insulate which affects their thermal regulation. The loss of thermal regulation creates high metabolic activity which causes exhaustion of metabolic energy and fat. It can also cause skin and eye lesions (Wake 2005).

7.3.4 Effect on Humans

All these crude oil pollutants have an ill effect on many living organisms. Some of these pollutants have mutagenic and carcinogenic effects. These pollutants have acute, chronic, genotoxic, carcinogenic, immunogenic, and teratogenic effects on humans (Paul et al. 2023; Raj and Das 2023). Acute health effects caused by hydrocarbons depend on their concentration and their type. It usually causes vomiting, eye irritation, nausea, and diarrhea but in severe cases, it causes skin allergies and skin irritation (Das and Mishra 2008; Das and Singh 2011). The human organs that have experienced significant historical harm from mono- and poly-aromatic hydrocarbon pollutants include the kidneys, lungs, and liver. Asthma and other respiratory disorders are also brought on by it. Redness and skin inflammation may be caused by abnormalities in cataracts and contact with the skin (Dey et al. 2023). Red blood cells might disintegrate as a result of the high naphthalene content.

Tumors, mutations, developmental abnormalities, and cancer are caused by cell damage. When these compounds are subjected to reactions involving other substances, they can give rise to epoxides and dihydrodiols, potentially resulting in genetic abnormalities. The term mutation refers to the binding of DNA by the dihydrodiols and epoxides of arenes. Some PAHs have been shown to have genotoxic effects in both rat studies and using mammalian cell lines, including those derived from humans, and performed in an artificial environment (in vitro). Deletion, frameshift mutations, phase arrest, and other chromosomal alternations are caused by base pair substitution in DNA caused by PAHs. Quinines, diol epoxides, and conjugated hydroxyalkyl are examples of PAH compounds that can attach to the nucleophilic centers of macromolecules. The model chemical for studying the toxicity of PAHs is benzo[a]pyrene. The waste from petroleum refineries contains aromatic hydrocarbon contaminants, which are proven carcinogens. They attach to DNA or cellular proteins, causing the growth of tumors and cancer. They lead to harmful consequences and cancer when they are converted into epoxides and dihydrodiols. Skin, lung, bladder, and gastrointestinal cancers are brought on by aromatic hydrocarbon pollutants including benzene and PAHs. According to reports, the presence of an aromatic amine in the structure of PAH compounds increases their carcinogenicity. It has been observed that the aromatic hydrocarbon pollutants PAHs inhibit the immune system, which either directly or indirectly causes cancer. In rodents, the presence of PAHs inhibits the immunological response. Infectious illnesses are associated with immune suppression. It increases cytokines secretion in the body which causes inflammation. In some conditions, it causes hypersensitivity and autoimmunity. According to certain experimental research, immunotoxicity is caused by consuming contaminated food containing PAHs.

Teratogenicity is defined as a deformity in the process of an embryo or fetus maturing and growing over time. Aromatic hydrocarbons like benzene, naphthalene, benzene, benzo[a]pyrene, and benzo[a]anthracene among others, are recognized to be extremely hazardous chemicals. A high amount of benzo[a]pyrene is responsible for congenital abnormality and lower weight of the body in the offspring. If PAHs enter a pregnant woman's body during pregnancy, they cause low weight at the time of birth, early premature delivery of the offspring, cognitively abnormal children, and cardiac defects. PAHs are the reason for early childhood respiratory problems such as asthma, reduced Intelligence Quotient at the early age of 2–3, and greater behavioral issues in the age period of 6–8. Baby's cord blood DNA is harmed by exposure to PAHs if they are exposed to them.

7.4 Sustainable Remediation

Petroleum hydrocarbon (PHC) pollution of soil poses a substantial hazard to people and other living things in polluted places and has a significant detrimental effect on the environment. Massive amounts of solid mine wastes and petroleum effluents have been produced as a result of substantial mining, petrochemical industry waste, and other human activities. The extraction of petroleum has significantly increased the release of harmful pollutants into the soil environment (Cordes et al. 2016). Surface oil deposits can be treated using a variety of methods, however, remediating subsurface oil is more challenging. Low-pressure water flushing is one technique, although caution must be exercised to prevent erosion.

The use of microorganisms in the biodegradation and removal of environmental pollutants has been applied to land treatment, bio pile/composting, and bio-slurry. These methods are of low cost, low energy consumption, and have the potential to accommodate large volumes of petroleum and environmental wastes (Mishra et al. 2020, 2021a, b, c; Mishra and Das 2023). It is generally known that bioremediation speeds up PWHC breakdown and many other environmental pollutants in aerobic environments (at the surface) (Mohanta et al. 2023; Kanungo et al. 2023). The number and diversity of the local hydrocarbon-degrading microbial community decline under anoxic environments (Sahoo et al. 2022; Singh et al. 2020b).

In some situations, burning oiled vegetation might be an alternative, however, burning could raise the hydrocarbon content of the underlying sediment. Cutting oily vegetation is an option, especially if keeping oil there could endanger wildlife, but it would require planning to operate in soft soil and remove and dispose of oiled waste. Oily soil remediation is a current research hotspot, and bioremediation offers benefits such as simplicity of application, economic viability, and lack of secondary contamination. The restoration of petroleum-contaminated soil is a very important and crucial subject that has drawn significant and growing research interest. PHC contaminants, which include saturated hydrocarbon (SHC), aromatic hydrocarbon (AHC), and non-hydrocarbon (NHC) chemicals, make up the great majority of substances and by-products that are categorised as priority environmental pollutants. The carbon–carbon and carbon-hydrogen bonds that make up the SHC molecular structure are easily broken down. Additionally, the SHC can gradually leave the soil through photosynthesis and volatilization because of its low boiling point. The difficulty of being removed from the soil is substantially increased by the complicated benzene ring and its higher boiling point. The incorrect management of petroleum effluents has led to the utilization of oil waste into the environment due to which the soil and groundwater get contaminated, and numerous other major environmental issues will arise and cause a significant loss of biodiversity (Lahiri et al. 2021; Mishra et al. 2021a, b, c). The majority of PHCs released into the environment are eventually broken down by different types of local microorganisms. These microorganisms can not only survive the physiological challenges caused by these substances, but they also use them as their only supply of carbon for growth and reproduction (Das and Ghosh 2018; Lahiri et al. 2021; Mohanty et al. 2018). Microorganisms' biological activity, stability, and capability to break down petroleum pollutants are all improved by the addition of charcoal, nutrients, and plants (Behera et al. 2021). Utilizing a combination of three microbiological techniques presents numerous undeniable benefits. Following restoration, these techniques will make the physical, chemical, and biological features of the soil ecosystem better. Additionally, without introducing secondary pollution, they may convert organic pollutants (such as carbon dioxide and water) into completely non-polluting inorganic compounds. Due to its benefits of safety, no secondary contamination, and economics, bioremediation is characterized as a clean soil treatment technology. Microbial remediation, plant-microbial remediation, and phytoremediation are the major methods used to remediate soil that has been polluted by petroleum (Ghosh et al. 2016a, b; Mohanty et al. 2017). Different methods of processing oil sludges can minimize toxicity, reduce sludge volume, and recover petroleum products. Bioremediation techniques involve bioaugmentation, biodegradation, bio-stimulation, rhizoremediation, landfarming, bio-venting, bio-sparging, composting, and phytoremediation processes (Das and Mishra 2010). These methods of treatment have been shown to decrease the toxic components in petroleum sludge and their detrimental impact on the environment and human health. One of the main methods of bioremediation used to clean up petroleum-contaminated soil is called "biostimulation," which involves changing the environment's temperature, moisture content, pH, redox potential, aeration, and mineral nutrition to promote the growth and metabolic activity of naturally occurring microbial populations that are responsible for soil degradation (Behera et al. 2022). The microorganisms used in biostimulation practice can use hydrocarbons as carbon sources for growth and are tolerant to a variety of hydrocarbons.

Biological augmentation is simpler and more cost-effective than physical and chemical cleaning techniques. Lipophilic microorganisms can be used to boost biological activities. Oleophilic bacteria may be found in a wide range of petroleum-contaminated environments, including saltwater, coastlines, muck, and soil. They could only be able to eat hydrocarbons while destroying or mineralizing toxic and hazardous petroleum pollution. Different kinds of contaminated settings contain different kinds of degrading microorganisms. DNA-based stable isotope probing (DNA-SIP) technique is employed to ascertain the varieties and activities of soil organisms (Ghosh et al. 2016a, b; Mohanty et al. 2016; Sanket et al. 2016). The exploitation of environmental waste such as low-grade waste has become urgent to combat the situation of decreasing high-grade ore reserves and increasing demand for the specific microorganisms capable of recovery of low-grade ores as well the recovery of petroleum products have been well recognized in biotechnology and many such bacteria have been reported (Bal et al. 2018; Das et al. 2012, 2015a; Ghosh et al. 2017; Ghosh and Das 2017).

Polycyclic aromatic hydrocarbon-contaminated soil frequently contains the phylum Actinomycetes. In the Philippines, contaminated soil was discovered to contain *Acidovorax, Rhodoferax, Hydrogenophaga, and Polaromonas. Acidobacteria* are present in contaminated soil. Numerous bacteria, including *Rhodococcus* species, *Pseudomonas* species, *Stenotrophomonas* sp., *Scedosporium boydii*, Brevibacterium, and bacterial consortium have been shown in studies that have the ability to break down petroleum pollutants (Ghosh et al. 2020; Ghosh and Das 2020; Behera et al. 2020; Mishra and Das 2021). Through aerobic mechanisms, the various bacteria break down hydrocarbons and other environmental contaminants, and hydrocarbon catabolism is frequently enhanced when oxygen acts as an electron acceptor (Mishra et al. 2023a, b). The processes of oxidation, reduction, hydroxylation, and dehydrogenation mediate the degradation in aerobic mode. Table 7.1 shows the petroleum hydrocarbons degrading bacteria and their preferred degradation substrates. Enzymes including monooxygenase, dioxygenase, cytochrome P450, peroxidase, hydroxylase, and dehydrogenase aid in the biodegradation of hydrocarbons (Mishra et al. 2009, 2020).

In a study by Jagaba et al. (2022), the effects of composting and the addition of sewage sludge on the biodegradation of oily sludge were examined. For the contaminated soil to sewage sludge ratio of 1:0.5, sewage addition led to rapid biodegradation of oil and grease (65.6%) during 9 weeks under low-temperature settings. The remaining 34% was labelled as recalcitrant because microorganisms were unable to access it. After all, it was tightly adsorbed to soil particles and/or remaining in soil

Table 7.1 Petroleum hydrocarbon-degrading bacteria and their preferred degradation substrates

Sl. No	Petroleum hydrocarbon components	Bacterial species	References
1	Aliphatic	*Pseudomonas* Species, and *Geobacillus thermodenitrification*	Abbasian et al. (2015)
2	Aromatics	*Sphingomonas*	Xu et al. (2018)
3	Asphaltenes	*Acinetobacter* species	Abbasian et al. (2015)
4	Alicyclic	*Mycobacterium, Brevibacterium* species	Xu et al. (2018)
5	Paraffin	*Thallassolituus, Cycloclasticus*	Adlan et al. (2020)
6	Naphthalene	*Agmenellum quadruplicatum, Cyanobacteria*	Cerniglia et al. (1983)
7	Cumene	*Streptobacillus, Streptococcus*	Abbasnezhad et al. (2011)
8	9-H Flourene	*Acinetobacter*	Das and Chandran (2011)
9	Phenanthrene	*Pseudomonas monteilii*	Spini et al. (2018)
10	Benzo(a)pyrene	*Mycobacterium vanbaalenii*	Moody et al. (2004)

macro and micropores. According to research by Saum et al. (2018), adding BC to soils with petroleum contamination did not speed up the process of bioremediation. The phytoremediation of petroleum-contaminated soil was enhanced by the use of plants (mesquite tree seedlings), compost, and combinations of these treatments.

7.5 Conclusion

A significant source of fatal air pollutants, such as benzene, toluene, ethylbenzene, and xylene, are produced by petroleum refineries. There are known carcinogens released by this site, including particulate debris, gases including nitrogen oxides (NOx), carbon monoxide (CO), hydrogen sulfide (H_2S), sulfur dioxide (SO_2), and other volatile PH compounds. Despite these well-known negative impacts, petroleum is nonetheless regarded as the cornerstone of contemporary civilization because of its wide range of applications and necessity for societal economic development. The petroleum industries generate enormous volumes of waste that need to be properly disposed of and valued during the processes of exploration, extraction, development, and production. For sustainable development, the reuse of this waste in diverse uses such as the building industry should be taken into consideration. Recent studies have examined the cooperation of various microorganisms in the degradation and detoxification of industrial pollutants like petroleum hydrocarbons, dyes, and textile wastewater, showing that the development of microbial consortia is a method that enhances and optimizes bioremediation processes. Bioremediation is an ecologically friendly and cost-effective method for eliminating petroleum-related soil pollutants. A number of bioremediation strategies, such as phytoremediation, rhizoremediation, bio-stimulating, bioaugmentation, and others, have been applied to remove petroleum hydrocarbons from soils that are polluted. In the bioremediation process, it is preferred to use plants and bacteria that are highly effective at degrading petroleum. A trend for the future is the development of various functional bacterial consortiums or genetically engineered bacteria, as well as the prospective use of integrated bioremediation methods for the bioremediation of petroleum wastes.

References

Abbasian F, Lockington R, Mallavarapu M, Naidu R (2015) A comprehensive review of aliphatic hydrocarbon biodegradation by bacteria. Appl Biochem Biotechnol 176(3):670–699. https://doi.org/10.1007/S12010-015-1603-5

Abbasnezhad H, Gray M, Foght JM (2011) Influence of adhesion on aerobic biodegradation and bioremediation of liquid hydrocarbons. Appl Microbiol Biotechnol 92(4):653–675. https://doi.org/10.1007/S00253-011-3589-4

Adlan NA, Sabri S, Masomian M, Ali MSM, Rahman RNZRA (2020) Microbial biodegradation of paraffin wax in Malaysian crude oil mediated by degradative enzymes. Front Microbiol 11. https://doi.org/10.3389/FMICB.2020.565608/FULL

Al-Rubaye AH, Jasim DJ, Ameen HFM, Al-Robai HA, Al-Assal JR (2023) The impacts of petroleum on the environment. IOP Conf Ser: Earth Environ Sci 1158(3). https://doi.org/10.1088/1755-1315/1158/3/032014

Ambaye TG, Chebbi A, Formicola F, Prasad S, Gomez FH, Franzetti A, Vaccari M (2022) Remediation of soil polluted with petroleum hydrocarbons and its reuse for agriculture: recent progress, challenges, and perspectives. Chemosphere 293:133572. https://doi.org/10.1016/J.CHEMOSPHERE.2022.133572

Bal B, Ghosh S, Das AP (2018) Microbial recovery and recycling of manganese waste and their future application: a review. Geomicrobiol J 36(1):85–96. https://doi.org/10.1080/01490451.2018.1497731

Behera ID, Basak G, Kumar RR, Sen R, Meikap BC (2020) Treatment of petroleum refinery sludge by petroleum degrading bacterium Stenotrophomonas pavanii IRB19 as an efficient novel technology. J Environ Sci Health Part A 56:1–13. https://doi.org/10.1080/10934529.2020.1866924

Behera ID, Nayak M, Biswas S, Meikap BC, Sen R (2021) Enhanced biodegradation of total petroleum hydrocarbons by implementing a novel two-step bioaugmentation strategy using indigenous bacterial consortium. J Environ Manag 292:112746. https://doi.org/10.1016/J.JENVMAN.2021.112746

Behera ID, Nayak M, Mishra A, Meikap BC, Sen R (2022) Strategic implementation of integrated bioaugmentation and biostimulation for efficient mitigation of petroleum hydrocarbon pollutants from terrestrial and aquatic environments. Mar Pollut Bull 177:113492. https://doi.org/10.1016/j.marpolbul.2022.11349

Bhattacharjee J, Mishra S, Das AP (2021) Recent advances in sensor-based detection of toxic dyes for bioremediation application: a review. Appl Biochem Biotechnol 2021:1–20. https://doi.org/10.1007/S12010-021-03767-7

Cerniglia CE, Freeman JP, Althaus JR, van Baalen C (1983) Metabolism and toxicity of 1- and 2-methylnaphthalene and their derivatives in cyanobacteria. Arch Microbiol 136(3):177–183. https://doi.org/10.1007/BF00409840

Cordes EE, Jones DOB, Schlacher TA, Amon DJ, Bernardino AF, Brooke S, Carney R, DeLeo DM, Dunlop KM, Escobar-Briones EG, Gates AR, Génio L, Gobin J, Henry LA, Herrera S, Hoyt S, Joye M, Kark S, Mestre NC, Metaxas A, Pfeifer S, Sink K, Sweetman AK, Witte U (2016) Environmental impacts of the deep-water oil and gas industry: a review to guide management strategies. Front Environ Sci 4:207190. https://doi.org/10.3389/FENVS.2016.00058/BIBTEX

Das AP, Mishra S (2008) Hexavalent chromium (VI): health hazards and environmental pollutant. J Environ Res Dev 2(3):386–392

Das AP, Mishra S (2009) Hexavalent chromium [Cr (VI)]: yellow water pollution and its remediation. Sarovar Saurabh ENVIS Newsl Wetl Ecosyst 5(2)

Das A, Mishra S (2010) Biodegradation of the metallic carcinogen hexavalent chromium Cr(VI) by an indigenously isolated bacterial strain. J Carcinog 9:6. https://doi.org/10.4103/1477-3163.63584

Das N, Chandran P (2011) Microbial degradation of petroleum hydrocarbon contaminants: an overview. Biotechnol Res Int 2011:1–13. https://doi.org/10.4061/2011/941810

Das AP, Singh S (2011) Occupational health assessment of chromite toxicity among Indian miners. Indian J Occup Environ Med 15(1):6. https://doi.org/10.4103/0019-5278.82998

Das AP, Swain S (2013) Algal Biosorption of toxic dye Methylene blue. In: A potential source of food, feed, biochemicals, biofuels and biofertilizers, international conference on algal biorefinery, vol 13. Indian Institute of Technology

Das AP, Ghosh S (2018) Bioleaching of manganese from mining waste materials. Mater Today: Proc 5(1):2381–2390. https://doi.org/10.1016/J.MATPR.2017.11.459

Das A, Ghosh S (2022) Role of microorganisms in extenuation of mining and industrial wastes. Geomicrobiol J 39:173–175. https://doi.org/10.1080/01490451.2022.2038953

Das AP, Sukla LB, Pradhan N, Nayak S (2011) Manganese biomining: a review. Biores Technol 102(16):7381–7387. https://doi.org/10.1016/J.BIORTECH.2011.05.018

Das AP, Sukla LB, Pradhan N (2012) Microbial recovery of manganese using *Staphylococcus epidermidis*. Int J Nonferrous Metall 2012(02):9–12. https://doi.org/10.4236/IJNM.2012.12002

Das AP, Ghosh S, Mohanty S, Sukla LB (2015a) Advances in manganese pollution and its bioremediation. In: Sukla LB, Pradhan N, Panda S, Mishra BK (eds) Environmental microbe biotechnology, 1st edn. Springer, Cham, pp 313–328. https://doi.org/10.1007/978-3-319-19018-1_16

Das AP, Ghosh S, Mohanty S, Sukla LB (2015b) Consequences of manganese compounds: a review. Toxicol Environ Chem 96(7):981–997. https://doi.org/10.1080/02772248.2015.1005428

Dey S, Tripathy B, Kumar MS, Das AP (2023) Ecotoxicological consequences of manganese mining pollutants and their biological remediation. Environ Chem Ecotoxicol. https://doi.org/10.1016/J.ENCECO.2023.01.001

Ghosh S, Das AP (2017) Bioleaching of manganese from mining waste residues using Acinetobacter sp. Geol Ecol Landsc 1(2):77–83. https://doi.org/10.1080/24749508.2017.1332847

Ghosh S, Das AP (2020) Microbial metagenomics: current advances in investigating microbial ecology and population dynamics. In: Nayak SK, Mishra BB (eds) Frontiers in soil and environmental microbiology, 1st edn. CRC Press, pp 247–254

Ghosh S, Mohanty S, Akcil A, Sukla LB, Das AP (2016a) A greener approach for resource recycling: manganese bioleaching. Chemosphere 154:628–639. https://doi.org/10.1016/J.CHEMOSPHERE.2016.04.028

Ghosh S, Mohanty S, Nayak S, Sukla LB, Das AP (2016b) Molecular identification of indigenous manganese solubilising bacterial biodiversity from manganese mining deposits. J Basic Microbiol 56(3):254–262. https://doi.org/10.1002/JOBM.201500477

Ghosh S, Bal B, Das AP (2017) Enhancing manganese recovery from low-grade ores by using mixed culture of indigenously isolated bacterial strains. Geomicrobiology 35(3):242–246. https://doi.org/10.1080/01490451.2017.1362080

Ghosh S, Kumar MS, Bal B, Das AP (2018) Application of bioengineering in revamping human health. Synth Biol: Omics Tools Appl 21–37. https://doi.org/10.1007/978-981-10-8693-9_2

Ghosh S, Gandhi M, van Hullebusch ED, Das AP (2020) Proteomic insights into Lysinibacillus sp.-mediated biosolubilization of manganese. Environ Sci Pollut Res 28(30):40249–40263. https://doi.org/10.1007/S11356-020-10863-4

Jafarinejad S (2017) Pollutions and wastes from the petroleum industry. In: Petroleum waste treatment and pollution control. Elsevier, pp 19–83. https://doi.org/10.1016/b978-0-12-809243-9.00002-x

Jagaba AH, Kutty SRM, Lawal IM, Aminu N, Noor A, Al-dhawi BNS, Usman AK, Batari A, Abubakar S, Birniwa AH, Umaru I, Yakubu AS (2022) Diverse sustainable materials for the treatment of petroleum sludge and remediation of contaminated sites: a review. Clean Waste Syst 2:100010. https://doi.org/10.1016/J.CLWAS.2022.100010

Kanungo J, Sahoo S, Bal M, Behera ID (2023) Performance of bioremediation strategy in waste lubricating oil pollutants: a review. Geomicrobiol J. https://doi.org/10.1080/01490451.2023.2245395

Lahiri D, Nag M, Dey A, Sarkar T, Joshi S, Pandit S, Das AP, Pati S, Pattanaik S, Tilak VK, Ray RR (2021) Biofilm mediated degradation of petroleum products. Geomicrobiol J 39:1–10. https://doi.org/10.1080/01490451.2021.1968979

Loya Y, Rinkevich B (1980) Effects of oil pollution on coral reef communities, vol 3

Mallick K, Sahu A, Dubey NK, Das AP (2023) Harvesting marine plastic pollutants-derived renewable energy: a comprehensive review on applied energy and sustainable approach. J Environ Manag 348:1–13

Mishra S, Das AP (2008) Hexavalent chromium reduction and 16S rDNA identification of bacteria isolated from a Cr (VI) contaminated site. Internet J Microbiol 7(1)

Mishra S, Das AP (2021) Current treatment technologies for removal of microplastic and microfiber pollutants from wastewater. Wastewater Treat: Cut-Edge Mol Tools Techn Appl Aspects 237–251. https://doi.org/10.1016/B978-0-12-821881-5.00011-8

Mishra S, Das AP (2022) Treatment of the wastewater polluted with synthetic microfiber released from washing machine. In: Das BB, Hettiarachchi H, Sahu PK, Nanda S (eds) Lecture notes in civil engineering, vol 207). Springer Science and Business Media Deutschland GmbH, pp 109–117. https://doi.org/10.1007/978-981-16-7509-6_9/COVER

Mishra S, Das AP (2023) Microbial remediation of synthetic microfiber contaminated wastewater. In: Shah MP (ed) Microbial technologies in industrial wastewater treatment. Springer, Singapore, pp 337–350. https://doi.org/10.1007/978-981-99-2435-6_17

Mishra S, Das AP, Seragadam P (2009) Microbial remediation of hexavalent chromium from chromite-contaminated mines of Sukinda Valley, Orissa (India). J Environ Res Dev 1122–1127

Mishra S, Rath C, Das AP (2019) Marine microfiber pollution: a review on present status and future challenges. Mar Pollut Bull 140:188–197. https://doi.org/10.1016/J.MARPOLBUL.2019.01.039

Mishra S, Singh RP, Rath CC, Das AP (2020) Synthetic microfibers: source, transport and their remediation. J Water Process Eng 38:101612. https://doi.org/10.1016/J.JWPE.2020.101612

Mishra A, Siddiqi H, Kumari U, Behera ID, Mukherjee S, Meikap BC (2021a) Pyrolysis of waste lubricating oil/waste motor oil to generate high-grade fuel oil: a comprehensive review. Renew Sustain Energy Rev 150:111446. https://doi.org/10.1016/J.RSER.2021.111446

Mishra S, Rout PK, Das AP (2021) Emerging microfiber pollution and its remediation. In: Prasad R (ed) Environmental and microbial biotechnology. Springer, Singapore, pp 247–266. https://doi.org/10.1007/978-981-15-5499-5_9

Mishra S, Swain S, Sahoo M, Mishra S, Das AP (2021c) Microbial colonization and degradation of microplastics in aquatic ecosystem: a review. Geomicrobiol J. https://doi.org/10.1080/01490451.2021.1983670

Mishra S, Dash D, Das AP (2022) Detection, characterization and possible biofragmentation of synthetic microfibers released from domestic laundering wastewater as an emerging source of marine pollution. Mar Pollut Bull 185:114254. https://doi.org/10.1016/J.MARPOLBUL.2022.114254

Mishra S, Dash D, Das AP (2023a) Aquatic microbial diversity on plastisphere: colonization and potential role in microplastic biodegradation. Geomicrobiol J 1–12. https://doi.org/10.1080/01490451.2023.2209750

Mishra S, Ghosh S, van Hullebusch ED, Singh S, Das AP (2023b) A critical review on the recovery of base and critical elements from electronic waste-contaminated streams using microbial biotechnology. Appl Biochem Biotechnol 2023:1–30. https://doi.org/10.1007/S12010-023-04440-X

Mohanty S, Ghosh S, Nayak S, Das AP (2016) Isolation, identification and screening of manganese solubilizing fungi from low-grade manganese ore deposits. Geomicrobiological J 34(4):309–316. https://doi.org/10.1080/01490451.2016.1189016

Mohanty S, Ghosh S, Nayak S, Das AP (2017) Bioleaching of manganese by Aspergillus sp. isolated from mining deposits. Chemosphere 172:302–309. https://doi.org/10.1016/J.CHEMOSPHERE.2016.12.136

Mohanty S, Ghosh S, Bal B, Das AP (2018) A review of biotechnology processes applied for manganese recovery from wastes. Rev Environ Sci Bio/Technol 17(4):791–811. https://doi.org/10.1007/S11157-018-9482-1

Mohanta S, Pradhan B, Behera ID (2023) Impact and remediation of petroleum hydrocarbon pollutants on agricultural land: a review. Geomicrobiol J 1–5. https://doi.org/10.1080/01490451.2023.2243925

Moody JD, Freeman JP, Fu PP, Cerniglia CE (2004) Degradation of benzo[a]pyrene by Mycobacterium vanbaalenii PYR-1. Appl Environ Microbiol 70(1):340–345. https://doi.org/10.1128/AEM.70.1.340-345.2004.PMID:14711661;PMCID:PMC321301

Nelson TP (2013) An examination of historical air pollutant emissions from US petroleum refineries. Environ Prog Sustain Energy 32(2):425–432. https://doi.org/10.1002/ep.11713

Paul A, Dey S, Ram DK, Das AP (2023) Hexavalent chromium pollution and its sustainable management through bioremediation. Geomicrobiol J 1–11. https://doi.org/10.1080/01490451.2023.2218377

Pradhan G, Tripathy B, Ram DK, Digal AK, Das AP (2023) Bauxite mining waste pollution and its sustainable management through bioremediation. Geomicrobiology J 1–10. https://doi.org/10.1080/01490451.2023.2235353

Raj K, Das AP (2023) Lead pollution: Impact on environment and human health and approach for a sustainable solution. Environ Chem Ecotoxicol 5:79–85. https://doi.org/10.1016/J.ENCECO.2023.02.001

Sahoo PP, Singh S, Rout PK, Mishra S, Das AP (2022) Microbial remediation of plastic pollutants generated from discarded and abandoned marine fishing nets. Biotechnol Genet Eng Rev. https://doi.org/10.1080/02648725.2022.2152629

Sanket AS, Ghosh S, Sahoo R, Nayak S, Das AP (2016) Molecular identification of acidophilic manganese (Mn)-solubilizing bacteria from mining effluents and their application in mineral beneficiation. Geomicrobiol J 34(1):71–80. https://doi.org/10.1080/01490451.2016.1141340

Saum L, Jiménez MB, Crowley D (2018) Influence of biochar and compost on phytoremediation of oil-contaminated soil 20(1):54–60. https://doi.org/10.1080/15226514.2017.1337063

Singh RP, Mishra S, Das AP (2020a) Synthetic microfibers: pollution toxicity and remediation. Chemosphere 257:127199. https://doi.org/10.1016/J.CHEMOSPHERE.2020.127199

Spini G, Spina F, Poli A, Blieux AL, Regnier T, Gramellini C, Varese GC, Puglisi E (2018) Molecular and microbiological insights on the enrichment procedures for the isolation of petroleum degrading bacteria and fungi. Front Microbiol 9:2543. https://doi.org/10.3389/FMICB.2018.02543/FULL

Tang J, Lu X, Sun Q, Zhu W (2012) Aging effect of petroleum hydrocarbons in soil under different attenuation conditions. Agric Ecosyst Environ 149:109–117. https://doi.org/10.1016/j.agee.2011.12.020

Tripathy B, Dash A, Das AP (2022) Detection of environmental microfiber pollutants through vibrational spectroscopic techniques: recent advances of environmental monitoring and future prospects. Crit Rev Anal Chem 1–11. https://doi.org/10.1080/10408347.2022.2144994

Vandana, Priyadarshanee M, Mahto U, Das S (2022) Mechanism of toxicity and adverse health effects of environmental pollutants. Microb Biodegrad Bioremediation: Techn Case Stud Environ Pollut 33–53. https://doi.org/10.1016/B978-0-323-85455-9.00024-2

Wake H (2005) Oil refineries: a review of their ecological impacts on the aquatic environment. Estuar Coast Shelf Sci 62(1–2):131–140. Academic. https://doi.org/10.1016/j.ecss.2004.08.013

Xu X, Liu W, Tian S, Wang W, Qi Q, Jiang P, Gao X, Li F, Li H, Yu H (2018) Petroleum hydrocarbon-degrading bacteria for the remediation of oil pollution under aerobic conditions: a perspective analysis. Front Microbiol 9:425106. https://doi.org/10.3389/FMICB.2018.02885/BIBTEX

Chapter 8
Microbial Remediation of Plastic Hydrocarbon Contaminants from Marine Ecosystem

G. Koteswara Reddy, Ch. Kavya, and K. Himabindu

Abstract In this situation, plastics around the world are used most popular. Plastic takes longer to break down and might sometimes take many years to break down in the surroundings. Waste plastic is extremely damaging to the terrestrial and marine environment. Numerous billions of marine species are dying in the ocean alone because of consuming plastic garbage, The least expensive and most environmentally beneficial option is bioremediation. Research has recently focused on the bioremediation of synthetic polymers, and numerous bacteria have been identified as potential degraders of these particles. This review article focuses on the marine environment's plastic contamination and how bioremediation can offer a remedy. In this procedure, contaminated areas are decontaminated using living creatures like bacteria and microorganisms. The revocation of. contaminants, pollutants, and toxins from soil, water, and other habitats also use this method.

Keywords Polymers · Degradation · Marine environment · Bioremediation

8.1 Introduction

Synthetic plastics have grown significantly in importance since the early 1950s because of their amazing physical and chemical characteristics. They are become an essential component of our life. Every year, tonnes of plastic are manufactured, and nearly half of them are intended for one use only. Humanity has overused them throughout time, and as a result, plastic debris is now everywhere (Andrady 1998). Every year, almost 300 million tonnes of plastic trash are produced. Marine scientists believe that plastic waste acts as a geological indication of the Anthropocene era. Although moving to biodegradable polymers could solve this problem, it is still

G. Koteswara Reddy (✉) · Ch. Kavya · K. Himabindu
Department of Biotechnology, Koneru Lakshmaiah Education Foundation (Deemed to be University), Green Fields, Vaddeswaram 522502, India
e-mail: koteswarareddy@kluniversity.in

I. D. Behera and A. P. Das (eds.), *Impact of Petroleum Waste on Environmental Pollution and its Sustainable Management Through Circular Economy*, Environmental Science and Engineering, https://doi.org/10.1007/978-3-031-48220-5_8

necessary to address the current negative impacts of synthetic plastic waste (Auta et al. 2018).

We conducted research and exploration to close this knowledge gap out of concern for the environment's protection and public safety. Physical, thermal, and chemical ways of deterioration have all been investigated as different modes of plastic degradation thus far. Researchers have recently become interested in the biodegradation of synthetic polymers, and a variety of microbes have been identified as possible degraders of these plastics (Barnes et al. 2009). Since various bacteria, fungi, and algae use petroleum hydrocarbons as the main source of carbon and energy for their growth and metabolic processes, microorganisms are the most effective degraders of hydrocarbon pollution (Behera et al. 2020, 2021). It has been discovered that microorganisms are able to adapt to their surroundings and change their catabolic pathways so they can either use plastic waste directly as a source of carbon or they can produce by-products that attack polymer structures. The present study is concerned with marine plastic pollution and possible solutions (Catania et al. 2020).

8.2 The Marine Environment and Plastics

Plastic is a synthetic organic polymer derived from petroleum that has properties that make it ideal for a variety of uses, including packaging, building, sports and household equipment, vehicles, electronics, and agriculture. (Danso et al. 2019; Behera et al. 2022). About half of the 300 million metric tonnes of plastic generated each year goes towards producing items with a single use, such as straws, cups, and shopping bags. Inappropriate disposal of plastic trash has the potential to affect the ecology and biodiversity (Debroas et al. 2017).

14 million tonnes or more of plastic enter the ocean annually. Plastic is currently the most common type of marine litter, accounting for 75% of all marine trash found in sediments from deep sea to surface waters. Plastic debris is strewn throughout the coastlines of every continent, with tourist sites and densely populated areas having the largest amounts (Kumari et al. 2019; Lucas et al. 2008; Mabrouk and Sabry 2001). The majority of terrestrial activities that contribute to plastic debris in the ocean include construction, tyre abrasion, urban and precipitation runoff, sewage overflows, rubbish, waste mismanagement, industrial operations, and illegal dumping. The biggest contributors to ocean plastic pollution are fishing, boating, and aquaculture. Systematic garbage management and disposal. A "plastic leak" occurs into rivers and oceans as a result. Ecosystems whose waste management systems are insufficient to reduce plastic waste can be harmed by the legal and illicit global trade in plastic trash (Delacuvellerie et al. 2019) (Figs. 8.1 and 8.2).

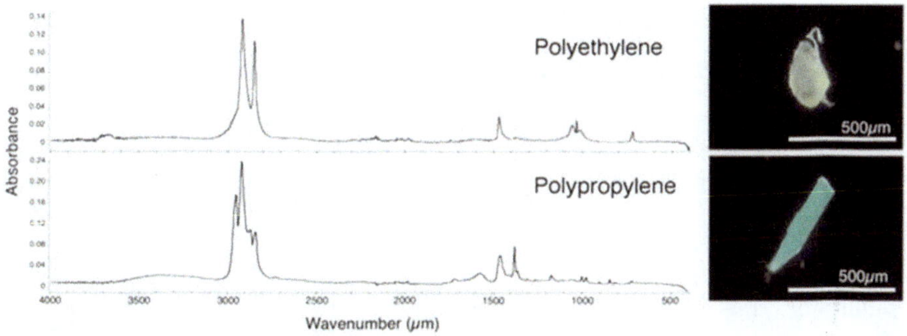

Fig. 8.1 Graphical representation of plastic fragments (Gewert et al. 2015)

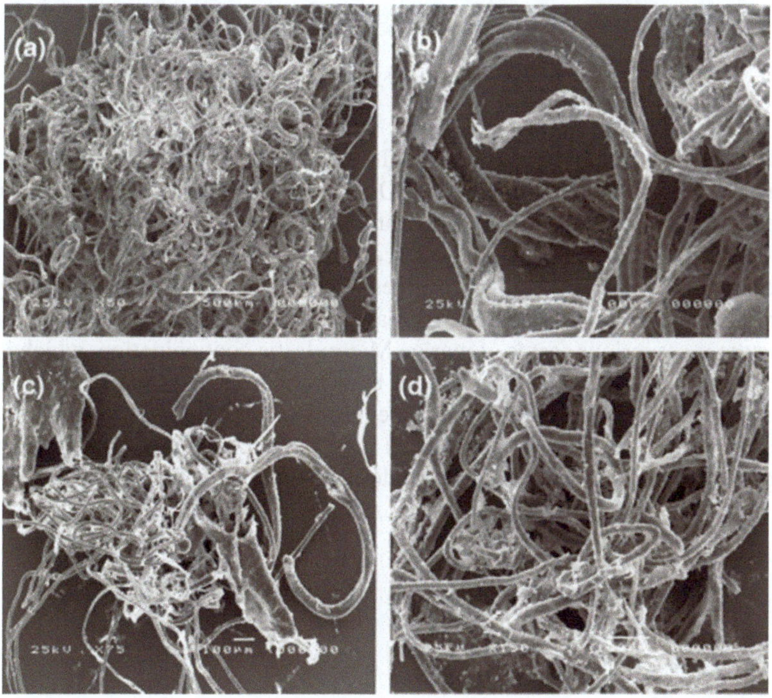

Fig. 8.2 Examining plastic fragments with scanning electron microscopy (SEM) (Harshvardhan and Jha 2013)

8.3 Problems with Small Particles

Since the advent of mass-produced plastic 60 years ago, humans have made more than 8 billion metric tonnes of plastic. Only 9% were reclaimed and the remaining 12% of this was carbonized. Over 80% of the residual plastic that has ever been created

Fig. 8.3 Accumulation of plastic in the marine system (Hung et al. 2016)

is eventually disposed of in landfills and the environment (Miraj Shaima et al. 2019; Mohanrasu et al. 2018; Nogi et al. 2014; O'Brine and Thompson 2010). The amount of plastic entering the oceans is worrisome; the ocean's biggest concentration of plastic trash is in the Pacific Ocean between California and Hawaii, estimated to be greater than that of France by three times (Howard et al. 2007). It is horrifyingly common. Pollution from plastic has increased a highly visible problem, nonetheless among the most serious forms of marine Microplastic contamination, which is difficult to detect. Plastic is not biodegradable, but it breaks down into smaller and smaller pieces, creating microplastics. Most of the billion of plastic trash in our oceans total tonnes are microplastics less than 5mm in size. Research on how these tiny particles impact our environment is yet mostly underdeveloped (Hung et al. 2016) (Fig. 8.3).

8.4 International Guidelines to Reduce Plastic Use and Waste

The various pollutants entering the seas frequently have similar causes and routes, including B. lack of access to wastewater treatment and sanitation or utilisation of natural resources inefficiently. Addressing These underlying factors may have an amplifying effect. For example, enhancing the management of large-scale wastewater in cities and regions reduces plastic emissions into the ocean, reduces nutrient loads, and improves fisheries and coral reef health (Palm et al. 2019; Peng and Fu 2020; Prabhat et al. 2013; Pu et al. 2018; Reddy et al. 2006). This means we could use the increased attention on plastic pollution to combat multiple marine contaminants simultaneously (Ishii et al. 2007). Enhancing wastewater management by designing and creating wastewater infrastructure that is sustainable for the three billion people who do not have access to managed wastewater treatment facilities. Untreated sewage includes various contaminants. It can cause major dangers to the health of people and the environment by exposing them to harmful substances, vector-borne illnesses, and eutrophication. Implementing rainwater and stormwater filtration will help to improve stormwater management. and garbage disposal at the estuary. Regulations and behavioral changes in using fertilizers and insecticides (Jayasiri et al. 2013).

8.5 Removal of Plastic Waste

Plastic is an abundant solid pollutant in the marine environment. Despite its durability, plastic breaks down into pieces under UV light and physical strain. Plastic, which are described as nano plastics smaller than five millimetres are accumulated because of all this. Inorganic processes that degrade plastics include photocatalysis, heating, dioxide, hydrolysed, and physical breakdown, depending on the type of causal factor (Jonathan et al. 2011). Polymers with heterocyclic rings in the primary polymeric chains, such as thermoplastics, polycaprolactone, and polychloroprene, as well as polyaddition or condensing monomers, are particularly susceptible to hydrolysed processes. Separation takes place because of mechanical stress in physical breakdown (Reisser et al. 2014; Restrepo-Florez et al. 2014; Sangeetha Devi et al. 2015). If polymers were available inside the aquatic organisms, those reactions can be triggered by storm and tide action, but they are also triggered by erosion from shore bed sediments. In maritime habitats, when saltwater serves as a barrier between sun energy and heat, values above 35 °C are also not typical (Kawai and Hu 2009) (Table 8.1).

Thus, only one polymer that could radiate heat whereby oxidising produces heat breakdown is trash that has deposited on top of the soil and has not interacted with water for a long time. Since polymer breakdown typically starts with exposure to ultraviolet radiation, a component of the solar spectrum, modelling of UV degradation is the technique that is most frequently used to examine the weathering of plastics (Sekiguchi et al. 2011; Sharon 2012; Sudhakar et al. 2008). To determine the impacts of Ultraviolet light, subject your model to weathering tests with UV lamps or direct sunlight. Additionally, most plastic products discovered on shores are single-use materials like polyethylene and polyester, which are either thrown there or float because they have a lower density than salt water (1.02 g/cm^3). When compared to polyethylene from open oceans, Ultraviolet light also substantially intensifies Ultraviolet damage (Kita et al. 1995).

Table 8.1 Marine environnement plastic pollution (Kumar et al. 2020)

Sl. NO	Bacteria	Plastic type	Biodegradability test conditions	Microbial degradation analysis
1	Bancroft's enteric bacteria	Polystyrene, phosphatidylethanolamine, positron emission tomography	Natural ions solution above ambient temperature, 1.25 months, 150 revolutions per minute	Decrease; *FTIR; **SEM
2	*Theilli* bacteria	Polystyrene, phosphatidylethanolamine, positron emission tomography, pancreatic polypeptide	Natural ions solution above ambient temperature, 1.25 months, 150 revolutions per minute	Loss of weight; *FTIR; **SEM
3	YP1 *Lactobacillus*	Positron emission tomography	Liquid carbon-free media at 30 °C for two months around 120 speeds with 1 g of monomer	Ductility; *FTIR; **SEM. and loss of weight
4	YT1 *E. coli asburia*	Positron Emission Tomography	120-speed, Liquid carbon-free media with 1 g of monomer, for two months at around 30 °C	Ductility; *FTIR; **SEM. and loss of weight
5	Native coastal microbes species	Positron emission tomography	Brining phosphate buffer for 6 months around 25 °C and 120 speed	Loss of weight; *FTIR; **SEM
6	Mixed culture of mesophilic bacteria (*Bacillus* and *Paenibacillus*)	Positron emission tomography	Basal medium containing 100 mg polymer at 30 °C for 2 months	Loss of weight; *FTIR; **SEM

* *Infrared Fourier transform, scanning electron microscope*

8.6 Biodegradation: Problem Solution

Since it is inexpensive and ecologically sustainable, microbial degradation appears to be a viable alternative. Marine generators contain waste material at all levels, such as the seafloor. Each of the above habitats contains organisms that are easily adaptable to plastic trash and can even create microbes on its top (Sugimoto et al. 2001; Syranidou et al. 2017; Tourova et al. 2020). The bioremediation method is significantly influenced by several elements. One of the most important of these are polymer properties and environmental conditions (Muller et al. 2005; Kanungo et al. 2023).

Marine particles are damaged via microbes in one or every of ways (Wilcox et al. 2016; Xanthos and Walker 2017):

Table 8.2 Organisms capable of degrading plastic that are isolated from a cold environment (Shibasaki et al. 2009)

S. No	Microbes	Origin	Type of plastic waste
1	*Psychrobacter*, Shewanella, several *Moritella* species, and bacterial infections	Silt from deep waters, the Kurile and Japan trenches	Posterior cruciate ligament
2	*Vibriosis parahemolyticus* and *Vibrio alginolyticus*	Aquatic ecosystems' benthic zones	PVA-LLDPE
3	*Rhodococcus* species, *Clonostachys rosea*, *Trichoderma* species, and *Pseudomonas* species	Arctic ground	PCL, is a readily available bag made of maize and potato starch
4	*Maritimum zalerion*	Ocean–atmosphere	Phosphatidylethanolamine
5	Fungal sp., Fungal Versicolor	Coastal Kovalam—500 m low price, at a depth of 5 m, in the Bay of Bengal	LDPE
6	*Falciparum* species	Tottori Prefecture's offshore areas and deep seas, including Toyama Bay	PCL
7	*Alcanivorax* species, *Tenacibaculum* species, and *Pseudomonas* species	A deep sea	Yarn PCL, PHB/V, and PBS fibres

With the aid of some essential catabolic enzymes, microorganisms use such chemicals as carbon resources. Byproducts from microorganisms damage polymer structures. The possibility for treating difficult and contaminated wastewater is increased by immobilised enzymes (Wei and Zimmermann 2017; Welden and Cowie 2017). The microbial enzymatic degradation mechanism includes the following steps: Microbial biofilm formation, initial deposition, and formation of plastispheres. Biodegradation: Effects of exomicrobial enzymes on polymers' structure, toxicological, and physical features. Enzyme depolymerisation of biological fragments to dimers, tri-, or monomers (Raghul et al. 2014). Further, with various microorganism species and their genetic mutation, synergistic effect, resistance, and higher adaptability in the toxic environment, indigenous microbial populations conduct better degradation to remove the pollutants (Mohanta et al. 2023; Behera et al. 2022) (Table 8.2).

8.7 Conclusion

Synthetic plastics have grown significantly in importance since the early 1950s because of their amazing physical and chemical characteristics. They are become an essential component of our life. Every year, tonnes of plastic are manufactured, and nearly half of them are intended for one use only. Humanity has overused them throughout time, and as a result, plastic debris is now everywhere. Recently, bioremediation of artificial polymers has attracted the glance of research, and many microorganisms have been found to potentially degrade these particles. This review document focuses on the plastic contaminants within the marine environment and how bioremediation might be one solution. In this procedure, contaminated areas are decontaminated using living creatures like bacteria and microorganisms. The revocation of. contaminants, pollutants, and toxins from soil, water, and other habitats also use this method.

References

Andrady AL (1998) Biodegradation of plastics: monitoring what happens. In: Pritchard G (eds) Plastics additives. Polymer science and technology series, vol 1

Auta HS, Emenike CU, Jayanthi B (2018) Growth kinetics and biodeterioration of polypropylene microplastics by *Bacillus sp.* and *Rhodococcus sp.* isolated from mangrove sediment. Mar Pollut Bull 127:15–21

Barnes DKA, Galgani F, Thompson RC (2009) Accumulation and fragmentation of plastic debris in global environments. Philos Trans R Soc B 364:1985–1998

Behera ID, Basak G, Kumar RR, Sen R, Meikap BC (2020) Treatment of petroleumrefinery sludge by petroleum degrading bacterium Stenotrophomonas pavanii IRB19 as an efficient novel technology. J Environ Sci Health Part A 56:1–13. https://doi.org/10.1080/10934529.2020.1866924

Behera ID, Nayak M, Biswas S, Meikap BC, Sen R (2021) Enhanced biodegradation of total petroleum hydrocarbons by implementing a novel two-step bioaugmentation strategy using indigenous bacterial consortium. J Environ Manage 292:112746. https://doi.org/10.1016/j.jenvman.2021.112746

Behera ID, Nayak M, Mishra A, Meikap BC, Sen R (2022) Strategic implementation of integrated bioaugmentation and biostimulation for efficient mitigation of petroleum hydrocarbon pollutants from terrestrial and aquatic environments. Mar Pollut Bull 177:113492. https://doi.org/10.1016/j.marpolbul.2022.11349

Catania V, Cascio Diliberto C, Cigna V (2020) Microbes and persistent organic pollutants in the marine environment. Water Air Soil Pollut 231(7):1–10

Danso D, Chow J, Streit WR (2019) Plastics: environmental and biotechnological perspectives on microbial degradation. Appl Environ Microbiol 85(19):e01095-e1119

Debroas D, Mone A, Ter Halle A (2017) Plastics in the North Atlantic garbage patch: a boat-microbe for hitchhikers and plastic degraders. Sci. Tot. Environ 599–600:1222–1232

Delacuvellerie A, Cyriaque V, Gobert S (2019) The plastisphere in the marine ecosystem hosts potential specific microbial degraders including *Alcanivorax borkumensis* as a key player for the low-density polyethylene degradation. J Hazard Mater 380:120899

Gewert B, Plassmann MM, Macleod M (2015) Pathways for degradation of plastic polymers floating in the marine environment. Environ Sci: Process Impacts 17(9):1513–1521

Harshvardhan K, Jha B (2013) Biodegradation of low-density polyethylene by marine bacteria from pelagic waters, Arabian Sea, India. Mar Pollut Bull 77(1–2):100–106

Howard GT, Mackie RI, Cann IK (2007) Effect of insertional mutations in the pueA and pueB genes encoding two polyurethanases in *Pseudomonas chlororaphis* contained within a gene cluster. J Appl Microbiol 103:2074–2083

Hung CS, Zingarelli S, Nadeau LJ (2016) Carbon catabolite repression and impranil polyurethane degradation in *Pseudomonas protegens* strain Pf-5. Appl Environ Microbiol 82:6080–6090

Ishii N, Inoue Y, Shimada KI, Tezuka Y, Mitomo H, Kasuya KI (2007) Fungal degradation of poly (ethylene succinate). Polym Degrad Stab 92(1):44–52
Jayasiri HB, Purushothaman CS, Vennila A (2013) Quantitative analysis of plastic debris on recreational beaches in Mumbai, India. Mar Pollut Bull 77(1–2):107–112
Jonathan RR, Huang J, Anand P (2011) Biodegradation of polyester polyurethane by endophytic fungi. Appl Environ Microbiol 77:6076–6084
Kanungo J, Sahoo S, Bal M, Behera ID (2023) Performance of bioremediation strategy in waste lubricating oil pollutants: a review. Geomicrobiol J. https://doi.org/10.1080/01490451.2023.2245395
Kawai F, Hu X (2009) Biochemistry of microbial polyvinyl alcohol degradation. Appl Microbiol Biotechnol 84:227–237
Kita K, Ishimaru K, Teraoka M (1995) Properties of poly(3-hy- droxybutyrate) depolymerase from a marine bacterium, *Alcaligenes faecalis* AE122. Appl Environ Microbiol 61(5):1727–1730
Kumar AG, Anjana K, Hinduja M (2020) Review on plastic wastes in marine environment – Biodegradation and biotechnological solutions. Mar Pollut Bull 150:110733
Kumari A, Chaudhary DR, Jha B (2019) Destabilization of polyethylene and polyvinylchloride structure by marine bacterial strain. Environ Sci Pollut Res 26(2):1507–1516
Lucas N, Bienaime C, Belloy C (2008) Polymer biodegradation: mechanisms and estimation techniques–a review. Chemosphere 73(4):429–442
Mabrouk M, Sabry S (2001) Degradation of poly (3-hydroxybutyrate) and its copolymer poly (3-hydroxybutyrate-co-3-hydroxyvalerate) by a marine *Streptomyces sp.* SNG9. Microbiol Res 156:323–335
Miraj Shaima SA, Parveen N, Zedan HS (2019) Plastic microbeads: Small yet mighty concerning. Int J Environ Health Res. Taylor & Francis
Mohanrasu.K, Premnath N, Prakash S (2018). Exploring multi potential uses of marine bacteria; an integrated approach for PHB production, PAHs and polyethylene biodegradation. J Photochem Photobiol B: Biol 185(2017)
Mohanta S, Pradhan B, Behera ID (2023) Impact and remediation of petroleum hydrocarbon pollutants on agricultural land: a review. Geomicrobiol J 1–5. https://doi.org/10.1080/01490451.2023.2243925
Muller RJ, Schrader H, Profe J (2005) Enzymatic degradation of PET: rapid hydrolyse using a hydrolase from *T. fusca*. Macromol Rapid Commun 26:1400–1405
Nogi Y, Yoshizumi M, Miyazaki M (2014) *Thalassospira povalilytica sp.* nov., a polyvinyl-alcohol-degrading marine bacterium. Int J Syst Evol Microbiol 64(PART 4):1149–1153
O'Brine T, Thompson RC (2010) Degradation of plastic carrier bags in the marine environment. Mar Pollut Bull 60(12):2279–2283
Palm GJ, Reisky L, Bottcher D (2019) Structure of the plastic-degrading *Ideonella sakaiensis* MHETase bound to a substrate. Nat Commun 1717:1–10
Peng L, Fu (2020) Micro- and nano-plastics in marine environment: source, distribution and threats—a review. Sci Total Environ 698:134254
Prabhat S, Bhattacharyya S, Vishal V (2013) Studies on isolation and identification of active microorganisms during degradation of polyethylene/starch film. Int Res J Environ Sci 2(9):83–85
Pu W, Cui C, Guo C, Wu ZL (2018) Characterization of two styrene monooxygenases from marine microbes. Enzym Microb Technol 112:29–34
Raghul SS, Bhat SG, Chandrasekaran M (2014) Biodegradation of polyvinyl alcohol-low linear density polyethylene-blended plastic film by consortium of marine benthic *vibrios*. Int J Environ Sci Technol 11(7):1827–1834
Reddy MS, Basha S, Adimurthy S, Ramachandraiah G (2006) Description of the small plastics fragments in marine sediments along the Alang-Sosiya ship-breaking yard, India. Estuar Coast Shelf Sci 68(3–4):656–660
Reisser J, Shaw J, Hallegraeff G (2014) Millimeter-sized marine plastics: a new pelagic habitat for microorganisms and invertebrates. PLoS ONE 9(6):1–11

Restrepo-Florez J-M, Bassi A, Thompson MR (2014) Microbial degradation and deterioration of polyethylene – a review. Int Biodeterior Biodegradation 88:83–90

Sangeetha Devi R, Rajesh Kannan V, Nivas D (2015). Biodegradation of HDPE by *Aspergillus spp.* from marine ecosystem of Gulf of Mannar, India. Mar Pollut Bull 96(1–2):32–40

Sekiguchi T, Saika A, Nomura K (2011) Biodegradation of aliphatic polyesters soaked in deep seawaters and isolation of poly(ε-caprolactone)-degrading bacteria. Polym Degrad Stab 96(7):1397–1403

Sharon M (2012) Studies on biodegradation of polyethylene terephthalate: a synthetic polymer biosynthesis of gold nanoparticles view project development of carbon solar cell view project

Shibasaki S, Kawabata A, Tanino T (2009) Evaluation of the biodegradability of polyurethane and its derivatives by using lipase-displaying arming yeast. Biocontrol Sci 14:171–175

Sudhakar M, Doble M, Murthy PS (2008) Marine microbe-mediated biodegradation of low- and high-density polyethylenes. Int Biodeterior Biodegrad 61(3):203–213

Sugimoto M, Tanabe M, Hataya M (2001) The first step in polyethylene glycol degradation by *Sphingomonas* proceeds via a flavo- protein alcohol dehydrogenase containing Flavin Adenine Dinucleotide. J Bacteriol 183:6694–6698

Syranidou E, Karkanorachaki K, Amorotti F (2017) Development of tailored indigenous marine consortia for the degradation of naturally weathered polyethylene films. PLoS ONE 12(8):1–21

Tourova T, Sokolova D, Nazina T (2020) Biodiversity of microorganisms colonizing the surface of polystyrene samples exposed to different aqueous environments. Sustainability 12(9):3624

Wei R, Zimmermann W (2017) Microbial enzymes for the recycling of recalcitrant petroleum-based plastics: how far are we? Microb Biotechnol 10(6):1308–1322

Welden NA, Cowie PR (2017) Degradation of common polymer ropes in a sublittoral marine environment. Mar Pollut Bull 118(1–2):248–253

Wilcox C, Mallos NJ, Leonard GH (2016) Using expert elicitation to estimate the impacts of plastic pollution on marine wildlife. Mar Policy 65:107–114

Xanthos D, Walker TR (2017) International policies to reduce plastic marine pollution from single-use plastics (plastic bags and microbeads): a review. Mar Pollut Bull 118(1–2):17–26

Chapter 9
Petroleum Hydrocarbon Waste Recycling, Reusing, Repairing, and Recovering Value Added Products

Anil Kumar Murmu, Lipika Parida, and Veda Prakash

Abstract Petroleum hydrocarbon waste is a common global concern that arises as a consequence of the extraction and refining procedures associated with fossil fuels. Improper disposal of untreated petroleum hydrocarbons presents a significant threat. The present chapter provides an in-depth analysis of the origins and nature of petroleum hydrocarbon waste that is commonly encountered in soils contaminated with such pollutants. Additionally, this study investigates the impacts of Petroleum hydrocarbon waste on soil microbiology, human well-being, and the degradation of terrestrial and aquatic ecosystems. The chapter furthermore presents a comprehensive examination of advanced biological approaches utilized in the treatment of soils that have been contaminated with petroleum hydrocarbons. The treatment methods encompass adsorption, membrane technology, thermal technologies, and biological treatment employing microorganisms.

Keywords Petroleum hydrocarbon waste · Phytoremediation · Biological treatment methods · Thermal treatment methods · Membrane technology

9.1 Introduction

Waste is described by the Environmental Quality Act of 1974 as "any matter in any physical form (solid, liquid, and gas) generated in such a large quantity that affects or alters the natural ecosystem" (Santiago et al. 2021). Any waste which comprises hydrocarbons is known as hydrocarbon waste. Hydrocarbons are the organic compounds made from the refinement or extraction of fossil fuels. Carbon and Hydrogen are mainly present in these organic compounds. These naturally occur in crude oil, which is created over time by the rapid oxidation of organic materials (Waples 2013; Tissot and Welte 1984). These wastes are produced throughout

A. K. Murmu · L. Parida (✉) · V. Prakash
Department of Chemical Engineering, Veer Surendra Sai University of Technology, Burla 768018, Odisha, India
e-mail: lparida_chemical@vssut.ac.in

I. D. Behera and A. P. Das (eds.), *Impact of Petroleum Waste on Environmental Pollution and its Sustainable Management Through Circular Economy*, Environmental Science and Engineering, https://doi.org/10.1007/978-3-031-48220-5_9

the gas and oil network at numerous points. The drilling for crude oil and gas on land and at sea generates these wastes. They are also created during unprocessed crude oil and gas transportation and storage.

Petroleum and its derivatives are the second most often utilized material in modern life, after water (Gardas et al. 2019). The level of living for humanity has improved and the scale of the global economy has grown since oil was first used as a source of energy in the 19th century (Inayat et al. 2020). Today, fossil fuels provide 85% of the world's energy, with oil and gas each contributing 35.3 and 20.5% (Mariyam et al. 2022). Our modern world is dominated by petroleum and products made from it. Petroleum is used in the production of almost everything we use daily, including plastics, oils, and almost everything else. But the petroleum-based goods we consume every day create a sizeable amount of hydrocarbon waste that needs to be dealt with. This waste is largely invisible to customers because it results from manufacturing processes and has the potential to pollute the environment. But hydrocarbon waste isn't only a by-product of production; it's all around us. Currently, oil and gas energy, together with related products, exert significant influence on global affairs. These products have undergone multiple upgrading and treatment steps before being made usable. The extraction and refining processes within the oil and gas industries result in the generation of a significant amount of toxic and hazardous wastes. These wastes exert a huge influence on the environment.

Hydrocarbons and their wastes are assemblages of various materials such as emulsion, solids, crude oil, and slurry. Petroleum and hydrocarbon waste exploration, manufacturing, and transportation may negatively impact the natural world, marine and terrestrial biota, and human health (Shahzad et al. 2020; Mohanta et al. 2023). The concentrations of hydrocarbons (35–55%), water (35–85%), and mineral particles (1–20%) differ from place to place in Petroleum Waste. These can be divided into two groups: hazardous waste and non-hazardous waste. Anything that is exploding, reactive, ignitable, hazardous, and corrosive is identified as hazardous. Solid hazardous waste like (sample bottles, batteries, catalysts, filters, fluorescent tubes, oily polluted soils), liquid harmful waste like (used lubes, lapsed chemicals, utilized acids), and soil/sludge are all examples of hazardous waste (Ng et al. 2021).

Non-hazardous waste refers to industrial waste which does not cause harm to humans or the ecology. However, it is subject to regulatory restrictions that prohibit its disposal in conventional trash containers or drainage systems. General waste includes (mixed garbage), automobile tires (tires of vehicles), paper (cardboard, newspapers and magazines, waste from offices, papers), glass (cold beverages, laboratory, bottles of reagent), garbage (cables, piping, packaging supplies, boxes, HDPE and LDPE bottles), and metallic substances (drums, pipes of steel and metal). Petroleum sludge (PS) is the petroleum industry's primary waste product (Srivastava et al. 2019).

Over 100,000 sites across the globe have been identified as dangerous for living organisms, especially humans, with most of these being a consequence of the improper disposal of unresolved petroleum hydrocarbons. Complete clean-up of these locations in a short duration has remained a problem, owing to the costs involved in disposal and processing. Another possibility is the absence of the most effective remediation approach, which leads to the degradation of various kinds of petroleum

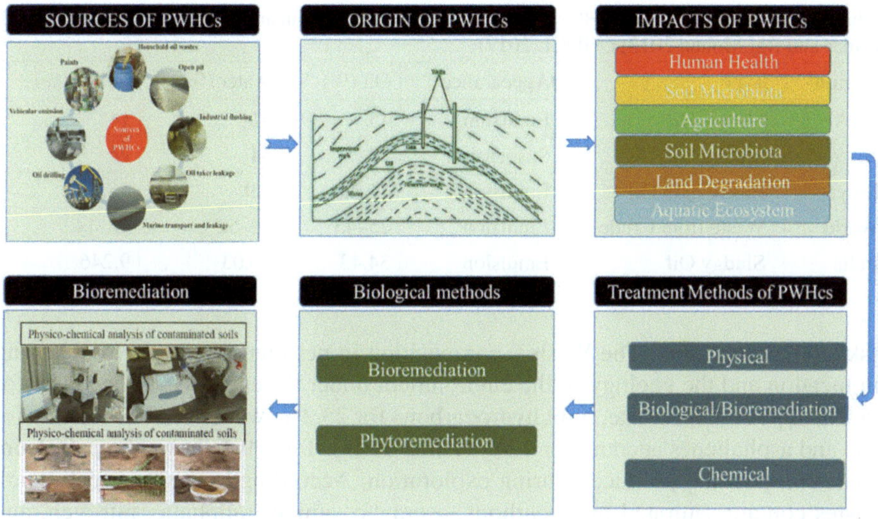

Fig. 9.1 Summary and overall plan of action for PWHCs from origin to remediation

waste hydrocarbons in a short span of time (Behera et al. 2022). The spilling and open discharge of fluid-based greasy garbage in the environment, along with their investigation and detection, represent significant concerns. The objective of this section is to (1) Look into the origin and chemical composition of petroleum waste present in soils that have been contaminated at disposal sites. (2) Assess the effects of petroleum waste on the wellness of humankind, microbiota of soil, deterioration of sites, and marine organisms. (3) To address the advanced biological processes that are currently accessible for the recuperation of polluted lands created due to the accumulation of Petroleum industry waste over a long span of time. A good petroleum waste treatment plan collects garbage from several sites. To remove hydrocarbon residues from the petroleum industry, this waste is analysed and treated. As shown in Fig. 9.1, these processes use biological, chemical, and physical methods.

9.2 Composition of Petroleum Waste

It has been widely observed that the process of crude oil exploration results in the generation of a significant amount of petroleum waste. This waste serves to provide invaluable information into the properties, behaviour, and content of source rocks (Hu et al. 2013; Ozdemir et al. 2020). Petroleum's chemical properties are almost equal to crude oil and, to a lesser extent, to rocks from where it is derived. Nevertheless, crude oil constitutes most of the petroleum, with trace amounts of water, emulsion, clay, and soil described in Table 9.1. The production of petroleum waste in the process of exploration, as well as its disposal through burial in large pits, poses potential

Table 9.1 Composition of PWHCs (Bidder and Hunt 1982; Mansur et al. 2015; Hu et al. 2013; Kumar and Raj Mohan 2013; OGDCL 2017)

Region	Type	Appearance	Oil (%)	Water (%)	Solids (%)
USA	Sloppy Oil	Emulsion	45	41	14
	Sludgy Oil	Suspension	22	54	24
Libya	Raw Crude Oil	Emulsion	41.08	3.9	55.02
World	Waste from Crude	Emulsion	25–45	32–52	12–14
India	Sludgy Oil	Emulsion	34.43	56.03	9.246

risks to human safety. The Waste's composition in petroleum may vary depending on location and the geology of the environment under the earth. Water accounts for 55.2% of petroleum waste, light hydrocarbons for 23.2%, waxes for 10.5%, clay for 9%, and asphaltenes account for 1.9% (Islam 2015). Table 9.1 shows the petroleum waste composition produced during exploration. According to the data, petroleum wastes contain saturated hydrocarbons, aromatic, amines, aliphatic, nitrogen, and oxides. Nevertheless, the composition varies by area, depending on the geologic and biogenic atmosphere. (Singh et al. 2017; Ozdemir et al. 2020) discovered that aromatic and aliphatic hydrocarbons dominate the world's petroleum reservoirs. The hydrocarbon composition in the range of 75–85% primarily consists of aromatic and aliphatic compounds. This composition mainly comprises petrol, kerosene, naphtha, diesel, crude oil, and bitumen. Additionally, trace amounts of nitrogen, sulphur, and oxygen are also present. Based on the available data, it has been observed that liquid petroleum wastes contain a significant presence of aliphatic hydrocarbons, including paraffin, alkanes, and alkenes, as well as aromatic hydrocarbons such as anthracene, naphthalene, pyrene, and chrysene. These hydrocarbons have the potential to cause substantial environmental pollution and disrupt the balance within an ecosystem (Connell et al. 2009; De Junet et al. 2013). To limit the environmental consequences of petroleum exploration and processing, environmental pollution caused by these activities must be remedied using advanced biological techniques.

9.2.1 Waste in the Oil and Gas Industry

From well extraction to petroleum products, the petroleum and natural gas industry generates trash that must be disposed of or recycled. These wastes need to be handled carefully because they contain hazardous and oily materials. In gas plants, flue gas (CO_2 and SO_2), Solid and organic wastes (wax, oil sludge, and plastic) as well as liquid wastes (oily chemicals, produced water, etc.) are the three types of waste that are produced most frequently. Process waste falls under the hazardous waste category, which also includes produced water, emissions of SO_X, CO_X, and oil sludge.

Table 9.2 Composition and Origin of Waste (Connell et al. 2009; Teng et al. 2021; Zhang et al. 2017; Guo et al. 2011; Liu et al. 2018)

Source	Composition
Produced Water	
• Drilling Operation for oil • Drilling Operation for gas	• Ammonia, BTS, H_2S • Dissolved gases, CO_2, O_2
Solid Waste	
• Landing Sludge • Drilling Waste • Refinery Sludge • Tank Bottom	• Complex, Oil content very low (10–30%) • Salts, Chemical additives, and heavy metals • Emulsifier, flocculating agent • Asphaltenes, heavy hydrocarbons
Flue gas	
• Gas and Oil Wells • Sour gas (CO_2 and H_2S) • Gas and Oil refining	• S containing compounds 1–3% • 2–3% required in NG, CO_2 present (4–50%)

Plastic, wood, drilling fluids, metal shavings, and sand mud make up most of the non-hazardous waste. The composition and origin of waste in the oil and gas industry are reported in Table 9.2.

9.2.1.1 Produced Water

The primary waste stream generated during the process of oil and gas exploration is known as produced water (PW). PW is a mixture of water from sedimentary sources and saltwater from aquifers that is extracted to the surface during exploration activities. The efficacy of PW is significantly influenced by various factors, including its geological formation, geographical location, reservoir lifespan, and production methods. PW typically consists of a mixture of oil (in dissolved or dispersed form), minerals, solids, additives, and dissolved gases (Kanungo et al. 2023). The chemical composition of PW includes polycyclic aromatic hydrocarbons (PAHs), toluene, benzene, xylenes (collectively known as BTEX), ethylbenzene, aliphatic hydrocarbons, phenols, and carboxylic acid (Ossai et al. 2020).

9.2.1.2 Solid and Organic Waste

Both organic and inorganic waste are produced by the oil and gas sector. Inorganic wastes include non-combustible materials like sand, fluid used for drilling, and scrap metal from construction sites. These wastes are also generated during the drilling process, site development, refinery of petroleum, and transportation processes. On the other hand, Organic Waste which is combustible in nature is mainly hydrocarbon waste. These are the most crucial elements for the conversion of waste to energy. Organic waste includes materials like oil sludge or plastic waste. The cleaning process of machinery, reforming section, hydrocarbon splitting, storing tank, and the treatment of effluent all result in the production of oil sludge.

9.2.1.3 Emission of Flue Gases

A gaseous hydrocarbon known as Natural Gas (NG) is present in coal beds, oil fields, and gas fields. Methane is the main component, along with other hydrocarbon and non-hydrocarbon gases such as H_2S, N_2, CO_2 (acid gases), and O2 are generated as by-products during thermogenic, acidogenesis, methanogenesis, and biogenic operations. These operations involve the transformation of organic substances and other compounds originating from deep within the earth. In comparison to other combustible fuels like coal and crude oil, NG is regarded as a clean fuel. The NG bears 60–70% of C, 5–30% of H_2, and 0–0.5% of S. Pollutants like H_2S, helium, CO_2, and water are also present in the raw gas.

9.3 Treatment Technologies of PW

PW involves the presence of toxic contaminants with extended half-lives, thereby presenting a significant threat to ecological and environmental well-being. Therefore, PW must be properly treated to meet regulatory standards before being recycled reused, or released into the environment. Various chemical, physical, and biological techniques have been employed for the purification of PW, as elaborated upon in the subsequent sections (Ahmadun et al. 2009).

9.3.1 Physiochemical Methods

9.3.1.1 Hydrocyclones

Dispersed oil particles and suspended solids with densities below those of water are separated using *Hydrocyclones*. They separate solids from fluids based on gravity differences. The hydrocyclones are supplied with pressure to induce a centrifugal motion, which helps in increasing the size of small particles in the PW and removing them from the main fluid. The overflow outlet allows clean fluid to exit while the bottom discharges solid particles. *Hydrocyclones* are very good at removing suspended particles, but they are ineffective at getting rid of dissolved solids. They are portable, last a long time, use less energy, and are simple to operate because of their smaller size, They are ideal for application in offshore installations. They are widely used in the treatment of PW on an industrial scale.

9.3.1.2 Coagulation/Flocculation

Using flocculation or coagulation to eliminate TSS, COD, and TOC from a variety of wastewater, including PW, is an efficient method of wastewater treatment. It entails adding chemicals at a high blending speed (>4500 rpm) to break down components and minimize surface tension. After that particles are aggregated, and the flocs are increased in size, and settled by gravity. Due to its poor TDS removal performance, this method is used as a preliminary treatment stage before membrane filtration, adsorption, and biological treatment.

9.3.1.3 Electro-coagulation

Electrocoagulation is an alternative method of coagulation that employs an electrical current between anode and cathode electrodes to produce hydroxyl and hydrogen radicals. The hydrogen gas bubbles attach to the particles and aggregate them, and they're then scanned from the surface of the water. Using the electro-coagulation technique, (Abdelwahab et al. 2008) removed 100% of the phenol from petrochemical wastewater. The phenol removal rate in refinery wastewater was 97% (Abdelwahab et al. 2008). The performance of electrocoagulation is largely dependent on the cathode and anode used.

9.3.1.4 Adsorption

The adsorption process is a surface phenomenon activity where contaminants from the liquid phase are chemically or physically bonded to the surface of the solid (of the adsorbent). The capacity for adsorption or effectiveness of removal is affected by a variety of variables, such as pH, temperature, the quality of the feed, and, most importantly, adsorbent properties. The most important factors to consider while choosing an adsorbent are (a) excellent adsorption capacity, (b) quick adsorption, (c) recuperation capability, and (d) plentiful accessibility. A large range of adsorbents can be utilized for the purification of PW. Modifications of surface and/or the use of artificial adsorbents can both increase an adsorbent's adsorption capacity, through the manufacturing process, which is not cost-effective. It's also essential to regenerate an adsorbent for the efficient treatment of PW. The use of chemicals and heat during regeneration may reduce adsorption capacity and increase cost.

9.3.1.5 Membrane Technology

PW is effectively treated using membrane technology. It is small, exhibits high removal effectiveness, and only needs a small amount of chemical addition. There are four categories of membranes based on pore size: reverse osmosis (RO), nanofiltration (NF), and ultrafiltration (UF). Their permeability is MF > UF > NF > RO in that order. While UF is frequently used to remove oil from the PW, MF membrane is employed to eliminate SS. In contrast to RO, which is used to remove dissolved solids, NF membranes are utilized to eliminate divalent ions, including heavy metals. UF is the finest among these for addressing PW due to its excellent recovery of water and low energy consumption. Nanofiltration and reverse osmosis are effective process water treatments. Small pores make them prone to clogging and energy-intensive, limiting their effectiveness.

9.3.1.6 Oxidation

Hydrogen peroxide, Ozone, chlorine dioxide, and UV radiations are a few examples of oxidizing agents that have been used (Hu et al. 2013). The oxidation method yields high water recovery, produces little sludge, and is easy to use. However, this technology's two biggest drawbacks are its high cost of chemicals and the production of toxic byproducts (Mansur et al. 2015). For the treatment of PW, Moraes et al., reported using Fe + H_2O_2 + UV and achieving 85% TOC removal (Hu et al. 2013). According to a study, ferrous ions in the oxidation reaction removed 70% of Total Organic Carbon (TOC) and 98% of Chemical Oxygen Demand (COD) from petroleum refinery effluent (Ossai et al. 2020).

9.3.2 Thermal Methods

Thermal treatments are recognized as reliable for PW treatment. Thermal technologies may be divided into three types based on variations in temperature and pressure: multiple-phase flash distillation (MSF), multi-effect distillation (MED), and vapor compression distillation (VCD).

MSF is used to boil PW, which is then transformed into steam in a low-pressure vessel. *MSF* bears a prolonged lifespan of 15–20 years and is economical., The formation of scale on heating surfaces, however, is a significant disadvantage when utilizing this technology, necessitating regular acid clearing or the application of scale blockers. Water is initially transformed to vapor and then cooled in *MED*. Due to the use of a variety of evaporators, it offers recovery of larger amounts of water (up to 65%) and is highly energy efficient. This is very helpful for PW which has a large TDS content. Still, scaling is a significant barrier to its widespread use. Water vapors are first produced by VCD, which is followed by thermal compression to raise the vapor's temperature and pressure. It uses less energy and has a less serious scaling problem than *MED* and *MSF* because it can work at lower temperatures (<70 °C). Utilizing a joint unit of *MED* and *VCD* technologies has been the subject of recent research. In numerous nations, the combined units are run on a large scale.

9.3.3 *Biological Treatment*

The organic element in PW is biodegraded by microorganisms in this operation. PW's quality is raised by the microorganisms using the organic carbon as a substrate. Meanwhile, a suitable harvesting technique is used to separate the generated microbial biomass from the aqueous medium. For the treatment, various microorganisms are used depending on the PW's characteristics. Aerobic and anaerobic biological treatments fall into these two categories. The use of oxygen is the primary distinction. Anaerobic treatment involves the utilization of an organic or inorganic substance, excluding oxygen, as an electron acceptor. On the other hand, aerobic treatment relies on oxygen as the terminal electron acceptor.

9.3.3.1 Aerobic Treatment

To deteriorate the contaminants in PW, aerobic microorganisms (algae, fungi, and bacteria) are used (Behera et al. 2020, 2021). The availability of oxygen is crucial for the growth and metabolic activity of aerobic bacteria as well as the degradation of organic pollutants. The treatment of PW requires the utilization of different systems, for example, the activated sludge process, biological aerated filters (BAFs), and aerobic treatment employing algae as a medium. The most basic form of aerobic therapy is known as BAF (Basic Aerobic Fitness). The process involves the cultivation of aerobic bacteria on a porous substrate. The microorganisms are nourished with PW and employ either adsorption or biosorption mechanisms to eliminate dissolved and suspended solids. This method has been successful in removing phenol, TOC, COD, oil, and BTEX from the PW.

9.3.3.2 Anaerobic Treatment

Higher treatment efficiency and portability have increased the use of anaerobic membrane bioreactors (AnMBRs). It regulates hydraulic and sludge retention times with membranes within anaerobic reactors to efficiently remove pollutants (OGDCL 2017). Various wastewater kinds have used AnMBR extensively (OGDCL 2017; Ossai et al. 2020). Their application for PW, however, has not received much attention.

9.3.3.3 Phytoremediation

The method of phytoremediation is effective in the management of PW. Plants do this by removing or degrading organic pollutants from the PW. Contaminants can be broken down directly in plant tissues or through interactions between active soil microorganisms in the rhizosphere and plant roots. The presence of organic acids, sugar, and oxygen derived from plant roots facilitates microbial growth and the

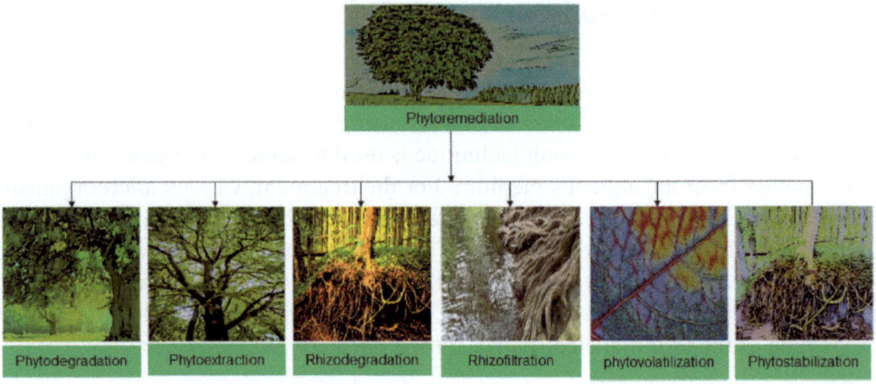

Fig. 9.4 Various mechanisms are adopted by plants during phytoremediation

decomposition of contaminants. Additionally, the roots of the plants exude several enzymes, including laccase and dehalogenase, which promote microbial development and the breakdown of contaminants. Phytoremediation systems are affected by the plant species used, the microbial community, the soil, the geographic location, the soil amendment methods, and the wastewater being treated. The effectiveness of treatment can be increased by looking into these factors. The addition of plants improved the microorganism-based remediation of oil wastes even further. Bioremediation is the term used to describe the microbial treatment of petroleum waste, while phytoremediation is the term used to describe the treatment of oil waste using plants. Consequently, the petroleum waste in contaminated soil can be reduced using either of the two methods, phytoremediation, or bioremediation. The process of phytoremediation involves several mechanisms, some of which are listed in (Fig. 9.4).

Phytodegradation

It refers to the process of pollutant degradation facilitated by plant enzymes. Phytodegradation has the capability to eliminate various contaminants from the environment, such as solvents in groundwater, petroleum, and aromatic components in soil. Dehalogenase, peroxidase, nitrilase, and nitroreductase are such few enzymes that are used by plants to break down organic pollutants.

Phytoextraction

Phytoextraction, also referred to as phytoaccumulation or phyto-absorption, is a process in which plant roots uptake contaminants from soil or water. These contaminants are then transported to and accumulated in the above-ground biomass of the plants.

Rhizodegradation

Rhizodegradation is the term for the breakdown of organic contaminants in soil by rhizosphere microorganisms. Plants may increase microbial activity in the rhizosphere by roughly 10–100 times by secreting exudates that contain flavonoids, amino acids, and sugars.

Rhizofiltration

Rhizofiltration refers to the process of adsorbing contaminants present in a solution within the root zone, also known as the rhizosphere, onto the roots of plants. It can also involve the absorption of contaminants into the roots of plants. Rhizofiltration is a remediation technique employed to cleanse groundwater. In greenhouses, plants are cultivated hydroponically, using water as a growing medium instead of traditional soil. The plants undergo acclimatization to the environment by utilizing water that is contaminated from the site. The plants are subsequently positioned in close proximity to a groundwater source that has been contaminated, facilitating the absorption of both water and contaminants by their roots. Once the roots have been saturated with the contaminant, the plants, including the roots, are harvested.

Phytovolatilization

The phenomenon referred to as phytovolatilization of petroleum hydrocarbons occurs when plant roots absorb pollutants, particularly volatile organic compounds (VOCs), from soil. These pollutants are subsequently released into the atmosphere through the leaves via evapotranspiration. The pollutants undergo a transformation process, converting them into a gaseous state before being subsequently released. The process described by Abdelwahab et al. (2008) is commonly employed in areas with marshy conditions or for the construction of wetlands (Fig. 9.4).

Phytostabilization

Phytostabilization involves the establishment of a plant cover on contaminated sites. The objective is to minimize the movement of contaminants within the vadose zone by accumulating or immobilizing them in the root zone, thereby reducing contamination in surrounding areas. The process of transcription and root growth effectively immobilizes pollutants by reducing leaching, controlling erosion, creating an aerobic environment in the root zone, and incorporating organic matter into the substrate, thereby binding the contaminants.

9.4 Solid Hydrocarbon Waste

One of the significant types of solid waste is oil sludge (OS), which is characterized as a dense, viscous, and partially solid combination of hydrocarbons, sand, clay, and certain metals (Jasmine and Mukherji 2014). The operating system (OS) is generated at multiple stages within the oil and gas sector, encompassing transportation vehicles, refining operations, and storage containers (Bujang et al. 2013; Jasmine and Mukherji 2014). Many processes, including incineration, solidification, oxidation, and biodegradation, are used to dispose of OS (Cloern et al. 2002). In contrast, a minimal amount of waste is recycled and subsequently transformed into usable products (Bidder and Hunt 1982).

9.4.1 Treatment Technologies of Solid Waste

9.4.1.1 Incineration

The process involves the complete combustion of waste sludge from the petroleum industry in the presence of ample air and auxiliary fuel. Rotary kilns and fluidized beds represent the prevailing categories of incinerators. The combustion temperatures observed in rotary kiln incinerators typically fall within the range of 980 to 1,200°C, while the residence time for this process is approximately 30 min. On the other hand, combustion temperatures in fluidized beds generally range from 730 to 760 °C, and the residence time for this method is measured in days. The fluidized bed treatment method is highly efficient in terms of mixing, fuel flexibility, pollutant emissions, and combustion efficiency. As a result, it is considered the optimal choice for treating low-quality sludge.

9.4.1.2 Solidification

Solidification is a process used to encapsulate or seal waste using an adhesive. The primary objective of this method is to prevent the waste from leaching into the environment, either through physical or chemical mechanisms. Additionally, solidification aims to transform the resulting products into environmentally friendly construction materials or non-hazardous waste suitable for disposal in landfills. The solidification process of hazardous waste by types of cement comprises three primary stages: I. Correcting chemical pollutants involves the chemical reaction between the hydration products of cement and the contaminants themselves. II. The physical absorption of contaminants that are found on the surface of hydrated cement products. III. Encapsulation of Contaminated Waste or Soil.

9.4.1.3 Oxidation

Oxidation treatment refers to the process of utilizing chemical and other oxidation agents to degrade organic contaminants. During the oxidation process, reactive agents are employed to introduce chemical reactions within the oily sludge. This facilitates the conversion of the organic compounds present in the sludge into either carbon dioxide or water, or into non-hazardous substances. Several oxidation reagents have been employed in the treatment of oily sludge in this process. These include Fenton's reagent, hypochlorite, ozone, ultrasonic irradiation, permanganate, and persulfate. The effectiveness of Ultrasonic irradiation in the process of oxidizing oily sludge has been demonstrated.

9.4.1.4 Biodegradation

Microorganisms have been utilized in the biodegradation and elimination of environmental pollutants through various methods such as land treatment, bio pile/composting, and bio-slurry.

Land Treatment

The land farming treatment method involves the biological, chemical, and physical degradation of oily sludge contaminants through the process of mixing them with soil. Land treatment is considered a favorable option for waste disposal due to its cost-effectiveness, minimal energy requirements, ability to handle large quantities of sludge, and straightforward operational procedures. Nonetheless, the process is characterized by a significant time investment and necessitates a substantial land area. Moreover, its efficacy may be limited in colder regions.

Bio-pile/Composting

Bio pile is a treatment procedure that converts petroleum waste into piles that marine microorganisms digest. The treatment technology above could replace land treatment, which requires a lot of land area. Composting uses organic resources to improve system efficiency. Total petroleum hydrocarbon (TPH) content decreased significantly after bio-augmentation with crude manure and straw. Oily sludge composting in piles for 56 days reduced this by 46–53%. The non-bio-augmented control piles only reduced TPH by 31%.

Bio Slurry Treatment

This treatment uses 5–50% w/v sludge-associated solids and water. This technique dissolves contaminants into the aqueous phase, resulting in a lot of solubilized pollutants. Microbial degradation reduces toxicity or converts contaminants into carbon dioxide and water. A study by Ossai et al. 2020 found that bio-slurry treatment on oily sludge significantly reduced total petroleum hydrocarbons. TPH decreased from 40.7 to 53.2% in two weeks, according to the study. After six weeks, the TPH decrease reached 63.7–85.5%.

9.5 Flue Gases

Cleaning and processing natural gas (NG) removes pollutants and contaminants before they may be piped or stored in cylinders. Acid gases like H_2S and CO_2 are the main contaminants in crude oil and gas. The concentration of these two gases needs to be minimized as per the standards set by the regulatory authority. The Natural Gas contains 60–80% C, and 2–20% H, along with impurities like elements Sulphur. Some important processes for the removal of H_2S and CO_2 are mentioned here.

9.5.1 Treatment Technologies of Effluent Flue Gases

9.5.1.1 Hydrodesulfurisation (HDS)

In this method, Hydrogen sulfide (H_2S) is formed from the hydrogen by mixing with sulfur which is initially present in oil and gas. The process takes place within a temperature range of 300 to 400 degrees Celsius and under a pressure range of 50 to 150 bar. This method helps in improving the properties of the fuel. This unit can easily be installed in the oil and gas refining process. Other impurities like N_2gas can also be removed with the help of desulfurization operation. It is an energy-intensive and expensive process. This process needs tail gas treatment. For this process, a large area and volume are required.

9.5.1.2 Flue Gas Desulfurization

Flue Gas Desulfurization (FGD) is a highly efficient method for removing sulfur dioxide (SO_2) from natural gas or flue gas. This method uses commercial services including wet and dry spray scrubbers, sorbent injection, and regeneration. Regenerating and reusing resources is 95% efficient with the procedure. Since there is no waste, the method is cost-effective. The spray dry scrubber is affordable.

Regeneration and spray drying consistently reduce sulfur levels, especially below 3.5%.

9.5.1.3 *CO_2 Emission*

The main components of acid gas are CO_2 and H_2S. CO_2 reduces natural gas (NG) calorific value and causes corrosion when mixed with water, making its removal crucial., NG contains 4–50% CO_2, depending on its origin and location. For pipeline and cylinder transportation, CO_2 content is regulated between 2 and 4%. In different regions, these regulations may vary and are determined by local authorities.

9.5.1.4 Chemical Absorption

A new substance is formed when CO_2 reacts with a sorbent in chemical absorption. After stripper or downstream unit regeneration, CO_2 is stored and transferred for further use. CO_2 absorption uses solvents with amines and carbonates at 40°C and low partial pressure. Sorbent regeneration occurs at 100–140 °C. CO_2 loading capacity decreases with active sorbent materials, affecting separation limit and capacity. Amine-based chemicals, derivatives of ammonia, are the most effective and ubiquitous commercial solvents. The amine chemicals MEA, DEA, and DGA are used in many commercial operations.

9.5.1.5 Adsorption

Adsorption is a common commercial method for removing CO_2, H_2S, water, and other pollutants from NG and flue gas. Certain gas components adhere selectively to surface active areas during this process. The interaction between gas and adsorbent material, which affects CO_2 recovery, is affected by pressure, temperature (in thermodynamic equilibrium), size, shape, surface area, and selectivity. The dispersion of gas particles on active sites and adsorbent material regeneration affect separation. Adsorption usually occurs at high pressure and low temperature. The most energy- and cost-intensive gas purification process is regenerating adsorbents, making this technique crucial.,

9.5.1.6 Membrane Separation

Membrane technology is economically viable and feasible due to several factors. The above include reduced energy intensity, phase shift-free separation, and procedural simplicity. Permeability and selectivity determine membrane quality. Membrane permeability measures volume. Selectivity is the ability to separate a natural gas element from a larger mixture. Polymeric (organic), inorganic, and mixed matrix membranes are used in diverse applications. The polymeric membrane, popular and profitable, is used in offshore and onshore operations. Commercial applications utilizing NG and other industrial processes favour the polymeric membrane due to its permeability and selectivity.

9.6 Conclusions

Modern society is concerned about petroleum waste hydrocarbon (PWHC), which is mostly produced through crude oil extraction, gas drilling, and transportation. Toxic and dangerous residues are classified as hazardous and non-hazardous waste. Due to improper disposal of unresolved petroleum hydrocarbons, over 100,000 areas worldwide are harmful to living species, including humans. These locations' disposal and processing costs are a major issue. The lack of effective cleanup technologies has also caused petroleum-derived hydrocarbon waste to degrade. A successful petroleum waste treatment strategy involves collecting waste from various locations and using biological, chemical, and physical methods to remove hydrocarbon residues from the petroleum industry. Physiochemical treatments include Hydrocyclones, flocculation/flocculation, electro-coagulation, adsorption, membrane technology, oxidation, and heat. Biological treatment uses microorganisms to biodegrade petroleum waste's organic components. In this case, aerobic and anaerobic treatments break down contaminants. Incineration, solidification, oxidation, biodegradation, and fuel gas treatment are used to handle solid hydrocarbon waste like oil sludge. Membrane separation is essential in the oil and gas sector because it efficiently separates oily sludge from other sources.

References

Abdelwahab O, Amin NK, El Ashtoukhy ESZ (2008) Electrochemical removal of phenol from oil refinery wastewater. J Hazard Mater 163:711–716. https://doi.org/10.1016/j.jhazmat.2008.07.016

Ahmadun FIR, Pendashteh A, Abdullah LC, Biak DRA, Madaeni SS, Abidin ZZ (2009) Review of technologies for oil and gas-produced water treatment. J Hazard Mater 170(2–3):530–551. https://doi.org/10.1016/j.jhazmat.2009.05.044

Behera ID, Basak G, Kumar RR, Sen R, Meikap BC (2020) Treatment of petroleum refinery sludge by petroleum degrading bacterium Stenotrophomonas pavanii IRB19 as an efficient novel technology. J Environ Sci Health Part A 1–13. https://doi.org/10.1080/10934529.2020.1866924

Behera ID, Nayak M, Biswas S, Meikap BC, Sen R (2021) Enhanced biodegradation of total petroleum hydrocarbons by implementing a novel two-step bioaugmentation strategy using indigenous bacterial consortium. J Environ Manag 292:112746. https://doi.org/10.1016/j.jenvman.2021.112746

Behera ID, Nayak M, Mishra A, Meikap BC, Sen R (2022) Strategic implementation of integrated bioaugmentation and biostimulation for efficient mitigation of petroleum hydrocarbon pollutants

from terrestrial and aquatic environments. Mar Pollut Bull 177:113492. https://doi.org/10.1016/j.marpolbul.2022.11349
Bidder WL, Hunt RG (1982) Industrial resource recovery practices: petroleum refineries and related industries. US Environmental Protection Agency, Office of Solid Waste
Bujang M, Ibrahim NA, Rak AE (2013) Biodegradation of oily wastewater by pure culture of Bacillus cereus. ARPN J Agric Biol Sci 8(2):1–8
Cloern JE, Canuel EA, Harris D (2002) Stable carbon and nitrogen isotope composition of aquatic and terrestrial plants of the San Francisco Bay estuarine system. Limnol Oceanogr 47(3):713–729. https://doi.org/10.4319/lo.2002.47.3.0713
Connell D, Miller G, Farrington J (2009) Petroleum hydrocarbons in aquatic ecosystems—behavior and effects of sublethal concentrations: part 1. Crit Rev Environ Sci Technol 11(1):37–104. https://doi.org/10.1080/10643388009381685
De Junet A, Basile-Doelsch I, Borschneck D, Masion A, Legros S, Marol C, Balesdent J, Templier J, Derenne S (2013) Characterization of organic matter from oregano-mineral complexes in an Andosol from Reunion Island. J Anal Appl Pyrol 99:92–100. https://doi.org/10.1016/j.jaap.2012.10.020
Gardas BB, Raut RD, Narkhede B (2019) Determinants of sustainable supply chain management: a case study from the oil and gas supply chain. Sustain Prod Consum 17:241–253. https://doi.org/10.1016/j.spc.2018.11.005
Guo S, Li G, Qu J, Liu X (2011) Improvement of acidification on dewaterability of oily sludge from flotation. Chem Eng J 168(2):746–751. https://doi.org/10.1016/j.cej.2011.01.070
Hu G, Li J, Zeng G (2013) Recent development in the treatment of oily sludge from petroleum industry: a review. J Hazard Mater 261:470–490. https://doi.org/10.1016/j.jhazmat.2013.07.069
Inayat A, Inayat M, Shahbaz M, Sulaiman SA, Raza M, Yusup S (2020) Parametric analysis and optimization for the catalytic air gasification of palm kernel shell using coal bottom ash as catalyst. Renew Energy 145:671–681. https://doi.org/10.1016/j.renene.2019.06.104
Islam B (2015) Petroleum sludge, its treatment and disposal: a review. Int J Chem Sci 13(4):1584–1602
Jasmine J, Mukherji S (2014) Evaluation of bioaugmentation and biostimulation effects on the treatment of refinery oily sludge using 2(n) full factorial design. Environ Sci Process Impacts 16(8):1889–1896. https://doi.org/10.1039/c4em00116h
Kanungo J, Sahoo S, Bal M, Behera ID (2023) Performance of bioremediation strategy in waste lubricating oil pollutants: a review. Geomicrobiol J. https://doi.org/10.1080/01490451.2023.2245395
Kumar B, Raj Mohan B (2013) Petroleum oily sludge and the prospects of microwave for its remediation. Int J Eng Res Technol 2:359–370. https://doi.org/10.17577/IJERTV2IS110125
Liu C, Zhang Y, Sun S, Huang L, Yu L, Liu X et al (2018) Oil recovery from the tank bottom
Mansur AA, Pannirselvam M, Al-Hothaly KA, Adetutu EM, Ball AS (2015) Recovery and characterization of oil from waste crude oil tank bottom sludge from Azzawiya oil refinery in Libya. J Adv Chem Eng 5(1):27200–27227. https://doi.org/10.4172/2090-4568.1000118
Mariyam S, Shahbaz M, Al-Ansari T, Mackey HR, McKay G (2022) A critical review on co-gasification and co-pyrolysis for gas production. Renew Sustain Energy Rev 161:112349. https://doi.org/10.1016/j.rser.2022.112349
Mohanta S, Pradhan B, Behera ID (2023) Impact and remediation of petroleum hydrocarbon pollutants on agricultural land: a review. Geomicrobiol J 1–5. https://doi.org/10.1080/01490451.2023.2243925
Ng JY, Wong DL, Kutty SRM, Jagaba AH (2021) Organic and nutrient removal for domestic wastewater treatment using bench-scale sequencing batch reactor. AIP Conf Proc 10(1063/5):0045224
OGDCL (2017) Oil and Gas Development Company Limited (OGDCL), Pakistan, annual report
Oil recovery from water-based drilling fluid waste. Fuel 237:335–343. https://doi.org/10.1016/j.fuel.2018.10.007

Ossai IC, Ahmed A, Hassan A, Hamid FS (2020). Remediation of soil and water contaminated with petroleum hydrocarbon: a review. Environ Technol Innov 17. https://doi.org/10.1016/j.eti.2019.100526

Ozdemir A, Karatas A, Palabiyik Y, Yaşar E (2020) Oil and gas exploration in Seferihisar Uplift (Western Turkey) containing an operable-size gold deposit: geochemical evidence for the presence of a working petroleum system. Geomech Geophys Geo-Energy Geo-Resour 6(1):1–22. https://doi.org/10.1007/s40948-020-00152-2

Santiago YC, Gonzalez AM, Venturini OJ, Maya DMY (2021) Assessment of the energy recovery potential of oil sludge through gasification aiming for electricity generation. Energy 215. https://doi.org/10.1016/j.energy.2020.119210

Shahzad A, Siddiqui S, Bano A et al (2020) Hydrocarbon degradation in oily sludge by bacterial consortium assisted with alfalfa (Medicago sativa L.) and maize (Zea mays L.). Arab J Geosci 13:879 .https://doi.org/10.1007/s12517-020-05902-w

Singh P, Jain R, Srivastava N, Borthakur A, Pal D, Singh R, Madhav S, Srivastava P, Tiwary D, Mishra PK (2017) Current and emerging trends in bioremediation of petrochemical waste: a review. Crit Rev Environ Sci Technol 47(3):155–201. https://doi.org/10.1080/10643389.2017.1318616

Srivastava M, Srivastava A, Yadav A, Rawat V (2019) Source and control of hydrocarbon pollution, hydrocarbon pollution and its effect on the environment. IntechOpen. https://doi.org/10.5772/intechopen.86487

Teng Q, Zhang D, Yang C (2021) A review of the application of different treatment processes for oily sludge. Environ Sci Pollut Res 28:121–132. https://doi.org/10.1007/s11356-020-11176-2

Tissot P, Welte D (1984) Petroleum formation and occurrence. Springer. WHO, International Program on Chemical Safety. World Health Organization

Waples DW (2013) Geochemistry in petroleum exploration. Springer Science & Business Media

Zhang Q, Hu J, Lee D-J, Chang Y, Lee Y-J (2017) Sludge treatment: current research trends. Bioresour Technol 243:1159–1172. https://doi.org/10.1016/j.biortech.2017.07.070

Chapter 10
Remediation Technologies for Petroleum Hydrocarbons from the Environment

Alisha Kakkar and Sudesh Kumar

Abstract Petroleum hydrocarbons (PHs) are widely utilized as substrates for industrial applications and as principal energy resources. However, over-utilization of PHs can contaminate the soil, posing considerable dangers to the ecosystem. To counter this, different physical, chemical, and biological approaches are employed to remediate contaminated soil. Nonetheless, many chemical approaches are costly, environmentally detrimental, and not effective. Consequently, researchers are focusing on developing new strategies that address soil pollution in an ecologically friendly approach. Bioremediation, which employs microorganisms, emerges as an environmentally conscious technique for PH degradation. Advanced technologies like Next-generation sequencing (NGS) have revolutionized monitoring practices at polluted regions, providing perspectives on biodegradation processes facilitated by microorganisms through selective ribosomal RNA (rRNA) targeting. The chapter examines the sustainable management of PHs from the environment by novel innovative remediation technologies, along with its challenges and future directions. Nanotechnology holds great potential for the degradation of pollutants; however, its viability is contingent on economic efficiency. To this end, the chapter proposes the application of microbial electrochemical methods that leverage reactive microbes to bioremediate and biodegrade PH-contaminated soil. Such an approach aligns with environment-friendliness and encourages the development of sustainable restoration strategies.

Keywords Petroleum hydrocarbons · Technologies · Microorganisms · Degradation · Chemical · Soil · Strategies · Remediation

---

Alisha Kakkar · S. Kumar (✉)
Department of Chemistry, Banasthali Vidyapith, Tonk, Rajasthan 304022, India
e-mail: kusudesh@banasthali.in

I. D. Behera and A. P. Das (eds.), *Impact of Petroleum Waste on Environmental Pollution and its Sustainable Management Through Circular Economy*, Environmental Science and Engineering, https://doi.org/10.1007/978-3-031-48220-5_10

10.1 Introduction

Petroleum hydrocarbons (PHs) play a vital role in many industries for making various products. The consumption of PH is estimated to reach 106.6 million barrels by 2030. Their usage varies by region, with Europe and Asia accounting for 32% and the Middle East for 53%. PHs, like polycyclic aromatic hydrocarbons and alkanes, are harmful to the environment because they stay in the environment for a long. The complex mixture of crude oil, containing aromatic hydrocarbons, alkanes, and resins, poses a high level of hazard to living organisms. It's crucial to understand how these pollutants behave in the environment and characterize their impact to minimize exposure for both animals and humans. Oil spills and PH can have long-lasting effects on the environment. They disperse, biodegrade very slowly, dissolve, and undergo photooxidation. This poses harmful impacts on the natural ecosystem, including humans, animals, and aquatic species. PH-contaminated soil can also alter the physico-chemical properties of the soil, affecting crop growth and productivity (Song et al. 2021). To minimize the harmful impacts of PH, it is necessary to use strategies and approaches to mitigate those harmful impacts based on risk assessment. These strategies involve various in-situ or ex-situ treatment methods, such as acoustic, ultrasonic, heat, electromagnetic, thermal, electric, biological, physio-chemical, and chemical approaches. However, many of these methods are costly, expensive, harmful to the environment, and not so efficient. This poses a significant health hazard to every living species. That's why researchers are actively working on developing innovative techniques to clean up the polluted environment. This chapter examines the sustainable management of PHs from the environment by novel innovative remediation technologies, along with its challenges and future directions.

10.2 Technologies for Removal of pH from Soil

Petroleum hydrocarbons have adverse effects on humans as well as the environment. It is crucial to develop new technologies and strategies to remediate the PH-polluted environment. Selecting the most effective remediation strategies to eliminate toxic PH from the natural ecosystem requires careful consideration of factors such as environmental conditions, contaminant composition, microbes, methods, time, cost, and regulations. It's also crucial to focus on implementing environment-friendly strategies to mitigate the risks associated with PH-contaminated soil. Different chemical, biological, and physical methods are explored to address this issue.

10.2.1 Physical Methods

Several methods have been reported to physically eliminate toxic PH from the soil. These methods are soil washing, soil flushing, thermal desorption, solidification, vitrification, stabilization, vapor extraction, and incineration. To restore and purify the polluted soil, there are two common methods: in-situ and ex-situ approaches. In-situ methods are done on-site, while ex-situ methods are done off-site, typically in a laboratory. For lighter molecules like BTEX, techniques such as air sparging, air extraction, ignition, or combinations of these methods are utilized. These combined approaches can enhance the efficiency of the remediation process. Ex-situ methods include destruction technologies, detoxification, and excavation to eliminate PH from both polluted soils (Abdel-Shafy and Mansour 2016).

10.2.1.1 Excavation

Excavation is a traditional method used to physically eliminate PH from contaminated soil. It involves the process of digging up the polluted soil and then transporting it to designated treatment or disposal locations. However, it's important to note that this method can present challenges, such as the potential risks associated with handling and transporting the polluted soil. To mitigate these issues and to ensure safe excavation, it is important to accurately assess and define the boundaries of the site beforehand. Additionally, Finding suitable locations for disposing of the excavated material can be costly and pose logistical challenges. Despite these considerations, excavation is often seen as a temporary measure for removing contaminants to restore environmental quality, as it evaluates and assesses potential risks associated with contamination (Sohoo et al. 2021).

10.2.1.2 Surface Capping

Surface capping is a technique that uses a physical barrier to protect soil from PH. It protects PH-contaminated soil from potential sources of contamination. It avoids the infiltration of rainfall water into PH-contaminated soil. It utilizes robust and impermeable barriers to prevent the movement of PH., preventing water and soil pollution (Zhang et al. 2022).

10.2.1.3 Use of Physical Barriers

Physical barriers are implemented to restrict the flow of PH from contaminating groundwater. These barriers help control the movement of pollutants and prevent their spread to nearby water sources. To prevent the lateral movement of petroleum hydrocarbons, physical barriers are placed in the down-gradient direction of the contaminated area. A novel approach involves using vertical barriers, such as curtains, slurry walls, steel shoring, plastic cut-off walls, and cement layered on top of the soil (Wagner and Yarmak 2017). In the case of more pliable soil, slurry walls are commonly utilized as physical barriers to control the movement of PH, which consists of a trench filled with cement, hard clay soil, or bentonite. Sheet piling cut-off walls are also employed to retain water and are inserted into the affected soil at depths ranging from 12–30 m to minimize the lateral migration and leakage of PH.

10.2.1.4 Washing of Soil Method

Soil washing is an essential method used outside of the contaminated site, that involves the separation and extraction of PH from the polluted soil through a leaching process. PH has the ability to attach itself to soil particles, including silts and clay (Tran et al. 2022) due to their chemical properties, leading to increased water pollution. Surfactants are often used to enhance the process. For soil remediation, washing solutions such as acetone ethyl acetate-acetone, hexane, and chlorinated organic compounds are commonly applied.

10.2.1.5 Vitrification

This method entails the utilization of high temperatures from 1600 to 2000 °C, to facilitate the transformation of various pollutants like polluted soil, as well as a range of organic substances, into glassy substances comprising crystals, massive glass, and frit. The technique achieves the immobilization and degradation of major pollutants through thermal decomposition, with the majority of pollutants being evaporated. The pollutants which are left are transformed into stable, unreactive, crystals and glass-like products. The liquified polluted soil after cooling, solidifies to become exceptionally robust, exhibiting properties akin to basalt rock, and is approximately ten-fold harder in comparison to solidified composite material. The elevated temperatures effectively annihilate organic substances and pollutants, yielding minimal side products (Vidonish et al. 2016). The vitrification of pollutants comprises majorly three processes: thermal, electrical, and plasma. All of these methods are energy-intensive. In electrical processes, high-voltage electric energy is derived from electrodes of graphite embedded in the polluted media (soil). Electric energy is supplied to raise the temperature of the soil to a range of 1400–2000 °C. Treatment typically extends to a depth of around 20 feet and depends on the concentration of alkali metal oxide content to heighten conductivity, heating, and point of liquefaction. When the soil contains adequate silica, elevated temperature causes the conversion of polluted soil into a glass-like substance. Inorganic contaminants are transformed within the resultant solid structure with glass-like characteristics. An exhaust hood captures emitted gases and directs them toward a treatment process consisting of a cooling unit for cooling, cleanser, activated charcoal, and oxidizing agent.

10.2.1.6 Incineration

This method entails the complete eradication of pollutants like PHs, dioxins, polychlorinated biphenyls (PCBs), and other organic substances through high-temperature ignition or burning, typically ranging from 870 to 1200 °C. Incineration occurs within various systems including circulating bed combustors, rotary kilns, liquid injection incinerators, and fluidized bed reactors. In the incineration process, the oxygen flow is regulated at nearly 10% for the combustion of volatile organic substances. Both the oxygen content and soil limit must be taken into account, along with the pollutant's minimum explosive concentration, to guarantee secure combustion. The emissions and vaporous by-products are cleansed through cleaning devices and electrostatic precipitators. Catalytic converters convert gases like carbon monoxide, and oxides of nitrogen and sulfur into less harmful counterparts such as carbon dioxide (CO_2), water (H_2O), and nitrogen gas (N_2) to mitigate environmental pollution and the release of harmful contaminants (Ivshina et al. 2015). Additionally, the residual gaseous and soil compounds that remained after the procedure can also be transformed into non-harmful products, while fly ash is often discarded in waste disposal sites. Following treatment, the soil contaminants are re-hydrated to control dust before they are employed in various applications.

10.2.1.7 Thermal Methods

Low-temperature thermal desorption (LTTD) is an effective method for removing PHs from contaminated soil. LTTD utilizes heat to physically segregate and scatter pollutants, effectively eliminating them from polluted soil. The soil is heated to approximately 1200 °F, causing the organic components to evaporate. The hydrocarbons are subjected to a catalytic oxidation process within a specialized chamber for treatment. The soil is subsequently cooled, stabilized, and made ready for proper disposal or potential reuse. Different soil types require specific treatment methods, such as LTTD, to effectively remove petroleum pollutants. The success of LTTD depends on factors like pollutant characteristics, site conditions, and regulatory and economic factors (Riser-Roberts 2020).

10.2.1.8 Microwave Method

Microwave heating is an inexpensive and less time-consuming method for treating petroleum-polluted soil. It utilizes electromagnetic irradiation, such as radio and infrared frequencies, to uniformly heat petroleum-polluted soil. The mechanisms involve the interaction of irradiations with dielectric materials, resulting in heat generation through ionic conduction and bipolar rotation. Bipolar rotation involves the movement of charged particles in a substance due to the alternating electric field generated by electromagnetic irradiation and ionic conduction. On the other hand, involves the flow of ions through a material, contributing to the overall heating process in microwave heating. The effectiveness of microwave heating in removing hydrocarbons is influenced by the permeability of the soil and the size of the particles. In the process of ionic conduction, the movement of free ions through electric fields facilitates rapid heating (Falciglia et al. 2018). However, this method is difficult, energy-intensive, technologically complex, and limited societal acceptance.

10.2.2 Chemical Methods

Chemical processes play a crucial role in the degradation of petroleum hydrocarbons in polluted soils. These processes involve both biotic and abiotic changes. Abiotic changes, such as hydrolysis and oxidation–reduction, break down the molecules without the involvement of microbes. Biotic changes, on the other hand, involve biodegradation and biotransformation processes occurring under aerobic or anaerobic conditions. These processes can take place under aerobic conditions, where oxygenase enzymes help break down the hydrocarbons into carbon dioxide and water as shown in Fig. 10.1 (Ossai et al. 2019).

10.2.2.1 Oxidation–Reduction Method

In-situ chemical oxidation reactions offer advantages for polluted soil remediation. Different oxidizing agents oxidants like O_2, MnO_4^-, $S_2O_8^{2-}$, O_3, and Fe^{2+} are used to transform PHs into less harmful and unreactive substances. However, it's important

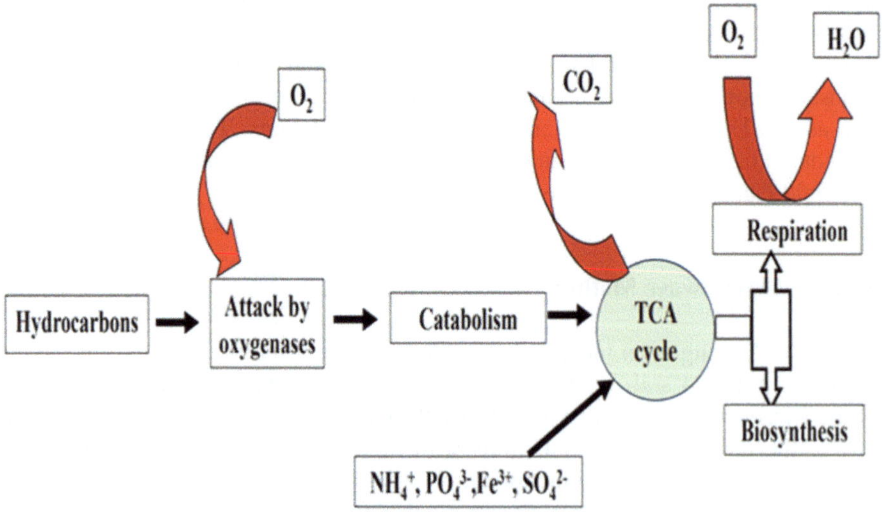

Fig. 10.1 Mechanism of chemical process

to note that these oxidants can be consumed by organic matter in the soil, which can hinder their effectiveness in treating contaminants.

10.2.2.2 Sonochemical Process

Sound waves consist of cyclic decompressions and compressions within all forms i.e., solid, liquid, and gas that possess elastic characteristics. These waves move horizontally through gas, liquid, and vacuum, while they travel perpendicular through solids substances. The sonochemical process includes subjecting heterogeneous and homogeneous solutions to more frequency ultrasonic waves to induce both chemical and physical changes through the formation of bubbles known as cavitation. The physical and chemical impacts of ultrasonic waves arise from the molecular response to wave interaction and the occurrence of cavitation, growth of the nucleus, and sudden contraction of acoustic bubbles within the liquid solution. The energetic contraction of acoustic bubbles generates light emission through acoustic collapse and the active species. The micro-bubbles accumulate energy from the ultrasonic waves and enlarge during the sound wave oscillation until they reach a maximum size, at which point they contract or collide forcefully. When acoustic bubbles are formed near a solid surface or large aggregates, the unequal implosion or contraction leads to the formation of solvent micro-jets, which interact with the solid with a significant force, leading to the creation of novel responsive surfaces of deterioration and abrasion.

10.2.2.3 Solidification Method

This technique also called as fixation of waste materials, plays a crucial role in confining and controlling the migration of pollutants by encapsulating them within a resilient matrix or converting them into stable forms. It involves the introduction of cement-like adhesive agents to the polluted media, resulting in the creation of an immobilized material such as a uniform solid mass or a clay-like structure, Various methods are employed for this purpose, including encapsulation, pozzolanic, thermoplastic, and organic polymer.

10.2.2.4 Photocatalytic-UV Oxidation

Photocatalytic oxidation and UV methods utilize oxidizing agents such as ozone and hydrogen peroxide (H_2O_2) and UV light to break down and convert toxic PHs. Intense UV light in combination with H_2O_2, helps break down and convert petroleum hydrocarbon compounds into salts, H_2O, and CO_2. When UV irradiation interacts with H_2O_2 in photocatalytic-UV oxidation, it triggers the activation of atomic bonds in it, makes the extremely reactive species, and generates highly reactive hydroxyl radicals (OH•). This reactive OH• then reacts with PH for further degradation. Gholami et al.

investigated the photocatalytic degradation rate of Zn-Co-layered double hydroxide (LDH) incorporated with biochar. It was observed that the degradation of Gemifloxacin was 92.7%. They also achieved an efficiency of 97.6% in degrading cefazolin sodium using Zn-Co-LDH incorporated with biochar. Additionally, a novel catalyst, WO_3, and N-Doped WO_3-x@Mesoporous Carbon (NM-WO_3-x@MC) composite showed degradation efficiencies of 90% for antibiotics and 76% for total organic carbon (Gholami et al. 2021). Moradi et al. also utilized the photocatalytic method in their study to degrade phenol in visible light using a composite of graphene oxide (Moradi et al. 2021).

10.2.2.5 Adsorption Method

Adsorption using activated carbon has been found to be effective in removing PHs from the soil. Activated carbon adsorption has been shown to decrease the potential impacts of organic contaminants. Activated carbon is known for its versatility and effective choice as an adsorbent for decontaminating PHs due to its hydrophobic nature, high porosity, and high specific surface area. Van der Waals forces play a crucial role in governing adsorption, while desorption takes place when equilibrium is achieved.

10.2.2.6 Electrochemical Method

The electrochemical method has been found to be effective in removing PHs from contaminated soil. In this system, electric current can act as either a source of electrons or a recipient of electrons to mitigate oil spills. Electromigration, electro-osmosis, and electrophoresis play a key role in dispersing and transporting toxic contaminants. In an electrochemical method, lightweight PHs can be volatilized using electromagnetic waves. Wang et al. carried out an investigation employing single-chamber microbial fuel cells to assess the efficiency of pyrene degradation. The results of their research indicated a notably high level of effectiveness in this degradation process (Wang et al. 2022).

10.2.3 Biological Methods

Removal of PHs by biological methods can be categorized into two groups: Bioremediation and phytoremediation. Bioremediation employs microorganisms like fungi and bacteria, while phytoremediation utilizes symbiont species and plants. These practices involve making changes to the soil and agricultural techniques. These methods involve various mechanisms such as phytodegradation, phytovolatilization, phytostimulation, rhizofiltration, phytoextraction, and phytostabilization (Ruley et al. 2022). The efficiency of the degradation mechanism depends on various factors like

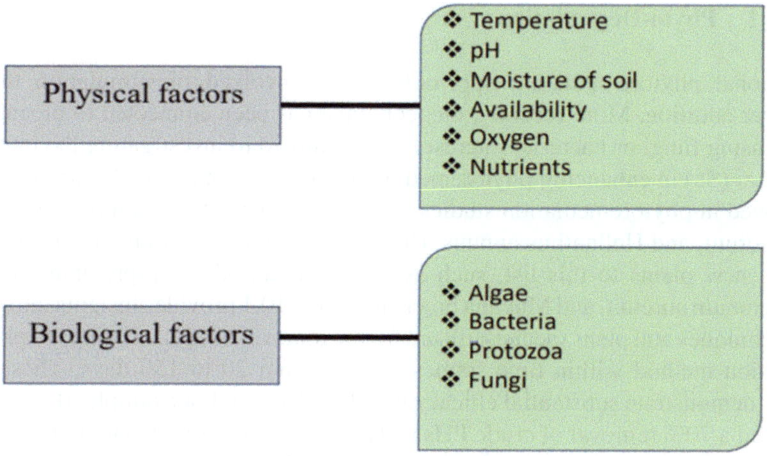

Fig. 10.2 Factors influencing the rate of biodegradation

amount, chemical composition, characteristics, and properties of PHs. Both processes i.e., chemical changes and biodegradation play important roles in the degradation and remediation of contaminated environments. The degradation rate of PHs is influenced by a range of biological and physical factors as depicted in Fig. 10.2.

Consequently, when formulating effective strategies for eliminating soil contaminated with PHs, it is imperative to account for the physicochemical properties and characteristics of the contaminants. For instance, aliphatic hydrocarbons exhibit a larger degradation rate in comparison to aromatic hydrocarbons, like BTEX and PAH compounds. This discrepancy arises from the structural simplicity and lesser weight of the aliphatic hydrocarbons, as an increase in the molecular weight and complexity of PHs results in the production of compounds that are less amenable to biodegradation. The use of biosurfactants can improve the degradation efficiency of sophisticated PHs. Various factors, including pH, temperature, and pressure, also play pivotal roles in this process. Moreover, under an anaerobic environment, the degradation rates of PAHs are dependent on the presence of terminal electron acceptors like O_2, SO_4^{2-}, and NO_3^-. Locations with limited oxygen availability experience reduced nutrient supply and diminished accessibility to terminal electron acceptors, crucial for microbial metabolic processes in the biodegradation method. Consequently, attempts have been made to provide alternative terminal electron acceptors into the soil, resulting in the enhancement of PH mineralization (Sun et al. 2022). Temperature, spanning a range from -10 to 55 °C, is another variable influencing the degradation rate of PHs. This temperature range affects physiological aspects, pollutant chemistry, and speed of degradation. Additionally, the solubility and rate of biodegradation reduce as temperatures decline.

10.2.3.1 Phyto-Degradation

Traditional phytoremediation approaches have involved biostimulation through fertilizer addition. More recently, the technique has been connected to bioaugmentation using fungi or bacteria. Interest has also turned to investigating the impact of selenium (Se) in enhancing phytoremediation efficiency in plants. Common species employed in phytoremediation studies include Zea mays, Sorghum bicolor, Lolium multiflorum, and Helianthus annuus. However, ongoing screening efforts are introducing new plants to this list, such as Azolla filicuilodes, Epipremnum aureum, Epipremnum aureum, and Mucuna bracteate. Table 10.1 provides insights into potential techniques and plant species utilized for the removal of PHs utilizing the phytoremediation method within time frames ranging from 20 to 150 days. These techniques demonstrate substantial efficacy in PHs reduction. For example, Baoune et al. achieved a 70% removal of crude PHs and a 61% removal of phenanthrene from the soil in just 14 days (Baoune et al. 2019). In general, phytoremediation encompasses a range of technologies capitalizing on the inherent capacities of plants to remediate PH-polluted soil. It has proven a highly efficient technique in the removal of organic contaminants, gaining attention for its large-scale capacity of pollutant degradation, ecological compatibility, and less expensive. Nevertheless, only a limited number of plants can efficiently uptake the organic compounds present in the soil, and most plants lack the necessary enzymes to degrade these organic chemicals. Noteworthy among these are a few plant species, such as members of the Cucurbita genus, which showcase remarkable capabilities to uptake substantial quantities of organic compounds present in the soil, Furthermore, researchers have pinpointed and extracted genes responsible for encoding enzymes with the capacity to degrade organic chemicals, sourced from microorganisms. The conceivable advancement of transgenic hyperaccumulator plants carrying these enzymatic codes holds the potential for advancing the phytoremediation of organic compounds (Malik et al. 2022).

10.2.3.2 Biodegradation of PHs Using Microbes

In recent decades, numerous researchers have extensively documented the significant contribution of microbes to the degradation of PHs in both water and soil. The vital role of microorganisms that degrade hydrocarbon is evident from their effective removal of PHs, facilitating the purification of polluted environments, encompassing soil, water, sediments, and various other polluted water bodies (Cabral et al. 2022). The utilization of oleophilic microorganisms to break down PHs in land–water environments is recognized for its environment-friendly nature and cost-effectiveness. Consequently, Microbial degradation plays a crucial role in upholding the balance of ecosystem well-being and environmental stability. A prominent degradation mechanism for PHs is biodegradation, a process that involves specific anaerobic and aerobic enzymes decomposing the substances (Imam et al. 2022). The utilization of oleophilic microorganisms to break down PHs in land–water environments is recognized for its

Table 10.1 Microbial degradation of various PHs

Microbial species	Polluted sample	Efficiency	References
Aspergillus sp. RFC-1	Various PHs	PHs were successfully removed from polluted sites with an efficiency ranging from 60.3 to 97.4%	Al-Hawash et al. (2019)
Aspergillus spp	Crude oil	Removal of 59–87.7%	Zhang et al. (2016)
Pseudomonas sp.	Gasoline, diesel, and used engine oil	The removal rate of 76% for gasoline, 83% for diesel, and 69% for used engine oil	Ruley et al. (2022)
Aspergillus ochraceus CBMAI 849	Benzo [a] pyrene	76.6–99.7% recovery from aqueous environments with high salinity	Passarini et al. (2011)
Candida tropicalis	Diesel	Removal rate of 83%	Adams et al. (2015)
Bacillus cereus D12, *Bacillus soreness* D11, and *Pseudomonas stutzeri* D13	Total Petroleum Hydrocarbons (TPH) in Diesel–Diesel Range Organics (DRO)	The Strains consortium achieved a petrol-diesel removal rate of up to 80%, with an extraction rate of 31.1 mg/g for TPH-DRO	Alsayegh et al. (2021)
Penicillium sp. RMA1 and RMA2	Crude oil	Removal rate by RMA1-57% and for RMA2-55%	Al-Hawash et al. (2018)
Ochrobactrum sp.	Crude oil	Removal rate of 83.49%	Varjani et al. (2015)

environment-friendly nature and cost-effectiveness. Consequently, Microbial degradation plays a crucial role in upholding the balance of ecosystem well-being and environmental stability. A prominent degradation mechanism for PHs is biodegradation, a process that involves specific anaerobic and aerobic enzymes decomposing the substances. Initially, the petroleum hydrocarbon undergoes an enzymatic oxidative attack, followed by the subsequent hydrocarbon ring cleavage. Polycyclic aromatic hydrocarbons (PAHs) can also be oxidized into derivatives of quinone and phenol through fungal action, involving peroxides (Koshlaf and Ball 2017). Through the aerobic transformation process, enzymes utilize the generated molecular oxygen to oxidize the molecular target of PAHs. This results in the creation of hydroxylated rings and a range of intermediary oxidized products. Enzymes utilize the molecular oxygen generated during aerobic transformation to oxidize polycyclic aromatic hydrocarbon substrates into intermediary hydroxylation rings and other forms of oxidized products. A compilation of potential microbes capable of PHs degradation is outlined in Table 10.1. While bacteria generally exhibit more rapid PH degradation rates, fungi require more time to adapt to environmental conditions and typically possess lower competing strengths. Consequently, fungal degradation relies on the specific capacities of each species and is primarily confined to the environment of soil, which can often contain toxic elements. An alternative degradation mechanism for PAHs and BTEX involves degradation in the absence of oxygen, characterized by compound carboxylation followed by breakage of the ring in the phenanthrene compound. This anaerobic route differs from the aerobic degradation process, highlighting the distinct approach to PHs degradation (Mohapatra et al. 2022). Despite these insights, the precise process of degradation of PHs remains incompletely comprehended, underscoring the need for further comprehensive research, particularly considering the intricate complexities associated with field-scale applications.

10.2.4 Other Biological Methods

10.2.4.1 Land Farming

In this method, soil that is contaminated with PHs is infused with air and evenly distributed in thin layers. This encourages the pollutant compounds' degradation while adding nutrients. Additionally, this approach involves immobilizing and decomposing the polluted soil through oxidation and microbial degradation. This treatment can be carried out in situ or ex-situ.

10.2.4.2 Composting

Composting is primarily employed to break down specific soil contaminants by combining mature organic matter with soil that is polluted. This approach involves the utilization of composted organic substances like used mushroom substrate, compost fishbone, and bark compost enriched with urea. Consequently, organic material compost with the addition of urea to the polluted soil offers carbon as well as nutrient supplements to microbes, bolstering their metabolic capacity and consequently fostering the degradation of organic matter present in the soil.

10.2.4.3 Bio-slurry Technology

This method stands as one of the most efficient off-site strategies for tackling soil pollution arising from PHs within regulated environments. To expedite the degradation of contaminants, a water-soil mixture is formed to make a slurry that constitutes around 10–60% in terms of weight and volume. This approach heightens the rate of transfer of mass and encourages interactions between nutrients, pollutants, and microbes existing within the slurry. Bioreactors tailored for slurry treatment, often referred to as slurry reactors, are generally operated in batch mode to simplify handling and enhance the rate of production of the process. Nevertheless, the accessibility of PHs to microbes for utilization as a source of carbon and energy highly impacts the pace of degradation. The persistence of PAHs within the soil and their limited availability due to physical barriers impact their rates of degradation. Additionally, the incomplete oxidation of PHs can generate harmful intermediate compounds, resulting in ineffective performance during microbial treatment processes. To enhance the solubility and availability of PHs, introducing a liquid state into bioreactors and integrating surfactants is pivotal. Further exploration is needed to fine-tune operating factors, encompassing factors such as oxygen concentrations, pH levels, nutrient availability, the diversity of native bacterial populations, and the type of supplement employed (Abena et al. 2019). These investigations should extend to larger-scale applications and the proper handling of waste streams. Recent research efforts focusing on mitigating polluted soils with PH employing bio-slurry techniques are concluded in Table 10.2.

10.2.4.4 Bioventing and Bio-sparging

In this method, an on-site method is used to eliminate contaminants present in the soil matrix through aerobic microbial degradation. This enhances the activity of microorganisms, typically leading to a reduction in water and soil pollution. Additionally, biosurfactants can be used as necessary to support the growth of microbial species. Constructed bioventing wells can be installed either vertically or horizontally, depending on the site's geological characteristics. Both bioventing and biosparging have the same operational mechanism and structure, but they exhibit disparity in their effective zone (Husaain et al. 2022).

Biosparging is primarily employed lower the level of groundwater level in the saturated region and is characterized by a higher flow rate of air injection in comparison to bioventing. Furthermore, in practical application, both bioventing and biosparging techniques are often integrated with different processes, including soil vapor extraction and permeable reactive barrier method.

10.2.4.5 Permeable Reactive Barriers

It's an underground wall designed to purify contaminated groundwater. This wall is "permeable," which means it permits groundwater to flow through it. To undergo treatment, water must pass through these reactive barriers. The materials within the wall, known as "reactive" materials, either capture or reduce the harmful effects of toxic contaminants. After treatment, the purified groundwater is released on the opposite side of the barrier. Compared to other conventional purification of groundwater methods, permeable reactive barriers offer several advantages: they have a low cost of maintenance, follow an in-situ method to remediate contaminants, and avoid the need for surface-level facilities. However, it's important to note that these reactive barriers are most suitable for shallow groundwater reservoirs and require a higher initial investment (Andrade and dos Santos 2020).

Table 10.2 PH remediation using bio-slurry technology

Polluted sample	Operational parameters and bacterial strains	Efficacy	References
Phenanthrene	Phenanthrene concentration of 75 mg/kg; microbial consortium	Removal efficiency of 100% in just 6 days	Woo et al. (2004)
PAHs	Soil with a PAH pollutant level of 10,973 mg/kg, mixed with a soil-to-water ratio of 30%; microbial consortium	Efficiency of 93.4% in only 12 days	Lewis (1993)
TPHs	Concentration of 39–41 g/kg and temperature: 25 ± 2 °C; microbial consortium	Removal efficiency of 74% in 210 days	Ramadass et al. (2018)
TPHs	Temperature of 27 °C, concentration: 9.45 g/kg soil; microbial consortium	Removal rate of 72.7% in 54 days	Bento et al. (2005)
TPHs	Concentration of TPHs-10g/kg; microbial consortium	Removal rate of 79% in 70 days	Suja et al. (2014)
TPHs	Neutral pH, concentration-42g/kg, temperature: 25 to 30 °C; microbial consortium	Removal rate of 76% in just 120 days	Farahat and El-Gendy (2007)
TPHs	Concentration: 38.08 g/kg and 25 °C temperature; microbial consortium	Removal rate of 79.8% in only 84 days	Hesnawi and Mogadami (2013)
PAHs	Concentration of pyrene-100 mg/L, temperature: 25 °C, soil to water ratio of 50%; microbial consortium	Removal rate of 77% in 60 days	Thavamani et al. (2012)
PAHs	Soil to water ratio-50%, temperature: 50 °C, concentration-3967 mg/L; microbial consortium	Removal rate of 90% in only 60 days	Kuppusamy et al. (2016)
PAHs	Concentration: 13g/kg, pH-7.5, temperature: 20 ± 2 °C, soil-to-water ratio of 25%; indigenous soil bacteria	Removal rate of 58.2% in 150 days	Fava et al. (2004)
PAHs	Concentration of PAHs-6.1 μg/g, temperature:30 °C, soil-to-water of 20%; natural soil bacteria	Removal rate of 40% for PAHs in 35 days	Beolchini et al. (2010)

Note TPHs: Total Petroleum Hydrocarbons; PAH: Polycyclic aromatic hydrocarbons

10.2.4.6 Natural Attenuation

In the natural attenuation process, native microbes decompose the PHs by relying exclusively on carbon, which is easily present in contaminated environments. When a region becomes polluted, native microorganisms undergo rapid multiplication and adjust to the recent PHs, leading to the complete biodegradation of PHs. This mechanism occurs organically in contaminated regions, especially those with lighter PHs, often in areas where alternative methods are not effective. Natural attenuation is particularly suited for remediating regions contaminated with PHs, as it requires no human assistance and minimal supervision. In biodegradation, this process effectively treats PH-contaminated regions. Globally, natural attenuation has been applied to manage 25% of soils contaminated with PHs. It proves a more efficient strategy than other remediation treatments, as it decreases the mass, harmfulness, movement, and amount of pollutants within the soil profile.

10.2.4.7 Dehalogenation and Dechlorination

In this method, PHs are eliminated through reduction. The dehalogenation and dechlorination procedures are utilized to reduce the levels of harmful compounds. These processes involve a gradual substitution of chlorine atoms with hydrogen, achieved through either partial vaporization or decomposition. Several studies have highlighted the involvement of bacteria, along with a subset of thermophilic and mesophilic methanogens, in reductive dehalogenation in the absence of oxygen. Halogen compounds can aggravate the adverse effects of organic compounds, increasing their toxic nature, mutagenic potential, and other detrimental properties (Zhu et al. 2022). Furthermore, these compounds can restrict treatment, as only specific halo-organic substances are biodegradable to a certain extent and may not degrade completely. Despite these challenges, only bacteria have cultivated a robust inclination to break down these substances through biodegradation.

10.3 Use of Nanomaterials in the Treatment of PHs

Nanomaterials possess distinctive attributes, including a large absorption tendency for ions, high stability, micro-porous structure, and large surface area. These characteristics position nanomaterials as strong candidates for environmental treatment applications. These characteristics make nanomaterials particularly suitable for in situ PH removal methods. Nanoparticles primarily rely on adsorption and oxidation as the main mechanisms to eliminate PHs present in polluted soil. Due to their high reactivity compared to other materials, nanomaterials offer certain advantages like less expensive and fast methods for PH removal. Researchers have demonstrated the potential of hybrid nanoparticles in soil remediation. Nanoparticles of manganese (Mn) and cobalt (Co) enabled the removal of PHs in 3 h from the soil matrix. Nanotechnology brings substantial advantages in the treatment of PH-polluted soil as shown in Table 10.3. However, concerns arise regarding its impact on living organisms. A thorough evaluation of their environmental risks and potential consequences is crucial. Understanding the potential dangers, accessibility, mobility, and possible risks linked to nanoparticles is crucial. The widespread application of

Table 10.3 Use of nanomaterials in the remediation of PHs

Pollutant type	Operating parameters	Nanomaterials	Removal rate (%)	References
PHs	Degradation of PHs occurs within 14 h and 5 days under conditions of pH 6.0 and with an oxidant concentration of 0.74 M	Microbially produced iron oxide	95	Bolade et al. (2021)
Diesel oil	Removal in just 2 h	Peroxymonosulfate-activated nanoscale zero-valent iron	61	Bajagain and Jeong (2021)
PHs	Removal in 3 days	Nanocomposite of chitosan and goethite	93	El-Sheshtawy et al. (2021)
PHs and PAHs n-alkanes	Removal in 7 days at pH 7 and temperature of 25 °C	Biochar-enhanced ferrous sulfate with activated persulfate	62	Xia et al. (2021)
PHs	Removal in 32 h at pH:9–109 and concertation of 94.20 mg/L	Iron nanoparticles enveloped in a coating derived from mortino berries	88	Murgueitio et al. (2018)
PHs	Removal of PHs in 7 days at a temperature of 20 °C	Nanoscale zero-valent iron coated with green mango peel extract	90	Desalegn et al. (2018)

Note: PHs: Petroleum hydrocarbon; PAHs: Polycyclic aromatic hydrocarbons

nanoparticles for broad-scale PH-contaminated soil treatment faces challenges due to the stability and persistence of nanoparticles (Usman et al. 2022). Before their extensive use, a deeper understanding of their impact on the environment is essential. Additionally, on a large scale, the applicability of nanotechnology relies on employing cost-effective, eco-friendly strategies in diverse fields like degradation using microorganisms, farming, and medicine.

10.3.1 Electro-Bioremediation (EB)

Electro-bioremediation (EB) is a technology that employs processes like electroosmosis or electromigration. It leverages a less electric voltage embedded in contaminated soil to enhance the in situ degradation of contaminants. This approach proves highly effective in contaminated soil having a large area and less conductivity. To address these limitations, ongoing studies are exploring the incorporation of surfactants, to enhance the accessibility of pollutants in the remediation process. They have disadvantages which include the absence of suitable terminal electron acceptors, slow kinetics, toxicity, incomplete biodegradation, and high cost of initial and maintenance. Therefore, more efficient and sustainable approaches are needed for the effective treatment of PHs from the ecosystem.

10.3.2 Bio-electrochemical Systems (BESs)

BES is an innovative advanced biotechnology that transforms organic substances present in polluted water into electricity for the treatment of wastewater. Notably, this technology boasts a unique trait: the microorganisms possess the ability to externally transfer electrons to the anode. This sets it apart from the traditional mechanism, which occurs inside the cells of microorganisms. Furthermore, compared to traditional inexpensive wastewater treatment methods, BES generates significantly reduced quantities of sludge while concurrently producing valuable substances like ethanol, polyhydroxyalkanoates, and methane. Over the past decade, BESs, including biological fuel cells, have been harnessed to treat wastewater from diverse sources such as sewage, the chocolate industry, beer production facilities, and pig farming. These systems showcase impressive removal rates for nitrates and sulfates, leading to a substantial 95% reduction in oxygen demand. BESs provides new avenues for employing this technology in treating PH-contaminated through the involvement of microorganisms. The solid anode electrode material of the biological fuel cell facilitates the oxidation of electron donors, such as PHs, and conveys electrons externally to the electrode's surface (Prathiba et al. 2022). Electrons are then directed to the negative terminal of the microbial fuel cell through an external pathway, where oxygen (O_2) is typically converted to water (H_2O). The solid anode electrode material of the biological fuel cell takes on the role of a never-depleting electron acceptor, facilitating the transfer of electrons from contaminated regions to regions rich in electron acceptors. Additionally, BESs have the potential to enhance the breakdown of PHs without necessitating additional energy or chemical inputs. This has the dual advantage of reducing functioning costs and generating electricity compared to other treatment methods. This current serves not only as a real-time indicator but also as a power source for wireless sensors to remotely monitor and track. Numerous research studies carried out over the past decade have addressed the issue of PHs contamination using BESs, with these findings succinctly concluded in Table 10.4. Consequently, there is an increasing need to further develop and implement these strategies on a broader scale, adhering to engineering procedures. Furthermore, biological fuel cells and BESs hold promise in enhancing the PHs remediation process, mainly saturated and aromatic compounds. Both in-situ and ex-situ approaches can utilize BESs and fuel cells for treating PH-contaminated soil. These technologies are regarded as biocompatible, resilient, adaptable, and environment-

friendly for degrading a range of contaminants. robust, versatile, and eco-friendly for remediating various pollutants. However, challenges such as sluggish microbial multiplication rate, limited source availability, expensive catalysts, and excess potential drop hinder their efficiency in large-scale processes. For instance, Zeppilli et al. noted that the use of a phosphate buffer at the electrode's anodic region could

Table 10.4 Removal of PH-contaminated soil using BESs

Pollutant sample	Microbial species	Experimental conditions	Removal efficiency (%)	References
Phenanthrene	Microbial consortium derived from the contaminated location	In batch sediment biofuel cells, phenanthrene serves as an electron donor, while oxygenated water functions as an electron acceptor	99	Yan et al. (2012)
Diesel oil	Anaerobic microbial community	Within tubular microbial fuel cells equipped with air cathodes, diesel functions as the electron donor, while oxygen serves as the electron acceptor	59	Wang et al. (2019)
PHs	Activated biomass	In single-chamber microbial fuel cells, PHs serve as the electron donor, while glucose, in combination with Tween 80, acts as the electron acceptor	45	Zhao et al. (2019)
Furfural	Mixed cultures	In batch single-chamber microbial fuel cells, furfural functions as the electron donor, while oxygen operates as the electron acceptor	95	Luo et al. (2010)
4-nitrophenol	Anaerobic sludge	In batch double-chambered microbial fuel cells, glucose serves as the electron donor, while 4-nitrophenol, in combination with H_2O_2, functions as the electron acceptor	100	Zhu and Ni (2009)
PHs	The microbial community combined and isolated from PH-contaminated soil	In batch U-tube single microbial fuel cells, PHs function as electron donors, while oxygen serves as the electron acceptor	15.2	Wang et al. (2012)
PHs	-	In continuous double microbial fuel cells, PHs serve as the electron donor, while oxygen acts as the electron acceptor	17	Li et al. (2019)

(continued)

Table 10.4 (continued)

Pollutant sample	Microbial species	Experimental conditions	Removal efficiency (%)	References
1,2-Dichloro ethane	Anaerobic microbial community	In continuous double microbial fuel cells, 1,2-dichloroethane functions as the electron donor, while oxygen serves as the electron acceptor	95	Pham et al. (2009)
Nitrobenzene	Mixed cultures	In continuous double microbial fuel cells, nitrobenzene functions as the electron acceptor and acetate acts as the electron donor	1.3	Mu et al. (2009)
PHs	–	In microbial fuel cells with bottle-type double reactors, PHs act as electron donors and oxygenated water serves as the electron acceptor	29	Wang et al. (2020)
Diesel oil	Heterogeneous microbial communities	In batch double microbial fuel cells, diesel serves as the electron donor, and oxygen functions as the electron acceptor	83	Cheng et al. (2017)

preserve the pH balance of the solution for brief periods but led to an excess potential at the anodic region, resulting in acidification and a decrease in microbial metabolism (Zeppilli et al. 2021). Other obstacles in the widespread application of BESs include current loss, less availability of sources, and a lack of a comprehensive understanding of microorganisms' electron transfer processes. To address these challenges, integrating microbial cathodes, surfactants, electroactive microorganisms, and redox agents holds the promise of mitigating electrical issues in treating PH-contaminated soil. However, the integration of these materials is still in its nascent stages and necessitates further dedicated research and exploration. The performance of BESs can be improved by concerning certain factors, as shown in Fig. 10.3.

10.4 Analysis of Remediation Techniques

The efficacy, consistency, and trustworthiness of treatment methods are fundamental parameters of their performance. These processes demand ongoing scrutiny for wider adoption. Additionally, robust treatment-checking procedures are imperative, encompassing duplication, accurate collection, and data analysis. A range of primary

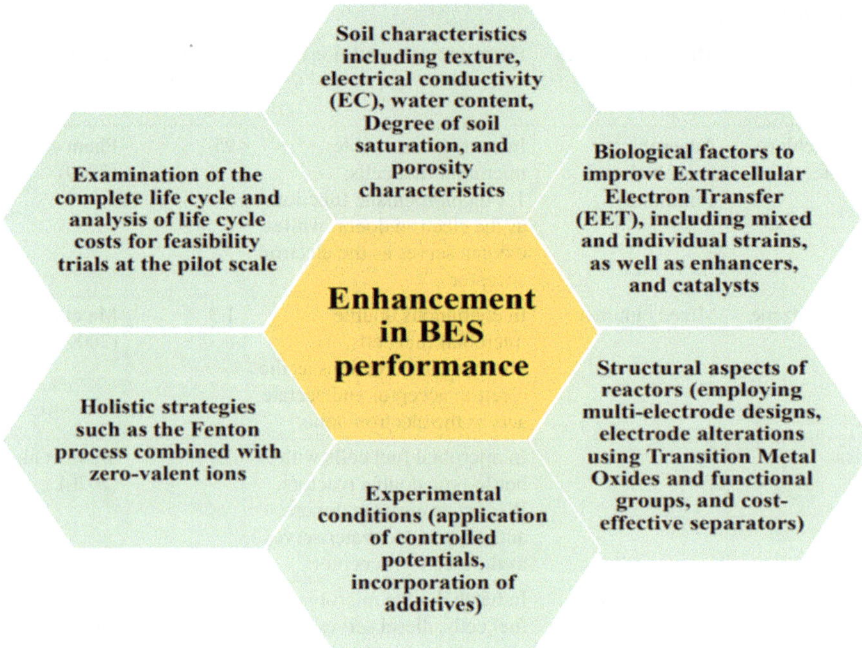

Fig. 10.3 Factors affecting Bio-electrochemical Systems (BES) performance

analytical and biological methods are currently used to oversee the soil treatment methods.

10.4.1 Biological Methods

Soil and water treatment techniques have gained significance on a large scale for rejuvenating polluted sites. To ensure the effective execution of such methods, it's crucial to develop efficient and precise approaches for seeing remediation processes. Over the course of time, several analytical methods have emerged to track the progress of environment remediation techniques. Nonetheless, these methods often have limitations such as limited effectiveness and difficulty distinguishing between biological and non-biological degradation processes. Therefore, biological methods have emerged as a valuable solution in distinguishing microorganisms. Bioremediation methods offer insights into the fate of compounds and majorly rely on microbial activity. There are two main categories of methods involved in this: culture-independent and culture-dependent (Daverey et al. 2019). Culture-dependent methods, such as enumeration on agar plates, carbon substrate utilization assessments, statistical estimations of microbial populations, transcriptomic analysis, and proteomic profiling, are utilized for distinguishing microorganisms and remediating pollutants. The immunochemical enzyme assay is the most adopted immunochemical method for distinguishing and tracking remediation processes. For example, according to Ko et al., their investigation into the analysis and identification of microcystin produced by cyanobacteria in the remediation of nutrient-rich water using immunoassay techniques demonstrates a notable efficacy in managing the remediation method (Ko et al. 2019). Culture-dependent methods have limitations in analyzing the complete microbial population in polluted regions, as only about 1% of microbes are cultivable. This constraint has led to the adoption of the culture-independent method, which bypasses traditional cultivation methods. The fingerprinting method offers comprehensive data of microbial populations without the need for culturing and provides large output, efficient and reproducible, like methods based on metagenomics.

Additionally, techniques like restriction fragment length polymorphism, polymorphic DNA random amplification, single-strand conformation polymorphism, sequence typing, fluorescence in situ hybridization, phospholipid fatty acid analysis, repetitive sequence-based PCR, reverse sample genome probing, and PCR can also be used to characterize and track the remediation technique involving microbes. However, molecular techniques utilized in intricate and persistently polluted soil and water often face challenges in terms of consistency and robustness. Another efficient approach to monitoring remediation techniques involves the use of enrichment methods. One such method involves incorporating a labelled nucleotide, bromodeoxyuridine, into the DNA of organisms with active metabolism. This facilitates the selective enrichment of microorganisms cultivated using xenobiotic substances. Another enrichment technique is stable-isotope probing, which involves the insertion of a substance enriched with carbon 13 isotope into the bacteria present in the soil. This method confirms that microbes that employ the 13C-labeled substrate result in a denser composition than the normal DNA with 12C, enabling the identification of cells displaying metabolic activity. The stable isotope probing technique is also used to track changes in the biological process. More recently, biofuel cells involving microorganisms and microorganism-based sensors have been used to evaluate remediation methods. Therefore, further research and exploration are needed in the fields of genetic modification and synthetic biology to improve the creation and advancement of microorganism-based sensors. These technologies hold promise for effective, precise tracking and evaluation of remediation processes and provide an eco-friendly approach to enhancing bioremediation techniques.

10.4.2 NGS Technology

Next-generation sequencing (NGS) technique offers many advantages in the field of PHs remediation. It provides deep knowledge of the microbial processes that affect contaminant degradation in soil environments. NGS technology facilitates the simultaneous examination of DNA sequences, allowing a comprehensive exploration of nucleic acids. This innovative technology is employed to assess the quality of the environment to gain insights into the genetic composition of various elements in the environment. It is utilized for the recognition of unique genetic elements, routes, and even shifts in microbial dynamics involved in the PHs biodegradation. Additionally, NGS technology offers insights into the crucial microorganisms engaged in the bioremediation and biodegradation of PHs.

Consequently, it has become an efficient method for tracking and data governance in polluted soil regions. The NGS process encompasses multiple steps, including the creation of libraries, decoding of bases, sequencing, genetic sequence alignment, and diverse annotation. The creation of a library begins with fragmenting DNA into multiple pieces, followed by enzymatic cleavage and sequence capture. However, NGS technology is often associated with higher costs and the need for more sophisticated equipment and chemicals compared to cutting-edge bioinformatics and statistical and utilities (Lopez 2020).

10.4.3 Analytical Methods

Analytical methods form the cornerstone of bioremediation and biodegradation, providing vital insights that are otherwise challenging to ascertain. On-site tracking of remediation through electrode material is the most used method, mainly in oxygen-depleted environments. This technique can identify both positively charged cations and negatively charged anions. However, its efficacy diminishes when dealing with water-repellent organic substances like PAHs. Contaminants and different intermediate substances can be characterized using techniques such as spectrometry and chromatography. These methodologies involve comprehensive physical and chemical analyses of specific substances and their intermediate compounds or end products, boasting high levels of remarkable selectivity, precision, and consistent reproducibility.

High-performance liquid chromatography (HPLC) and Gas chromatography (GC) are instrumental in monitoring the biodegradation and bioremediation of heavy metals, PHs, and lubricious slurry. The integration of mass spectrometry (MS) with GC excels in recognizing contaminants and their transitional forms, employing regulated and suitable techniques for refining the soil. Additionally, diverse spectroscopy technologies, including Raman spectroscopy, Nuclear magnetic resonance (NMR), Fourier-transform infrared (FTIR), Atomic absorption spectroscopy (AAS), 3D-fluorescence spectroscopy, and X-ray diffraction (XRD) are harnessed to analyze and track different contaminants like PHs, PAHS, and residual petroleum sediment (Patil et al. 2022). Among these, FTIR and AAS stand out for their low cost, reliability, responsiveness, and albeit with somewhat lower precision. Another approach to categorizing pollutants is by exploiting their chemical and physical characteristics through thermal processing. In the larger picture, petroleum products pose a substantial threat to both human well-being and nature. Researchers globally are focusing on degradation as a safe means to mitigate such pollution. Modern analytical techniques provide diverse solutions in this endeavour.

Through in-depth investigation, different analytical methods have emerged, underlining the critical role of appropriate specimen prepping in calculating the accuracy of quantitative hydrocarbon measurement outcomes. GC technique is a prominent choice for analyzing substances and their derivative compounds. Metabolite characterization, facilitated by GC–MS and mass spectrometry-liquid chromatography (MS-LC), elucidates degradation routes. FTIR, NMR, MS-LC, and HPLC are increasingly utilized in biodegradation techniques to evaluate contaminants of petroleum products.

10.5 Challenges

Choosing and implementing suitable eco-friendly bioremediation techniques relies on a range of parameters that impact the efficacy, applicability, and actual progress of the bioremediation process. These factors include both non-biological and biological considerations and non-scientific and scientific influences. Moreover, the lack of comprehensive empirical proof concerning the real-world application of these factors and their effects on bioremediation efficacy underscores the need for a thorough understanding. Consequently, it is essential to have a comprehensive grasp of various aspects such as physical and chemical attributes, characteristics of pollutants and environment, characteristics of living organisms, origin identification, age, categorization, composition, diversity, concentration levels, prevailing site conditions, procedural intricacies, geological and hydrological location, monitoring complexities, spatial demands, necessary extent of remediation criteria, remedial tactics, approach to mitigating risks life-cycle assessment, economic analysis ratio, anticipated outcomes, and the ability to meet prerequisites and regulatory criteria. These aspects collectively guide the selection and projection of eco-friendly mitigation technology for PH-contaminated soil.

Furthermore, in addition to the aforementioned considerations, the comprehensive implementation of eco-friendly bioremediation research necessitates a fundamental shift in how we approach the restoration of polluted areas. This entails recognizing economic, ecological, and societal dimensions as integral components in choice determination. Collaboration among academia, governmental bodies, and industries plays a pivotal role in advancing eco-friendly bioremediation practices and tackling existing challenges. This collaborative effort is vital for effectively integrating eco-friendly principles into remediation endeavours, fostering progress in both developed and under-developed nations. Technologies for the eco-friendly bioremediation and biodegradation of PH-contaminated soil are depicted in Fig. 10.4.

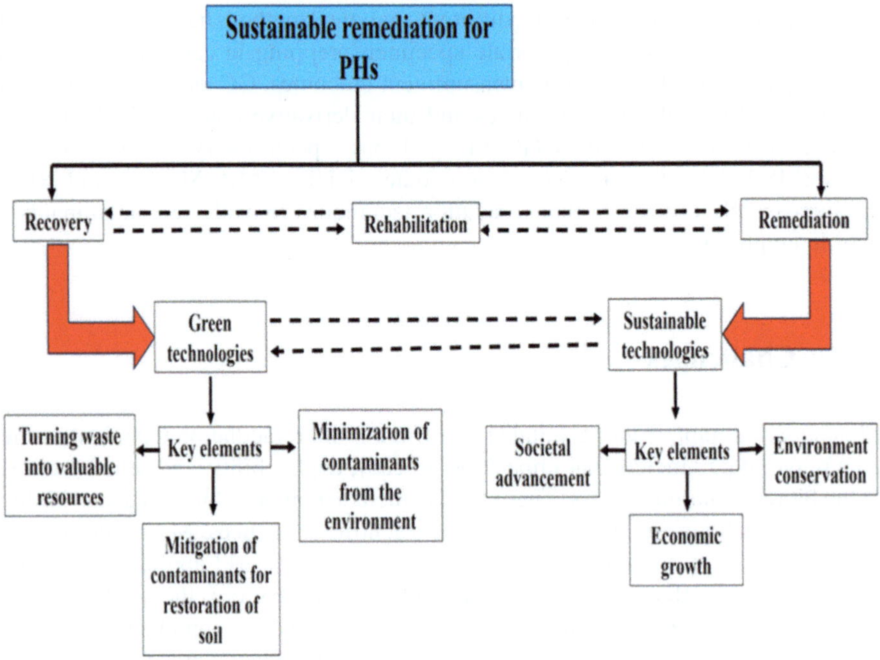

Fig. 10.4 Sustainable bioremediation of PH-contaminated soil

10.6 Conclusion and Future Outlook

This chapter explains the remediation techniques for PH-polluted soil both in laboratory and field applications. While a variety of methods are in practice, it's important to note that no single method can be universally used to mitigate PHs present in the soil matrix and other polluted regions. It's crucial to emphasize that the factors involving the bioremediation techniques have a significant effect on the overall efficiency of the bioremediation process. In the current landscape, there's a growing demand for sustainable methods and innovative mechanisms that mitigate PH-contaminated soil efficiently at a reasonable cost. Implementing innovative technologies for PHs management, combined with thoughtful decision-making processes, plays a pivotal role in achieving sustainability objectives over a long time for contaminants removal from polluted regions. In parallel, the establishment of tools for managing based on risk assessment, driven by databases, systems utilizing machine learning algorithms, life cycle assessment (LCA), and life cycle cost analysis (LCCA) studies, as well as the utilization of NGS technology, merits greater attention. Furthermore, exploring low-cost nanomaterials and biologically-driven electrochemical technologies offers promising avenues for addressing the challenges to the remediation of soil environments. Equally important is the emphasis on evaluating and analyzing methods to gauge the effectiveness of restoration, not only in creating novel in-situ techniques but also in creating novel freely available software tailored to diverse needs. Such comprehensive strategies can lead to the development of eco-friendly remediation techniques that can fit various conditions of the environment and diverse levels of pollutants.

References

Abdel-Shafy HI, Mansour MS (2016) A review on polycyclic aromatic hydrocarbons: source, environmental impact, effect on human health and remediation. Egypt J Pet 25(1):107–123. https://doi.org/10.1016/j.ejpe.2015.03.011

Abena MTB, Li T, Shah MN, Zhong W (2019) Biodegradation of total petroleum hydrocarbons (TPH) in highly contaminated soils by natural attenuation and bioaugmentation. Chemosphere 234:864–874. https://doi.org/10.1016/j.chemosphere.2019.06.111

Adams GO, Fufeyin PT, Okoro SE, Ehinomen I (2015) Bioremediation, biostimulation and bioaugmention: a review. Int J Environ Bioremediat Biodegrad 3(1):28–39. https://doi.org/10.12691/ijebb-3-1-5

Al-Hawash AB, Alkooranee JT, Abbood HA, Zhang J, Sun J, Zhang X, Ma F (2018) Isolation and characterization of two crude oil-degrading fungi strains from Rumaila oil field. Iraq Biotechnol Rep 17:104–109. https://doi.org/10.1016/j.btre.2017.12.006

Al-Hawash AB, Zhang X, Ma F (2019) Removal and biodegradation of different petroleum hydrocarbons using the filamentous fungus Aspergillus sp. RFC-1. Microbiology 8(1):e00619. https://doi.org/10.1002/mbo3.619

Alsayegh SY, Al-Ghouti MA, Zouari N (2021) Study of bacterial interactions in reconstituted hydrocarbon-degrading bacterial consortia from a local collection, for the bioremediation of weathered oily-soils. Biotechnol Rep 29:e00598. https://doi.org/10.1016/j.btre.2021.e00598

Andrade DC, dos Santos EV (2020) Combination of electrokinetic remediation with permeable reactive barriers to remove organic compounds from soils. Curr Opin Electrochem 22:136–144. https://doi.org/10.1016/j.coelec.2020.06.002

Bajagain R, Jeong SW (2021) Degradation of petroleum hydrocarbons in soil via advanced oxidation process using peroxymonosulfate activated by nanoscale zero-valent iron. Chemosphere 270:128627. https://doi.org/10.1016/j.chemosphere.2020.128627

Baoune H, Aparicio JD, Acuña A, El Hadj-khelil AO, Sanchez L, Polti MA, Alvarez A (2019) Effectiveness of the Zea mays-Streptomyces association for the phytoremediation of petroleum hydrocarbons impacted soils. Ecotoxicol Environ Saf 184:109591. https://doi.org/10.1016/j.ecoenv.2019.109591

Bento FM, Camargo FA, Okeke BC, Frankenberger WT (2005) Comparative bioremediation of soils contaminated with diesel oil by natural attenuation, biostimulation and bioaugmentation. Bioresour Technol 96(9):1049–1055. https://doi.org/10.1016/j.biortech.2004.09.008

Beolchini F, Rocchetti L, Regoli F, Dell'Anno A (2010) Bioremediation of marine sediments contaminated by hydrocarbons: experimental analysis and kinetic modeling. J Hazard Mater 182(1–3):403–407. https://doi.org/10.1016/j.jhazmat.2010.06.047

Bolade OP, Akinsiku AA, Oluwafemi OS, Williams AB, Benson NU (2021) Biogenic iron oxide nanoparticles and activated sodium persulphate for hydrocarbon remediation in contaminated soil. Environ Technol Innov 23:101719. https://doi.org/10.1016/j.eti.2021.101719

Cabral L, Giovanella P, Pellizzer EP, Teramoto EH, Kiang CH, Sette LD (2022) Microbial communities in petroleum-contaminated sites: structure and metabolisms. Chemosphere 286:131752. https://doi.org/10.1016/j.chemosphere.2021.133207

Cheng Y, Wang L, Faustorilla V, Megharaj M, Naidu R, Chen Z (2017) Integrated electrochemical treatment systems for facilitating the bioremediation of oil spill contaminated soil. Chemosphere 175:294–299. https://doi.org/10.1016/j.chemosphere.2017.02.079

Daverey A, Pandey D, Verma P, Verma S, Shah V, Dutta K, Arunachalam K (2019) Recent advances in energy efficient biological treatment of municipal wastewater. Bioresour Technol Rep 7:100252. https://doi.org/10.1016/j.biteb.2019.100252

Desalegn B, Megharaj M, Chen Z, Naidu R (2018) Green mango peel-nano zerovalent iron activated persulfate oxidation of petroleum hydrocarbons in oil sludge contaminated soil. Environ Technol Innov 11:142–152. https://doi.org/10.1016/j.eti.2018.05.007

El-Sheshtawy HS, Aman D, Nassar HN (2021) A novel bioremediation technique for petroleum hydrocarbons by bacterial consortium immobilized on goethite-chitosan nanocomposite. Soil Sediment Contam 31:1–24. https://doi.org/10.1080/15320383.2021.1916737

Falciglia PP, Roccaro P, Bonanno L, De Guidi G, Vagliasindi FG, Romano S (2018) A review on the microwave heating as a sustainable technique for environmental remediation/detoxification applications. Renew Sustain Energy Rev 95:147–170. https://doi.org/10.1016/j.rser.2018.07.031

Farahat LA, El-Gendy NS (2007) Comparative kinetic study of different bioremediation processes for soil contaminated with petroleum hydrocarbons. Mater Sci Res Int 4(2):269–278. https://doi.org/10.13005/msri/040206

Fava F, Berselli S, Conte P, Piccolo A, Marchetti L (2004) Effects of humic substances and soya lecithin on the aerobic bioremediation of a soil historically contaminated by polycyclic aromatic hydrocarbons (PAHs). Biotechnol Bioeng 88(2):214–223. https://doi.org/10.1002/bit.20225

Gholami P, Khataee A, Bhatnagar A, Vahid B (2021) Synthesis of N-doped magnetic WO3–x@ mesoporous carbon using a diatom template and plasma modification: visible-light-driven photocatalytic activities. ACS Appl Mater Interfaces 13(11):13072–13086. https://doi.org/10.1021/acsami.0c21076

Hesnawi RM, Mogadami FS (2013) Bioremediation of Libyan crude oil-contaminated soil under mesophilic and thermophilic conditions. APCBEE Proc 5:82–87. https://doi.org/10.1016/j.apcbee.2013.05.015

Hussain A, Rehman F, Rafeeq H, Waqas M, Asghar A, Afsheen N, Rahdar A, Bilal M, Iqbal HM (2022) In-situ, ex-situ, and nano-remediation strategies to treat polluted soil, water, and air–a review. Chemosphere 289:133252. https://doi.org/10.1016/j.chemosphere.2021.133252

Ossai IC, Ahmed A, Hassan A et al (2019) Remediation of soil and water contaminated with petroleum hydrocarbon: a review. Environ Technol Innov. https://doi.org/10.1016/j.eti.2019.100526

Imam A, Suman SK, Kanaujia PK, Ray A (2022) Biological machinery for polycyclic aromatic hydrocarbons degradation: a review. Bioresour Technol 343:126121. https://doi.org/10.1016/j.biortech.2021.126121

Ivshina IB, Kuyukina MS, Krivoruchko AV, Elkin AA, Makarov SO, Cunningham CJ, Peshkur TA, Atlas RM, Philip JC (2015) Oil spill problems and sustainable response strategies through new technologies. Environ Sci Process Impact R Soc Chem 17:1201–1219

Ko SR, Srivastava A, Lee N, Jin L, Oh HM, Ahn CY (2019) Bioremediation of eutrophic water and control of cyanobacterial bloom by attached periphyton. Int J Environ Sci Technol 16(8):4173–4180. https://doi.org/10.1007/s13762-019-02320-8

Koshlaf E, Ball AS (2017) Soil bioremediation approaches for petroleum hydrocarbon polluted environments. AIMS Microbiol 3(1):25. https://doi.org/10.3934/microbiol.2017.1.25

Kuppusamy S, Thavamani P, Megharaj M, Naidu R (2016) Biodegradation of polycyclic aromatic hydrocarbons (PAHs) by novel bacterial consortia tolerant to diverse physical settings–assessments in liquid-and slurry-phase systems. Int Biodeterior Biodegrad 108:149–157. https://doi.org/10.1016/j.ibiod.2015.12.013

Lewis RF (1993) Site demonstration of slurry-phase biodegradation of PAH contaminated soil. J Air Waste Manag Assoc 43(4):503–508. https://doi.org/10.1080/1073161X.1993.10467149

Li X, Li Y, Zhang X, Zhao X, Sun Y, Weng L, Li Y (2019) Long-term effect of biochar amendment on the biodegradation of petroleum hydrocarbons in soil microbial fuel cells. Sci Total Environ 651:796–806. https://doi.org/10.1016/j.scitotenv.2018.09.098

López ONB (2020) Role of modern innovative techniques for assessing and monitoring environmental pollution. In: Bioremediation and biotechnology. Springer, Cham, pp 75–91. https://doi.org/10.1007/978-3-030-35691-0_4

Luo Y, Liu G, Zhang R, Zhang C (2010) Power generation from furfural using the microbial fuel cell. J Power Sources 195(1):190–194. https://doi.org/10.1016/j.jpowsour.2009.06.057

Malik B, Pirzadah TB, Hakeem KR (2022) Phytoremediation of persistent organic pollutants (POPs). In: Phytoremediation. Academic, pp 415–436. https://doi.org/10.1016/B978-0-323-89874-4.00010-8

Mohapatra B, Dhamale T, Saha BK, Phale PS (2022) Microbial degradation of aromatic pollutants: metabolic routes, pathway diversity, and strategies for bioremediation. In: Microbial biodegradation and bioremediation. Elsevier, pp 365–394. https://doi.org/10.1016/B978-0-323-85455-9.00006-0

Moradi M, Vasseghian Y, Khataee A, Harati M, Arfaeinia H (2021) Ultrasound-assisted synthesis of FeTiO3/GO nanocomposite for photocatalytic degradation of phenol under visible light irradiation. Separ Purif Technol 261:118274. https://doi.org/10.1016/j.seppur.2020.118274

Mu Y, Rozendal RA, Rabaey K, Keller J (2009) Nitrobenzene removal in bioelectrochemical systems. Environ Sci Technol 43(22):8690–8695. https://doi.org/10.1021/es9020266

Murgueitio E, Cumbal L, Abril M, Izquierdo A, Debut A, Tinoco O (2018) Green synthesis of iron nanoparticles: application on the removal of petroleum oil from contaminated water and soils. J Nanotechnol. https://doi.org/10.1155/2018/4184769

Passarini MR, Rodrigues MV, da Silva M, Sette LD (2011) Marine-derived filamentous fungi and their potential application for polycyclic aromatic hydrocarbon bioremediation. Mar Pollut Bull 62(2):364–370. https://doi.org/10.1016/j.marpolbul.2010.10.003

Patil SM, Rane NR, Bankole PO, Krishnaiah P, Ahn Y, Park YK, Yadav KK, Amin MA, Jeon BH (2022) An assessment of micro-and nanoplastics in the biosphere: a review of detection, monitoring, and remediation technology. Chem Eng Technol 430:132913. https://doi.org/10.1016/j.cej.2021.132913

Pham H, Boon N, Marzorati M, Verstraete W (2009) Enhanced removal of 1, 2-dichloroethane by anodophilic microbial consortia. Water Res 43(11):2936–2946. https://doi.org/10.1016/j.watres.2009.04.004

Prathiba S, Kumar PS, Vo DVN (2022) Recent advancements in microbial fuel cells: a review on its electron transfer mechanisms, microbial community, types of substrates and design for bio-electrochemical treatment. Chemosphere 286:131856. https://doi.org/10.1016/j.chemosphere.2021.131856

Ramadass K, Megharaj M, Venkateswarlu K, Naidu R (2018) Bioavailability of weathered hydrocarbons in engine oil-contaminated soil: impact of bioaugmentation mediated by Pseudomonas spp. on bioremediation. Sci Total Environ 636:968–974. https://doi.org/10.1016/j.scitotenv.2018.04.379

Riser-Roberts E (2020) Remediation of petroleum contaminated soils: biological, physical, and chemical processes. CRC Press. https://www.routledge.com/Remediation-of-Petroleum-Contaminated-Soils-Biological-Physical-and-Chemical/Riser-Roberts/p/book/9780367400446

Ruley JA, Amoding A, Tumuhairwe JB, Basamba TA (2022) Rhizoremediation of petroleum hydrocarbon–contaminated soils: a systematic review of mutualism between phytoremediation species and soil living microorganisms. Phytoremediation 263–296. https://doi.org/10.1016/B978-0-323-89874-4.00008-X

Sohoo I, Ritzkowski M, Kuchta K, Cinar SO (2021) Environmental sustainability enhancement of waste disposal sites in developing countries through controlling greenhouse gas emissions. Sustainability 13(1):151. https://doi.org/10.3390/su13010151

Song X, Wu X, Song X, Shi C, Zhang Z (2021) Sorption and desorption of petroleum hydrocarbons on biodegradable and non-degradable microplastics. Chemosphere 273:128553. https://doi.org/10.1016/j.chemosphere.2020.128553

Suja F, Rahim F, Taha MR, Hambali N, Razali MR, Khalid A, Hamzah A (2014) Effects of local microbial bioaugmentation and biostimulation on the bioremediation of total petroleum hydrocarbons (TPH) in crude oil contaminated soil based on laboratory and field observations. Int Biodeterior Biodegrad 90:115–122. https://doi.org/10.1016/j.ibiod.2014.03.006

Sun J, Zhang Z, Wang H, Rogers MJ, Guo H, He J (2022) Exploration of the biotransformation of phenanthrene degradation coupled with methanogensis by metabolites and enzyme analyses. Environ Pollut 293:118491. https://doi.org/10.1016/j.envpol.2021.118491

Thavamani P, Megharaj M, Naidu R (2012) Bioremediation of high molecular weight polyaromatic hydrocarbons co-contaminated with metals in liquid and soil slurries by metal tolerant PAHs degrading bacterial consortium. Biodegradation 23(6):823–835. https://doi.org/10.1007/s10532-012-9572-7

Tran HT, Lin C, Hoang HG, Bui XT, Vu CT (2022) Soil washing for the remediation of dioxin-contaminated soil: a review. J Hazard Mater 421:126767. https://doi.org/10.1016/j.jhazmat.2021.126767

Usman M, Jellali S, Anastopoulos I, Charabi Y, Hameed BH, Hanna K (2022) Fenton oxidation for soil remediation: a critical review of observations in historically contaminated soils. J Hazard Mater 424:127670. https://doi.org/10.1016/j.jhazmat.2021.127670

Varjani SJ, Rana DP, Jain AK, Bateja S, Upasani VN (2015) Synergistic ex-situ biodegradation of crude oil by halotolerant bacterial consortium of indigenous strains isolated from on shore sites of Gujarat, India. Int Biodeterior Biodegrad 103:116–124. https://doi.org/10.1016/j.ibiod.2015.03.030

Vidonish JE, Zygourakis K, Masiello CA, Sabadell G, Alvarez PJJ (2016) Thermal treatment of hydrocarbon-impacted soils: a review of technology innovation for sustainable remediation. Engineering 2:426–437

Wagner AM, Yarmak E Jr (2017) Using frozen barriers for containment of contaminants. Cold Regions Research and Engineering Laboratory, US Army Engineer Research and Development Center Hanover United States. https://apps.dtic.mil/sti/citations/AD1039597

Wang X, Cai Z, Zhou Q, Zhang Z, Chen C (2012) Bioelectrochemical stimulation of petroleum hydrocarbon degradation in saline soil using U-tube microbial fuel cells. Biotechnol Bioeng 109(2):426–433. https://doi.org/10.1002/bit.23351

Wang H, Lu L, Mao D, Huang Z, Cui Y, Jin S, Zuo Y, Ren ZJ (2019) Dominance of electroactive microbiomes in bioelectrochemical remediation of hydrocarbon-contaminated soils with different textures. Chemosphere 235:776–784. https://doi.org/10.1016/j.chemosphere.2019.06.229

Wang H, Lu L, Chen H, McKenna AM, Lu J, Jin S, Zuo Y, Rosario-Ortiz FL, Ren ZJ (2020) Molecular transformation of crude oil contaminated soil after bioelectrochemical degradation revealed by FT-ICR mass spectrometry. Environ Sci Technol 54(4):2500–2509. https://doi.org/10.1021/acs.est.9b06164

Wang H, Chen P, Zhang S, Jiang J, Hua T, Li F (2022) Degradation of pyrene using single-chamber air-cathode microbial fuel cells: electrochemical parameters and bacterial community changes. Sci Total Environ 804:150153. https://doi.org/10.1016/j.scitotenv.2021.150153

Woo SH, Lee MW, Park JM (2004) Biodegradation of phenanthrene in soil-slurry systems with different mass transfer regimes and soil contents. J Biotechnol 110(3):235–250. https://doi.org/10.1016/j.jbiotec.2004.02.007

Xia C, Liu Q, Zhao L, Tang J, Wang L (2021) Enhanced degradation of petroleum hydrocarbons in soil by FeS@ BC activated persulfate and its mechanism. Separ Purif Technol 282:120060. https://doi.org/10.1016/j.seppur.2021.120060

Li YT, Zhang JJ, Li YH, Chen JL, Du WY (2022) Treatment of soil contaminated with petroleum hydrocarbons using activated persulfate oxidation, ultrasound, and heat: a kinetic and thermodynamic study. Chem Eng 428:131336. https://doi.org/10.1016/j.cej.2021.131336

Yan Z, Song N, Cai H, Tay JH, Jiang H (2012) Enhanced degradation of phenanthrene and pyrene in freshwater sediments by combined employment of sediment microbial fuel cell and amorphous ferric hydroxide. J Hazard Mater 199:217–225. https://doi.org/10.1016/j.jhazmat.2011.10.087

Zeppilli M, Paiano P, Torres C, Pant D (2021) A critical evaluation of the pH split and associated effects in bioelectrochemical processes. Chem Eng Sci 422:130155. https://doi.org/10.1016/j.cej.2021.130155

Zhang JH, Xue QH, Gao H, Ma X, Wang P (2016) Degradation of crude oil by fungal enzyme preparations from Aspergillus spp. for potential use in enhanced oil recovery. J Chem Technol Biotechnol 91(4):865–875. https://doi.org/10.1002/jctb.4650

Zhao L, Deng J, Hou H, Li J, Yang Y (2019) Investigation of PAH and oil degradation along with electricity generation in soil using an enhanced plant-microbial fuel cell. J Clean Prod 221:678–683. https://doi.org/10.1016/j.jclepro.2019.02.212

Zhu X, Ni J (2009) Simultaneous processes of electricity generation and p-nitrophenol degradation in a microbial fuel cell. Electrochem Commun 11(2):274–277. https://doi.org/10.1016/j.elecom.2008.11.023

Zhu X, Wang X, Li N, Wang Q, Liao C (2022) Bioelectrochemical system for dehalogenation: a review. Environ Pollut 293:118519. https://doi.org/10.1016/j.envpol.2021.118519

Chapter 11
Biodegradation of Synthetic Polyethylene Terephthalate (PET) into Bis-(2-Hydroxyethyl) Terephthalate (BHET)

G. Koteswara Reddy, T. Manas, and B. Devi Sri Siddhartha

Abstract The biodegradation of synthetic polyethylene terephthalate (PET) into bis-(2-hydroxyethyl) terephthalate (BHET) is an active area of research. PET, a durable and non-biodegradable plastic, accumulates in the environment as waste. Enzymes called PETases can break down PET into its monomers, with BHET being a crucial intermediate. BHET can then be further processed into terephthalic acid (TPA) and ethylene glycol (EG), which can be used to create new PET or other chemicals. Ongoing research aims to optimize the efficiency of PET-degrading enzymes, offering potential solutions to address plastic waste through the biodegradation of PET into environmentally friendly products. Additionally, researchers are exploring various strategies to enhance the biodegradation of PET. This includes the genetic engineering of microorganisms to improve their PET-degrading capabilities and the development of novel enzyme variants with enhanced activity and stability. Furthermore, efforts are being made to optimize the conditions for PET degradation, such as temperature, pH, and substrate concentration, to maximize efficiency and scalability. The goal is to establish cost-effective and sustainable biodegradation processes that can be implemented on a large scale to tackle the growing issue of plastic waste and reduce the environmental impact of PET.

Keywords Biodegradation · Polyethylene · Phytoremediation · Oxidation

G. Koteswara Reddy (✉) · T. Manas · B. Devi Sri Siddhartha
Department of Biotechnology, Koneru Lakshmaiah Education Foundation (Deemed to be University), Green Fields, Vaddeswaram 522502, India
e-mail: koteswarareddy@kluniversity.in

I. D. Behera and A. P. Das (eds.), *Impact of Petroleum Waste on Environmental Pollution and its Sustainable Management Through Circular Economy*, Environmental Science and Engineering, https://doi.org/10.1007/978-3-031-48220-5_11

11.1 Introduction

One approach to biodegrading PET involves the use of enzymes called PETases. PETases are enzymes produced by certain bacteria and fungi that can break down PET into its monomers. Through a process called hydrolysis, these enzymes cleave the ester bonds in PET, resulting in the formation of mono-(2-hydroxyethyl) terephthalate (MHET) (Narayan and Miyakawa 2017). Another enzyme called MHETase further acts on MHET, breaking it down into bis-(2-hydroxyethyl) terephthalate (BHET) monomers. BHET is a key intermediate in the degradation pathway of PET, and its formation is crucial for the complete breakdown of PET into its constituent monomers. Once BHET is produced, it can undergo further enzymatic or chemical processes to yield the ultimate monomers of PET: terephthalic acid (TPA) and ethylene glycol (EG) (Tournier et al. 2020). These monomers can then be utilized as building blocks to produce new PET or other useful chemicals. Research in the field of PET biodegradation is still ongoing, and scientists are continually exploring and optimizing the efficiency of enzymes involved in PET degradation (Chen et al. 2020). This field holds promise for developing sustainable solutions to address the issue of plastic waste by enabling the biodegradation of PET into environmentally benign products.

Polyethylene terephthalate (PET) is a commonly used synthetic polymer in the production of plastic bottles, packaging materials, and textiles (Tournier et al. 2020). Its durability and resistance to degradation contribute to its accumulation as waste in the environment (Tribedi and Sil 2013). However, researchers have been focusing on enhancing the biodegradation of PET (Balasubramanian and Natarajan 2013) to reduce its environmental impact (Webb et al. 2013a, b). One approach involves the use of enzymes known as PETases (Chen et al. 2020), which can break down PET into its monomers (Ghanbari et al. 2018) through hydrolysis (Lv et al. 2019). These monomers (Oda et al. 2020), such as terephthalic acid (TPA) (Hirth et al. 2016) and ethylene glycol (EG) (Lenz et al. 2019), can then be utilised to produce new PET or other beneficial chemicals. Ongoing research aims to optimise the efficiency of PET-degrading enzymes, offering promising solutions for addressing plastic waste.

11.1.1 Types of Synthetic Polymers

Some familiar household synthetic polymers include Nylons in textiles (Adrio et al. 2005) and fabrics (Awaja and Pavel 2005), Teflon in non-stick pans, Bakelite for electrical switches, polyvinyl chloride (PVC) in pipes, etc. Considering the environmental concerns associated with non-biodegradable synthetic polymers such as polyethylene used in plastic kits and covers, as well as polybutadienes used in tire manufacturing, there is a growing interest in exploring alternative options (Zimmermann et al. 2021). Bioplastics are being considered as potential substitutes, as they offer more sustainable alternatives to petroleum-based synthetic polymers. These biodegradable

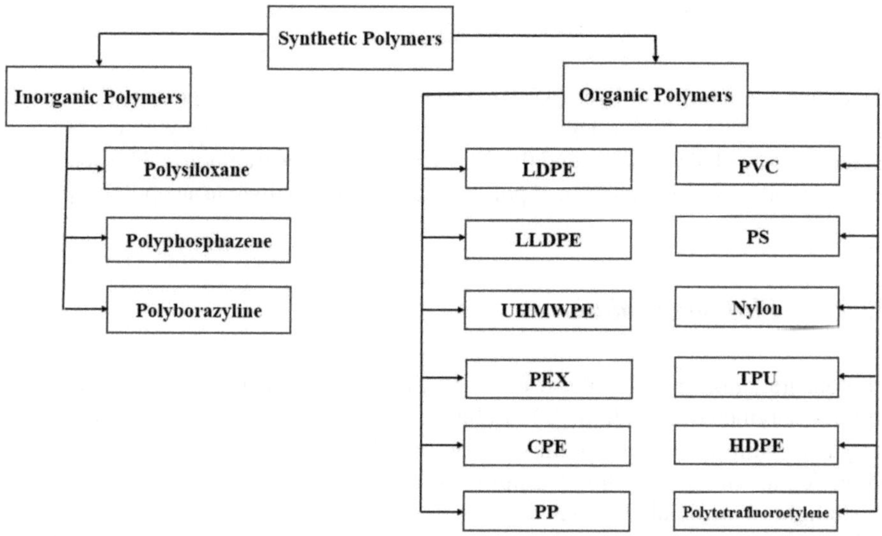

Fig. 11.1 Types of synthetic polymers (*Note* LDPE: Low-Density Polyethylene, LLDPE: Linear Low-Density Polyethylene, UHMWPE: Ultra-High Molecular Weight Polyethylene, PEX: Cross-Linked Polyethylene, CPE: Chlorinated Polyethylene, PP: Polypropylene, PVC: Poly Vinyl Chloride, PS: Polystyrene, TPU: Thermoplastic polyurethane, HDPE; High-Density Polyethylene)

materials have the potential to address the environmental issues caused by traditional polymers and reduce our reliance on non-renewable resources (Fig. 11.1).

11.2 Conventional Chemical Degradation Techniques

Conventional chemical degradation techniques refer to methods used to degrade or break down polymers through chemical means. These techniques involve the use of various chemical agents or processes to initiate polymer degradation. Here are some common conventional chemical degradation techniques:

11.2.1 Hydrolysis

Hydrolysis (Ronkvist et al. 2009) involves the use of water or aqueous solutions to break down polymers. It typically involves the cleavage of chemical bonds within the polymer chain through the addition of water molecules. This process can be accelerated using catalysts or elevated temperatures (Södergård et al. 2002).

11.2.2 Oxidation

Oxidative degradation utilizes oxidative agents such as hydrogen peroxide, ozone, or oxygen to initiate polymer degradation. The oxidative agents react with the polymer chain, causing chain scission and the formation of smaller fragments. This technique is often used for the degradation of polymers such as polyethylene and polypropylene.

11.2.3 Photodegradation

Photodegradation involves the use of light, typically ultraviolet (UV) radiation, to initiate polymer degradation. UV radiation can break the chemical bonds in the polymer chain, leading to chain scission and the formation of smaller fragments. Photodegradation is commonly employed for the degradation of polymers exposed to sunlight or artificial UV sources.

11.2.4 Thermal Degradation

Thermal degradation involves subjecting polymers to high temperatures, which can break the polymer chain. This process typically occurs through pyrolysis or thermal cracking, resulting in the formation of smaller molecules. Thermal degradation is often used for the controlled decomposition of polymers, such as in recycling processes (Ion et al. 2021).

11.2.5 Chemical Agents

Various chemical agents can be used to induce polymer degradation. For example, acids, bases, or organic solvents can be employed to degrade certain polymers. These agents can react with the polymer chain, causing cleavage or depolymerisation.

It is often important to note that conventional chemical degradation techniques may not always result in environmentally friendly (de Castro and Tavares 2021) or sustainable solutions, as they often involve the use of harsh chemicals or energy-intensive processes. There is ongoing research and development in the field of eco-friendly degradation techniques, including enzymatic degradation and biodegradation processes, which offer more sustainable alternatives to conventional chemical methods.

11.3 Biodegradation Techniques

Biodegradation techniques are used to break down organic materials, including polymers, into simpler compounds by the action of microorganisms such as bacteria, fungi, and enzymes (Yang et al. 2014). Here are some key uses and benefits of biodegradation (Ghosh et al. 2018) techniques:

11.3.1 Environmental Sustainability

Biodegradation techniques offer a more sustainable approach to managing organic waste (Khan and Anjum 2021), including biodegradable polymers. By facilitating the natural breakdown of materials through microbial action, these techniques contribute to reducing the accumulation of waste in landfills and ecosystems, leading to a cleaner and healthier environment (Kuroda et al. 2022).

11.3.2 Plastic Waste Management

Biodegradation techniques provide a potential solution for the degradation of synthetic polymers, including plastics, which are not readily biodegradable in nature. By utilizing enzymes or microorganisms capable of breaking down polymers, it becomes possible to convert plastic waste into simpler compounds, which can be further utilized or assimilated by natural systems.

11.3.3 Reduced Dependency on Fossil Fuels

Many conventional polymers, such as those derived from petrochemicals, require significant amounts of energy and resources for production. Biodegradable polymers offer an alternative that can be derived from renewable sources, such as plant-based feedstocks, reducing the reliance on fossil fuels and contributing to a more sustainable and circular economy.

11.3.4 Agricultural Benefits

Biodegradable polymers can find applications in agriculture, where they can be used as mulch films, seed coatings, or soil conditioners. These materials can provide temporary protection, aid in moisture retention, and improve crop yields. After use, they can be left in the field, where they undergo biodegradation (Montazer et al. 2021) without the need for removal or disposal.

11.3.5 Water and Soil Remediation

Biodegradation techniques can be employed for the remediation of polluted water bodies and contaminated soils (Quartinello et al. 2018). By introducing specific microbial strains or enzymes capable of degrading pollutants, such as hydrocarbons or pesticides, the techniques facilitate the breakdown of harmful substances, leading to the restoration of ecosystems (Su et al. 2020; Mohanta et al. 2023).

It's important to note that the efficacy of biodegradation techniques can vary depending on factors such as the type of polymer, environmental conditions, and the availability of appropriate microbial populations. Ongoing research and development aim to optimize and expand the application of biodegradation techniques to address biodegradation techniques.

11.4 Conventional Biodegradation of Synthetic Polyethylene

Conventional biodegradation of synthetic polyethylene, a highly durable and resistant polymer, is a complex process due to its inherent properties that make it resistant to breakdown by biological organisms. However, scientists and researchers have been exploring various strategies and approaches to enhance the biodegradation of polyethylene. In this comprehensive reply, we will delve into four key methods: microbial degradation, enzymatic degradation, the use of bio-additives, and physical and chemical treatments.

11.4.1 Microbial Degradation

Certain microorganisms, including bacteria, fungi, and consortium, have demonstrated the ability to break down polyethylene through their metabolic processes (Behera et al. 2020, 2021). For instance, species such as *Pseudomonas* and *Bacillus* bacteria, as well as *Aspergillus* and *Penicillium* fungi, have shown promise in degrading polyethylene. This discusses the mechanisms employed by microorganisms, the factors influencing degradation, and the potential applications and challenges associated with microbial plastic degradation (Webb et al. 2013a, b).

Polyethylene is a particularly challenging plastic to degrade due to its high molecular weight and low reactivity. However, several microorganisms have been identified that can degrade polyethylene, with bacterial species being the most extensively studied. The discovery and characterization of polyethylene-degrading bacteria have shown promising results in the degradation of polyethylene. Fungi have also shown potential in the degradation of polyethylene, with some species capable of producing lignin-modifying enzymes that can degrade polyethylene. Synergistic microbial communities have also been shown to enhance the efficiency of polyethylene degradation.

11.4.2 Enzymatic Degradation

Microorganisms produce enzymes responsible for breaking down various compounds, including polyethylene. Scientists have identified and isolated specific enzymes, such as polyethylene hydrolases and esterases, which can degrade polyethylene. These enzymes can be employed in treating polyethylene waste, either by direct application or by engineering microorganisms to produce them.

Enzymatic degradation involves the utilization of enzymes to catalyze the breakdown of polyethylene. Several types of enzymes have been identified as key players in the degradation process, including esterases, lipases, and peroxidases. Esterases and lipases cleave the ester bonds in polyethylene, while peroxidases initiate oxidative degradation by generating free radicals. The synergistic action of these enzymes enables the effective breakdown of polyethylene into smaller fragments.

11.4.3 Bio-additives

Another approach to enhancing the biodegradability of polyethylene is the incorporation of bio-additives during its production. Bio-additives, such as pro-oxidants or biodegradable polymers, play a crucial role in initiating and accelerating the degradation process, thereby making polyethylene more susceptible to microbial attack. Bio-additives used in polyethylene degradation play a crucial role in enhancing the biodegradability and eco-friendliness of this widely used plastic. In this report, we will explore the significance of bio-additives, their mechanisms, and their applications in polyethylene degradation.

Polyethylene (PE) is a synthetic polymer known for its non-biodegradable nature, leading to significant environmental pollution. However, the incorporation of bio-additives in polyethylene formulations can facilitate its degradation and contribute to a more sustainable waste management system.

In conclusion, bio-additives used in polyethylene degradation offer a sustainable approach to addressing the environmental challenges posed by this non-biodegradable plastic. Through their diverse mechanisms, including nucleation, plasticisation, and enzymatic degradation, bio-additives enhance the biodegradability of polyethylene and enable the production of more eco-friendly plastic products. The application of bio-additives in polyethylene formulations holds great promise for a more sustainable and responsible plastic waste management system.

11.4.4 Physical and Chemical Treatments

Physical and chemical treatments can also facilitate the biodegradation of polyethylene by modifying its structure and rendering it more accessible to microbial activity. Techniques like UV irradiation, thermal treatments, and oxidation have been shown to alter the structure of polyethylene and enhance its biodegradability. These treatments can be combined with microbial or enzymatic (Herrero Acero et al. 2011) degradation approaches to improve the overall biodegradation process (Pluta and Krasowska 2020).

It is essential to acknowledge that conventional biodegradation of polyethylene is typically a slow process, often taking years or even decades for complete degradation to occur. Moreover, the effectiveness of biodegradation can be influenced by several factors, including environmental conditions, the presence of oxygen, moisture content, temperature, and the specific microbial community present. Scientists and researchers are continuing to explore and develop more efficient and sustainable methods for the biodegradation of synthetic polyethylene (Musa et al. 2020) (Fig. 11.2).

11.5 Advanced Biodegradation Mediated Techniques

Biodegradation is the natural process by which microorganisms break down organic materials into simpler compounds, such as water, carbon dioxide, and other substances. Advanced biodegradation-mediated techniques are methods used to enhance and speed up the biodegradation process, particularly in the treatment of environmental contaminants such as pollutants, hazardous wastes, and industrial chemicals. Here are some examples of advanced biodegradation-mediated techniques:

11.5.1 Bioremediation

Bioremediation is the process of using microorganisms to break down environmental contaminants such as petroleum hydrocarbons, solvents, and heavy metals (Kanungo et al. 2023). Bioremediation can be carried out in situ, where the contaminated material is treated on-site, or ex-situ, where the contaminated material is removed from the site and treated elsewhere. Bioremediation is a cost-effective and environmentally friendly method of cleaning up contaminated sites (Al-Musawi et al. 2020).

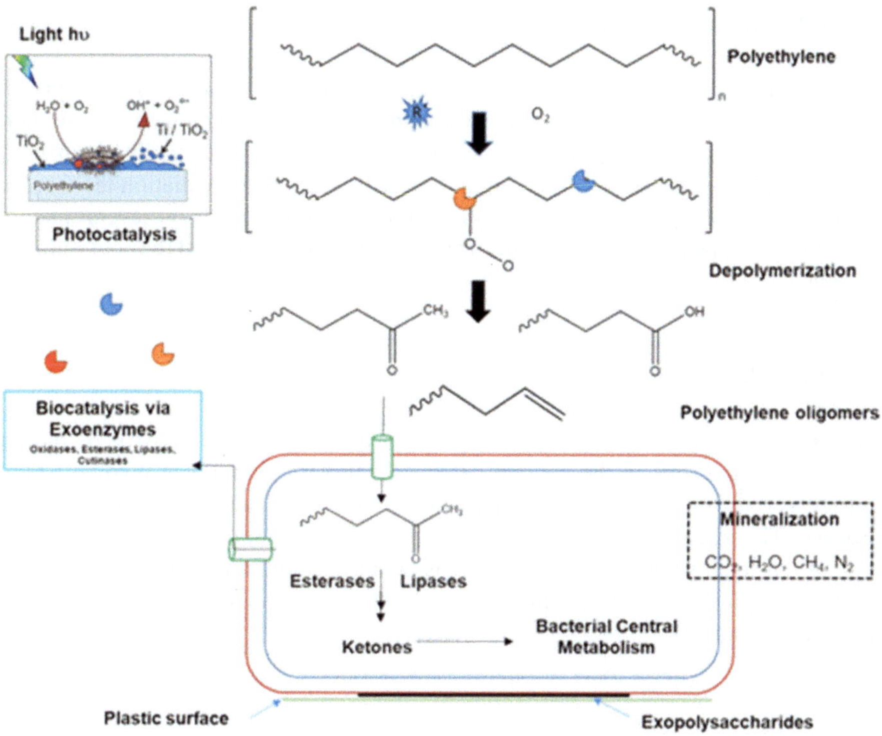

Fig. 11.2 Schematic representation of PE degradation (Ghatge et al. 2020)

The success of bioremediation depends on several factors, including the type of contaminants present, the environmental conditions, and the type and activity of the microorganisms used.

11.5.1.1 Natural Attenuation

Natural attenuation is a passive bioremediation technique that relies on the naturally occurring microorganisms in the environment to degrade contaminants (Kawai et al. 2014). This technique is typically used in situations where the contaminant concentration is low, and the site is stable. Natural attenuation works by using the existing microbial population to break down contaminants into less toxic compounds. The microbes involved in natural attenuation can be aerobic or anaerobic, depending on the type of contaminant present and the environmental conditions.

11.5.1.2 Bioaugmentation

Bioaugmentation is an active bioremediation technique that involves adding specific microorganisms to a contaminated site to enhance the biodegradation process (Behera et al. 2022). Microorganisms can be naturally occurring or genetically engineered. Bioaugmentation can be used in conjunction with other bioremediation techniques, such as biostimulation. The success of bioaugmentation depends on several factors, including the type of microorganisms used, the site conditions, and the contaminants present.

11.5.1.3 Biostimulation

Biostimulation is a technique that stimulates the growth and activity of indigenous microorganisms in the contaminated site. Biostimulation can involve adding nutrients, oxygen, or other substances to the site to enhance the biodegradation process. The success of biostimulation depends on several factors, including the type of nutrients added, the site conditions, and the contaminants present. Biostimulation is typically used in conjunction with other bioremediation techniques, such as bioaugmentation.

11.5.2 Phytoremediation

Phytoremediation is a technique that uses plants to remove, stabilize, or detoxify contaminants from the environment. This approach offers several advantages over traditional remediation techniques, such as low cost, minimal disturbance to the environment, and the potential for long-term sustainability. Phytoremediation also offers aesthetic and ecological benefits, such as improving soil quality, enhancing biodiversity, and reducing erosion. Phytoremediation can be used for soil, groundwater, and air remediation, and there are several different phytoremediation techniques that can be employed depending on the specific contaminant and site conditions.

11.5.2.1 Phytoextraction

It is a phytoremediation technique that involves using plants to absorb and accumulate contaminants from the soil into their above-ground tissues, which are then harvested and disposed of. This approach is particularly useful for removing heavy metals, radionuclides, and organic contaminants from the soil. Phytoextraction works by using hyperaccumulating plants, which can absorb and tolerate high concentrations of contaminants. These plants can be grown on the contaminated site and allowed to take up the contaminants through their roots. The plants are then harvested, and the contaminants are removed from the site along with the plant biomass. Phytoextractions The advantages of phytoextraction include low cost, minimal site disruption, and the ability to remove contaminants from the soil without having to excavate and dispose of the contaminated soil. However, the effectiveness of phytoextraction depends on several factors, such as the type and concentration of contaminants, the availability of hyperaccumulating plants, and the harvesting and disposal methods used.

11.5.2.2 Phyto Stabilization

Phytostabilization is a phytoremediation technique that involves using plants to stabilize contaminants in the soil, reducing their mobility and availability for uptake by plants and animals. This approach is particularly useful for stabilizing metals and metalloids in the soil, reducing the risk of contamination to groundwater and other environmental receptors. Phyto stabilization works by using plants that can secrete substances that can bind with the contaminants and immobilize them in the soil. These plants can be grown on the contaminated site and allowed to take up the contaminants through their roots. The contaminants are then bound to the plant tissues or the surrounding soil, reducing their mobility and availability for uptake by other plants and animals. The advantages of Phyto stabilization include low cost, minimal site disruption, and the ability to stabilize contaminants in the soil without having to remove them from the site. However, the effectiveness of phytostabilization depends on several factors, such as the type and concentration of contaminants, the availability of appropriate plants, and the long-term stability of the immobilized contaminants.

11.5.2.3 Phytodegradation

Phytodegradation is a phytoremediation technique that involves using plants to degrade organic contaminants in the soil or water. This approach is particularly useful for degrading organic compounds, such as petroleum hydrocarbons, pesticides, and herbicides, which are difficult to remediate using traditional techniques. Phytodegradation works by using plants that can produce enzymes that can break down organic contaminants into harmless byproducts. These plants can be grown on the contaminated site or in a hydroponic system, and the contaminants are degraded as they pass through the plant tissues. The advantages of phytodegradation include low cost, minimal site disruption, and the ability to degrade organic contaminants without generating toxic byproducts.

11.5.3 Composting

Composting is a natural process that converts organic materials, such as food waste and yard waste, into a nutrient-rich soil amendment that can be used to improve soil quality and plant growth. This is an environmentally friendly way to dispose of organic waste and can be done on a small scale in a backyard or on a larger scale in commercial operations. Composting works by providing the right conditions for microorganisms, such as bacteria and fungi, to break down the organic materials into a stable, humus-like substance. The microorganisms require oxygen, moisture, and a source of carbon and nitrogen to carry out the composting process.

In conclusion, composting is an effective and environmentally friendly way to dispose of organic waste and improve soil quality. Whether you are composting in a backyard or on a larger scale, the composting process requires attention to the C: N ratio, moisture content, and oxygen supply to ensure that the microorganisms can carry out the composting process efficiently. The benefits of composting include reducing waste, improving soil quality, and saving money.

11.5.4 Anaerobic Digestion

Anaerobic digestion is a natural biological process that breaks down organic matter in the absence of oxygen to produce biogas and a nutrient-rich fertilizer. The process is carried out by a diverse community of microorganisms, including bacteria, archaea, and fungi. The anaerobic digestion process typically involves four stages:

11.5.4.1 Hydrolysis

In the first stage, complex organic compounds are broken down into simpler molecules, such as sugars, amino acids, and fatty acids, by enzymes produced by bacteria.

11.5.4.2 Acidogenesis

In the second stage, the simpler molecules produced in the hydrolysis stage are broken down further into organic acids, such as acetic acid and butyric acid, by a different group of bacteria.

11.5.4.3 Acetogenesis

In the third stage, the organic acids produced in the acidogenesis stage are converted into acetate, hydrogen, and carbon dioxide by a group of bacteria called acetogens.

11.5.4.4 Methanogenesis

In the final stage, the acetate, hydrogen, and carbon dioxide produced in the previous stages are converted into methane and carbon dioxide by a group of archaea called methanogens. The biogas produced by the methanogens can be used as a renewable energy source for heating, electricity generation, or transportation. The nutrient-rich fertilizer produced during the anaerobic digestion process can also be used as a soil amendment to improve soil fertility. Anaerobic digestion can be carried out on a small scale, such as in a backyard digester, or on a larger scale in commercial operations. The process is particularly suited to organic wastes with high levels of biodegradable material, such as food waste, agricultural waste, and wastewater.

In conclusion, anaerobic digestion is an effective and environmentally friendly way to produce renewable energy and nutrient-rich fertilizer from organic waste. The process is carried out by a diverse community of microorganisms and requires attention to factors such as substrate type, temperature, and pH to ensure that the process is carried out efficiently.

11.6 Pros and Cons

Pros of Biodegradation of Synthetic Polyethylene Terephthalate (Yoshida et al. 2016) (PET) into Bis-(2-Hydroxyethyl) Terephthalate (BHET): Environmental benefits: PET waste is a significant contributor to plastic pollution (Pintus and Fantoni 2018) in the environment, and biodegradation of PET into BHET can reduce pollution, making it a more sustainable and eco-friendly option. Resource conservation (Gertler and Nägeli 2019): Recycling of PET into BHET reduces the demand for virgin materials used in the production of PET and its precursors, leading to the conservation of natural resources (Koshti et al. 2018). Cost-effective: BHET is a valuable compound that can have uses as a precursor for the synthesis of various chemicals, making PET recycling a cost-effective solution for PET waste management (Ragaert et al. 2017).

Energy-saving: Recycling of PET (Al-Musawi et al. 2002) into BHET requires less energy than the production of virgin materials, leading to reduced greenhouse gas emissions and energy consumption (Wei and Zimmermann 2017). Cons of Biodegradation of Synthetic Polyethylene Terephthalate (PET) into Bis-(2-Hydroxyethyl) Terephthalate (BHET): Limited scalability: Current methods for PET biodegradation into BHET are not yet scalable for large-scale industrial applications, making it difficult to implement on a commercial scale. Technical challenges: The process of PET biodegradation into BHET involves a very complex set of chemical reactions and requires specialized equipment and expertise, which can possibly increase the cost and technical challenges. Impurities: The BHET produced from PET biodegradation may contain impurities, such as catalyst residues, which can limit its applications in certain industrial processes. Competing recycling technologies: There are competing recycling technologies for PET, such as mechanical recycling, which have their very own pros and cons, and the choice of recycling technology depends on several factors, including the type of waste and end-use applications.

11.7 Future Scope

The biodegradation of synthetic polyethylene terephthalate (PET) into bis-(2-hydroxyethyl) terephthalate (BHET) (Shahanaz et al. 2020) has significant future scope in several areas, including:

Sustainable waste management: The biodegradation of PET into BHET provides a sustainable and environmentally friendly approach to PET waste management, reducing plastic pollution and conserving natural resources.

Industrial applications: BHET is a valuable compound that can be used as a precursor for the synthesis of various chemicals, such as polyester (Wei et al. 2014) resins and coatings, making it a potential replacement for petroleum-based raw materials. Bioplastics production: BHET can be further converted into bio-based polyesters, such as polybutylene terephthalate (Shirke et al. 2018) (PBT), which have applications in the production of bioplastics, reducing the reliance on fossil fuels (Ronkvist et al. 2010). Biodegradable plastics: BHET can be used as a building block for biodegradable polymers, such as polyhydroxyalkanoates (PHAs), which have applications in the production of biodegradable plastics. Technology development: The biodegradation of PET into BHET is still in the early stages of development, and further research is required to improve the efficiency and scalability of the process, which presents opportunities for technological advancements and innovation. In conclusion, the biodegradation of PET into BHET has a significant future scope in several areas (Sudhakar et al. 2008), including sustainable waste management, industrial applications, bioplastics production, biodegradable plastics, and technology development.

11.8 Conclusion

In the realm of sustainable microbial technology, there is a growing focus on utilizing it for the bioremediation of synthetic microfibers, as well as its potential contribution to the bioeconomy. This summary aims to provide an overview of sustainable microbial technology for synthetic microfiber bioremediation and its role in the bioeconomy, with a specific focus on the biodegradation of Synthetic Polyethylene Terephthalate textile (PET) into Bis-(2-Hydroxyethyl) Terephthalate (BHET).

These microfibers, commonly found in textiles and other materials, are known for their slow degradation and accumulation in various ecosystems. To combat this issue, researchers have explored the potential of microbial degradation as an environmentally friendly approach. One specific area of interest within sustainable microbial technology is the biodegradation of PET, a common type of synthetic microfiber. PET is widely used in the production of clothing, carpets, and packaging materials, among others.

In conclusion, sustainable microbial technology offers a promising avenue for the bioremediation of synthetic microfibers. The focus on PET biodegradation into BHET highlights the potential of microbial processes to address the environmental challenges posed by microfiber waste. Moreover, this technology aligns with the principles of the bioeconomy by converting waste into valuable materials. Continued research and optimization efforts in this field hold the key to unlocking sustainable solutions for microfiber waste management and the development of a circular economy.

References

Adrio LA, Rodríguez Couto S, Sanromán MÁ (2005) Ability of Geotrichum sp. strains for decolourisation and detoxification of textile dyes. Biotechnol Lett 27(14):1001–1005

Al-Musawi TJ, Al-Hajaj A, Al-Sabagh AM, Al-Hinai MA (2020) Recycling of polyethylene terephthalate (PET) waste to bis-(2-hydroxyethyl) terephthalate (BHET) and terephthalic acid (TPA). J Clean Prod 268:122131

Awaja F, Pavel D (2005). Recycling of PET. Eur Polym J 41(7):1453–1477.

Balasubramanian S, Natarajan K (2013) Biodegradation of polyethylene terephthalate (PET) by Aspergillus sp. and Bacillus sp. from marine environment. J Appl Polym Sci 128(2):1039–1045

Behera ID, Basak G, Kumar RR, Sen R, Meikap BC (2020) Treatment of petroleum refinery sludge by petroleum degrading bacterium Stenotrophomonas pavanii IRB19 as an efficient novel technology. J Environ Sci Health Part A 56:1–13. https://doi.org/10.1080/10934529.2020.1866924

Behera ID, Nayak M, Biswas S, Meikap BC, Sen R (2021) Enhanced biodegradation of total petroleum hydrocarbons by implementing a novel two-step bioaugmentation strategy using indigenous bacterial consortium. J Environ Manage 292:112746. https://doi.org/10.1016/j.jenvman.2021.112746

Behera ID, Nayak M, Mishra A, Meikap BC, Sen R (2022) Strategic implementation of integrated bioaugmentation and biostimulation for efficient mitigation of petroleum hydrocarbon pollutants from terrestrial and aquatic environments. Mar Pollut Bull 177:113492. https://doi.org/10.1016/j.marpolbul.2022.11349

Chen Z, Wang Y, Cheng Y, Wang X, Tong S, Yang H, Wang Z (2020) Efficient biodegradation of highly crystallized polyethylene terephthalate through cell surface display of bacterial PETase. Sci Total Environ 709:136138

de Castro AML, Tavares LBB (2021) Status and future trends in the recycling of polyethylene terephthalate (PET) waste. J Environ Manag 278:111560

Gertler C, Nägeli I (2019) Comparative life cycle assessment of PET bottle waste management options: recycling, waste-to-energy, landfilling, and littering. Resour Conserv Recycl 142:235–246

Ghanbari M, Rezaei S, Karimi K (2018) Biodegradation of polyethylene terephthalate in medium containing seawater and/or manganese salts by newly isolated bacteria from the Persian Gulf. J Polym Environ 26(5):2075–2086

Ghatge S, Yang Y, Ahn JH (2020) Biodegradation of polyethylene: a brief review. Appl Biol Chem 63:27

Ghosh S, Kumari A, Das P (2018) Biodegradation of polyethylene terephthalate fibers by Pseudomonas and Bacillus species. Appl Biochem Biotechnol 185(1):239–252

Herrero Acero E, Ribitsch D, Steinkellner G, Gruber K, Greimel KJ, Eiteljoerg I, Trotscha E, Wei R, Zimmermann W, Zinn M, Cavaco-Paulo A, Freddi G, Schwab H, Guebitz G (2011) Enzymatic surface hydrolysis of PET: effect of structural diversity on kinetic properties of cutinases from Thermobifida. Macromolecules 44(12):4632–4640

Hirth T, Schmid J, Sieber V, Zinn M (2016) Enzymatic hydrolysis of polyethylene terephthalate films in a membrane reactor. Green Chem 18(15):4208–4218

Ion S, Voicea S, Sora C, Gheorghita G, Tudorache M, Parvulescu VI (2021) Sequential biocatalytic decomposition of BHET as valuable intermediator of PET recycling strategy. Catal Today 366:177–184

Kanungo J, Sahoo S, Bal M, Behera ID (2023) Performance of bioremediation strategy in waste lubricating oil pollutants: a review. Geomicrobiol J. https://doi.org/10.1080/01490451.2023.2245395

Kawai F, Kawabata T, Oda M (2014) Current knowledge on enzymatic PET degradation and its possible application to waste stream management and other fields. Appl Microbiol Biotechnol 98(12):563–572

Khan SA, Anjum MN (2021) Advances in enzymatic hydrolysis of polyethylene terephthalate (PET): an eco-friendly approach for waste management. J Environ Manage 289:112502

Koshti R, Mehta L, Samarth N (2018) Biological recycling of polyethylene terephthalate: a mini review. J Polym Environ 26:3520–3529

Kuroda K, Narihiro T, Nakaya Y, Noguchi TQ, Maeda R, Nobu MK, Ohnishi Y, Kumaki Y, Aizawa T, Satoh H (2022) Elucidation of the biodegradation pathways of bis (2-hydroxyethyl) terephthalate and dimethyl terephthalate under anaerobic conditions revealed by enrichment culture and microbiome analysis. Chem Eng J 450:137916

Lenz R, Gardères J, Brzezinski R (2019) Hydrolysis of polyethylene terephthalate fibers by alkaline and enzymatic methods. J Polym Environ 27(2):267–276

Lv R, Chen Y, Liu X, Liu C, Du L (2019) Enhanced hydrolysis of polyethylene terephthalate by adsorption of PETase on functionalized graphene oxide. Biores Technol 273:391–397

Mohanta S, Pradhan B, Behera ID (2023) Impact and remediation of petroleum hydrocarbon pollutants on agricultural land: a review. Geomicrobiol J 1–5. https://doi.org/10.1080/01490451.2023.2243925

Montazer Z, Zhong C, Yeow JT, Tang Y (2021) Biodegradation of polyethylene terephthalate (PET) waste using Bacillus sp. enriched consortium. J Environ Manag 281:111887.

Musa N, Ghazali MN, Naim MN, Samsudin MW, Jamaludin NA (2020) Sustainable degradation of polyethylene terephthalate (PET) by UV-assisted photocatalysis and subsequent biodegradation by Pseudomonas sp. 2.9. J Environ Manag 262:110279

Narayan R, Miyakawa T (2017) Enzymatic hydrolysis of polyethylene terephthalate. Macromol Rapid Commun 38(3):1600623

Oda K, Yamagami Y, Yoshida S, Taniguchi I, Tanaka Y, Miyamoto K (2020) Enzymatic hydrolysis of polyethylene terephthalate using TPA-forming carboxylesterase mutants. Appl Microbiol Biotechnol 104(4):1595–1604

Pintus V, Fantoni GP (2018) Advances in the biodegradation of polyethylene terephthalate: a review. Environ Sci Pollut Res 25(14):13709–13718

Pluta M, Krasowska K (2020) Chemical recycling of polyethylene terephthalate (PET): a review. Recycling 5(3):22

Quartinello F, Luu W, Tosta G, Bernardi L, Campana Filho SP (2018) Biodegradation of poly (ethylene terephthalate) (PET) by thermobifida fusca. J Polym Environ 26(10):4135–4144

Ragaert K, Delva L, Van Geem K, Mechanical Recycling of PET (2017) State of the art and recent developments. Macromol Mater Eng 302(10):1700263

Ronkvist ÅM, Nordgren N, Albertsson AC (2010) Laccase-catalyzed surface modification of a polyester film by the phenolic radical from TEMPO. Macromolecules 43(24):10348–10354

Ronkvist ÅM, Xie W, Lu W, Gross RA (2009) Cutinase-catalyzed hydrolysis of poly (ethylene terephthalate). Macromolecules 42(6):5128–5138

Shahanaz K, Ghosal D, Ghosh SK (2020) Isolation and characterization of a novel polyethylene terephthalate (PET) degrading bacterium, Gordonia sp. strain IITR100. J Environ Manag 262:110328

Shirke AN, Basu B, Madras G (2018) Biodegradation of polyethylene terephthalate (PET) by different dimethyl terephthalate (DMT) oligomers and the corresponding enzymatic hydrolysis: a molecular simulation study. ACS Sustain Chem & Eng 6(2):2756–2764

Södergård A, Stolt M, Pitkänen P (2002) Degradation of polyethylene terephthalate at low temperatures under controlled composting conditions. J Environ Polym Degrad 10(3):137–1461

Su W, Zhang J, Zhang X (2020) Isolation and characterization of a polyester-degrading Pseudomonas sp. strain from the gut of a polyethylene terephthalate (PET)-eating worm. Front Microbiol 11:594012

Sudhakar M, Doble M, Murthy PS (2008) Biodegradation of polyester polyurethane by Aspergillus sp. and Bacillus sp. Int Biodeter

Tournier V, Topham CM, Gilles A, David B, Folgoas C, Moya-Leclair E, Kamionka E, Desrousseaux ML, Texier H, Gavalda S, Cot M, Guémard E, Dalibey M, Nomme J, Cioci G, Barbe S, Chateau M, André I, Duquesne S, Marty A, Pelloux-Leon N (2020) An engineered PET depolymerase to break down and recycle plastic bottles. Nature 580(7802):216–219

Tribedi P, Sil AK (2013) Cellulase-assisted biodegradation of polyethylene terephthalate. Int Biodeterior Biodegrad 85:248–256

Webb HK, Arnott J, Crawford RJ, Ivanova EP (2013a) Plastic degradation and its environmental implications with special reference to poly (ethylene terephthalate). Polymers 5(1):1–18

Webb HK, Crawford RJ, Ivanova EP (2013b) Plastic degradation and bioremediation of waste poly (ethylene terephthalate) with microbial cultures. Appl Environ Microbiol 79(8):2478–2484

Wei R, Zimmermann W (2017) Microbial enzymes for the recycling of recalcitrant petroleum-based plastics: how far are we? Microb Biotechnol 10(6):1308–1322

Wei R, Oeser T, Billig S, Zimmermann A, Zimmermann W (2014) A disulfide bridge in the calcium binding site of a novel thermostable polyesterase is essential for polyester hydrolysis. Appl Environ Microbiol 80(24):7325–7333

Yang Y, Yang J, Wu WM, Zhao J, Song Y, Gao L (2014) Biodegradation and mineralization of polystyrene by plastic-eating mealworms: part 1. Chemical and physical characterization and isotopic tests. Environ Sci & Technol 48(20):12080–12086

Yoshida S, Hiraga K, Takehana T, Taniguchi I, Yamaji H, Maeda Y, Toyohara K, Miyamoto K, Kimura Y, Oda K (2016) A bacterium that degrades and assimilates poly (ethylene terephthalate). Science 351(6278):1196–1199

Zimmermann W, Billig S, Kroutil M (2021) Biodegradation of polyethylene terephthalate. In: Handbook of environmental chemistry. Springer, pp 1–26

Chapter 12
Circular Economy Model for Petroleum Waste and Its Implementation in India

Lipika Parida, Anil Kumar Murmu, and Veda Prakash

Abstract Petroleum waste is a major issue in the petroleum industry, affecting land and water resources. It is produced through various processes, such as exploration, drilling, refinery, spent petroleum products, and oil spills. Proper management of waste can lead to environmental and health hazards, contaminating water sources, threatening animal life, and causing ecosystems to suffer. To manage petroleum waste effectively, strategies like improved drilling techniques, efficient refining processes, and promoting recyclability are implemented. The Circular Economy concept, first proposed in the 1960s, promises to maximize resource utilization while reducing waste output. It emphasizes sustainability and the importance of fulfilling the current generation's needs while safeguarding future generations' needs. The model encourages the use of clean technologies and renewable energy sources to mitigate environmental consequences and decrease dependency on non-renewable resources. In a developing and densely populated nation like India, the Circular Economy model can generate new business prospects, foster innovation, and facilitate job creation in sectors like recycling, renewable energy, and green technologies. It is particularly important for managing petroleum waste, which poses a significant environmental threat and requires resource conservation, waste reduction, pollution prevention, and energy recovery. Collaboration among governments, industries, research institutions, and communities is essential for developing policies, regulations, and technologies that facilitate waste reduction, resource recovery, and sustainable practices across the petroleum value chain.

Keywords Circular economy · Petroleum products · Recycling · Sustainable · India

L. Parida (✉) · A. K. Murmu · V. Prakash
Department of Chemical Engineering, Veer Surendra Sai University of Technology, Burla 768018, Odisha, India
e-mail: lparida_chemical@vssut.ac.in

I. D. Behera and A. P. Das (eds.), *Impact of Petroleum Waste on Environmental Pollution and its Sustainable Management Through Circular Economy*, Environmental Science and Engineering, https://doi.org/10.1007/978-3-031-48220-5_12

12.1 Introduction

Petroleum waste comprises all waste materials that are produced throughout the processes of petroleum exploration, production, refining, and utilization. The rate of generation has been experiencing an upward trend due to the growing energy demand. There are four main forms of waste in the petroleum industry, which vary depending on the stage of the petroleum lifecycle and the specific processes involved. These forms include exploration and drilling waste, refinery waste, spent petroleum products, and oil spills and leaks. Exploration and drilling waste comprise drilling muds, cuttings, and produced water that is known to harbor diverse contaminants, including heavy metals, hydrocarbons, and drilling additives. Refinery waste comprises various waste streams that are produced by petroleum refineries, including spent catalysts, sledges, and wastewater. These waste streams contain oil, chemicals, and heavy metals (Bhattacharyya and Shekdar 2003; Behera et al. 2020). Improper management of these waste materials can potentially result in environmental and health hazards. Spent petroleum products are classified as hazardous waste. These include utilized motor oil, lubricants, and various products derived from petroleum. Oil leaks and spills can occur as a result of accidental spills that happen during the transportation, storage, or extraction of petroleum. Improper disposal of waste has the potential to contaminate water sources, pose a threat to animal life, and result in enduring consequences for ecosystems (Behera et al. 2021). To prevent the contamination of land and water resources, it is crucial to effectively manage and dispose of these wastes properly.

Various strategies and practices are implemented to effectively manage petroleum waste and minimize its environmental impact. The reduction and prevention of waste in the petroleum lifecycle can be achieved through the implementation of various technologies and methods. These include the adoption of improved drilling techniques, the utilization of efficient refining processes, and the incorporation of product designs that promote recyclability. By implementing these measures, waste generation can be decreased or even avoided altogether. The process of recycling and recovering valuable materials from petroleum waste has the potential to decrease reliance on raw resources and mitigate environmental issues. The objective is achieved by implementing and promoting recycling programs targeting the disposal of used petroleum products, including motor oil and lubricants. The toxicity of petroleum waste can be mitigated or converted through the application of suitable treatment techniques before disposal. It is imperative to adhere to rigorous environmental protocols when disposing of waste to prevent the contamination of land and water sources. In the event of an oil spill or leak, quick response and effective cleanup protocols are of the utmost importance. The tasks encompassed within this scope involve the mitigation of oil spills, the extraction and treatment of recovered oil, and the restoration of ecosystems impacted by such incidents. The development of regulations and guidelines for petroleum waste management and disposal is a crucial responsibility of governments. The purpose of these guidelines is to ensure the safe handling, treatment, and disposal of exercises. Ongoing research and development endeavors are necessary to foster the creation of cutting-edge technologies and solutions aimed at reducing petroleum waste, improving recycling methods, and advancing sustainable alternatives to petroleum-derived products. To ensure the efficient management of petroleum waste, it is essential to employ a comprehensive approach that incorporates regulatory measures, industry best practices, and public awareness. This multifaceted strategy aims to mitigate environmental damage and facilitate the shift towards a more sustainable energy system.

The classic linear economic model involves the collection and transformation of raw resources into products that customers use until they discard them as waste, with little regard for the environmental imprint or effects. Since the Industrial Revolution, the linear economic model has been used in a variety of industrial sectors to prioritize economic gains over all other factors (CIASULLO et al. 2019). According to the literature, it has been found that this particular model has been associated with significant environmental issues. These issues include natural resource depletion, trash build-up, and climate change (Velenturf and Purnell 2021). However, initiatives to develop alternative economic models came up, resulting in the Circular Economy (CE) (Reike et al., 2018) to prevent such consequences. The idea of CE was proposed in the 1960s as a sustainable management approach. It can minimize waste generation and mitigate the effects of global warming (CIASULLO et al. 2019).

12.2 Circular Economy Model

The CE model is a framework for optimizing resource use and reducing waste output. It operates on the ideas of material reduction, reuse, recycling, and products throughout. The concept operates on the principle of implementing a closed-loop system, wherein products and materials are maintained in active use for an extended duration through the application of diverse strategies. This approach effectively prolongs the life cycle of products. In practical terms, it involves minimizing waste generation to the greatest extent possible. Whenever a product approaches the end of its functional lifespan, recycling is employed to retain its materials within the economy whenever feasible. The aforementioned items have the potential to be repeatedly utilized, resulting in the generation of additional value (Kalmykova et al. 2018; Ellen MacArthur Foundation (EMF) News 2015a, b). The Circular Economic approach represents a deviation from the conventional linear economic model that follows a take-make-consume-throw-away sequence. The model is dependent on abundant supplies of cost-effective and readily available materials and energy sources. Therefore, unlike the linear economy, the CE operates on the premise that waste is non-existent and instead regarded as a valuable resource, as depicted in Fig. 12.1 (United Nations Development Programme, NY, USA 2019). It emphasizes sustainability and the importance of fulfilling the current generation's needs while safeguarding future generations' needs.

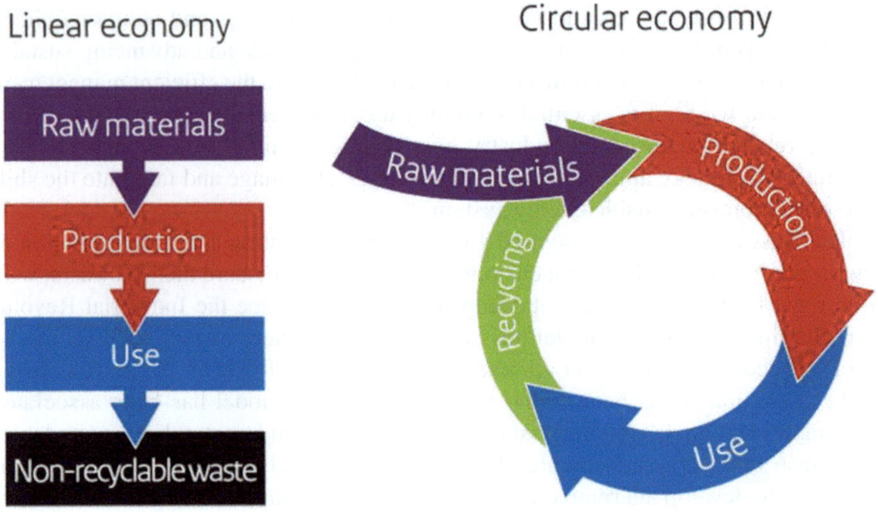

Fig. 12.1 Linear and CE Model (United Nations Development Programme, NY, USA 2019)

12.2.1 Fundamental Characteristics of a Circular and Sustainable Economic Model

The circular and sustainable economic model is characterised by several fundamental attributes. These characteristics are critical for ensuring the economic system's durability and environmental sustainability. The products are engineered with a focus on durability, ease of repair, and the use of recyclable materials that can be safely reintegrated into the environment. The model facilitates the optimization of resource utilization by minimizing waste and maximizing material value through various strategies such as recycling, remanufacturing, and refurbishment. The model encourages the utilization of renewable energy sources and the adoption of clean technology to reduce environmental implications and dependency on non-renewable resources. The primary objective of this initiative is to minimize waste generation by implementing enhanced production processes, optimizing product design, and influencing consumer behavior. Furthermore, it places significant emphasis on the implementation of appropriate waste management practices, encompassing essential aspects such as recycling, composting, and the secure disposal of hazardous materials. Manufacturers and producers are advised to assume accountability for the complete lifecycle of their products, encompassing disposal and recycling, by implementing measures like product take-back programs and eco-design requirements. The circular and sustainable economy model entails the collaboration of businesses, governments, consumers, and other stakeholders to foster innovation, exchange knowledge, and establish supportive policies and regulations. The model acknowledges that the implementation of a circular and sustainable economy has the potential to generate new business prospects, foster innovation, and facilitate job creation in sectors such as recycling, renewable energy, and green technologies (Annarelli et al. 2016). The spreading of knowledge and the provision of education to individuals and communities regarding the advantages and methodologies of a circular and sustainable economy are of utmost importance to facilitate its widespread acceptance and encourage changes in behavior. By embracing a circular and sustainable economy model, the shift from the linear "take-make-waste" approach can be achieved. This shift will result in the establishment of a resilient and environmentally conscious economic system that promotes long-term prosperity and ensures the preservation of the planet for future generations.

12.2.2 Benefits of the CE Model

12.2.2.1 Protects the Environment

CE has the capacity to significantly reduce the utilization of natural resources, minimize disruptions to landscapes and habitats, and actively promote the protection of biodiversity by implementing product reuse and recycling practices. Furthermore, the establishment of a CE framework can significantly lower yearly greenhouse gas emissions. In order to address the environmental impacts, it is recommended to prioritize the development of products that are efficient and sustainable right from the beginning. This approach is effective in reducing energy and resource consumption, as studies indicate that more than 80% of the overall environmental effect of a product is established during the design stage (Diaz et al. 2021). In addition, the adoption of durable products that can be reused, upgraded, and repaired would have a significant impact on reducing waste generation. The packaging industry is currently undergoing substantial growth. The primary goal is to tackle the problem of excessive packaging and improve its design to promote reuse and recycling (Coelho et al. 2020).

12.2.2.2 Reduces Raw Material Dependence

The global population is experiencing steady growth, leading to an increase in raw materials demand. There are limitations on the supply of necessary raw materials. The limited availability of resources also results in countries relying on external sources for their raw materials. The process of recycling raw materials helps to address and reduce the potential risks associated with supply, including price fluctuations, limited availability, and dependence on imports. This is particularly important in the case of key raw resources that are required for the manufacturing of technologies such as batteries and electric motors. These technologies are crucial to meeting climate-change targets (Malinauskaite et al. 2021).

12.2.2.3 Creates Jobs and Saves Consumers Money

The implementation of a CE model has the ability to enhance competitiveness, drive innovation, foster economic growth, and generate employment opportunities. According to a press release from the European Commission (European Commission—Press release 2010), the development of robust CE measures in Europe has the potential to generate a 0.5% growth in the European Union's GDP by the year 2030. Additionally, this initiative could result in the creation of approximately 700,000 new job opportunities. The adoption of CE principles in the redesign of materials and products has the potential to stimulate innovation in various sectors of the economy. Consumers will benefit from the availability of enhanced, long-lasting products that incorporate innovative features, thereby improving their overall quality of life and delivering long-term cost savings.

12.2.3 Need for CE Model for Petroleum Waste

Petroleum waste poses a substantial environmental threat, necessitating the implementation of a CE model to effectively manage and mitigate its adverse effects. The adoption of the CE model offers several avenues to address and reduce petroleum waste. Figure 12.2 illustrates the potential applications of petroleum industry sludge for sustainable development (Jagaba et al. 2022).

Resource Conservation: Petroleum is a finite and valuable resource that must be conserved. The adoption of CE principles enables the reclassification of petroleum waste from a discarded waste material to a valuable input. This process facilitates the retrieval, reutilization, and recycling of petroleum-derived materials, thereby prolonging their durability and diminishing the demand for new resources. This model conserves natural resources while reducing the environmental effect of petroleum production. The various applications of petroleum waste are summarized in Fig. 12.2.

Waste Reduction and Pollution Prevention: Petroleum waste, including but not limited to oil spills, industrial by-products, and used oil, poses significant risks to both human health and ecosystems (Kanungo et al. 2023). The implementation of a CE approach can effectively reduce waste generation and prevent pollution. This is achieved through the promotion of efficient utilization of petroleum products, the adoption of appropriate disposal and recycling methods, and the encouragement of cleaner alternative development.

Energy Recovery: The presence of substantial energy within petroleum waste presents an opportunity for harnessing this energy through the utilization of diverse technologies. The CE approach prioritizes energy recovery from petroleum waste instead of solely considering it as a disposal issue. Various techniques, including thermal conversion, anaerobic digestion, and pyrolysis, can be utilized to transform waste into valuable energy sources. This not only decreases dependence on fossil fuels but also encourages the adoption of sustainable energy practices.

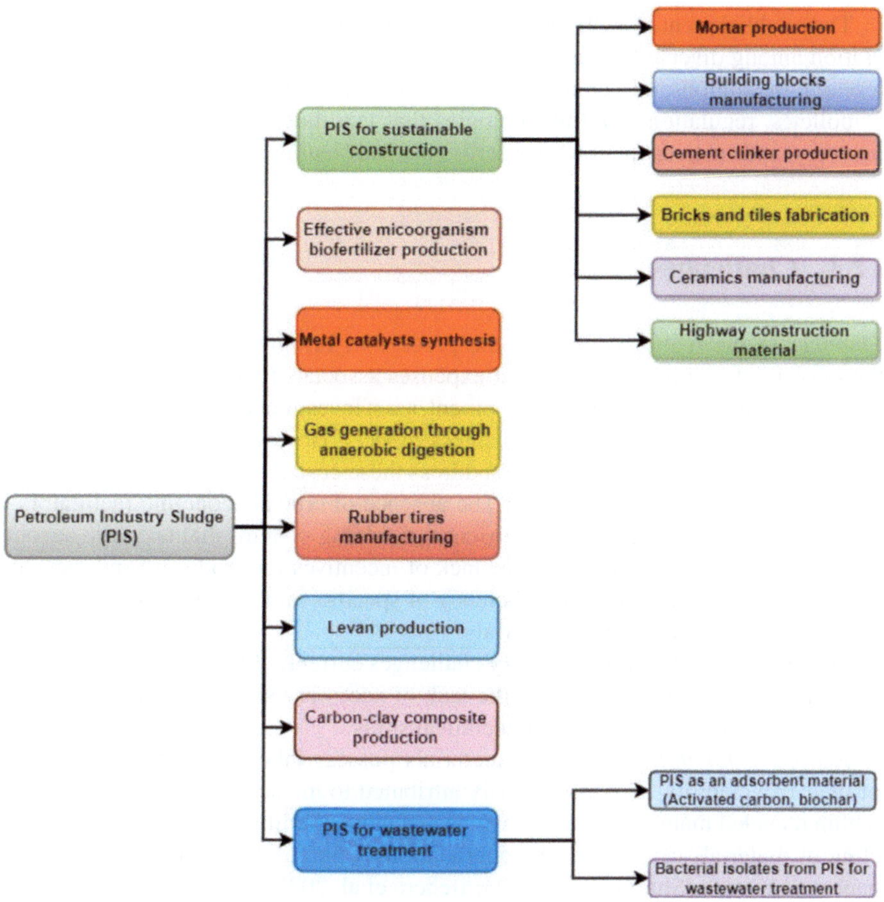

Fig. 12.2 Possible utilisations of petroleum industry waste for sustainable development (Jagaba et al. 2022)

Economic Opportunities: The implementation of a CE approach for managing petroleum waste has the potential to generate novel economic prospects. The stimulation of job growth, the fostering of technological advancements, and the establishment of a green economy are possible through the development of innovative recycling and recovery technologies, the establishment of waste management and recycling facilities, and the promotion of sustainable practices.

Climate change mitigation: This involves addressing the impact of petroleum extraction, production, and disposal on greenhouse gas emissions and subsequent climate change. The implementation of CE principles, such as waste reduction, recycling, and energy recovery, can effectively mitigate emissions related to petroleum waste. This is in line with worldwide initiatives to combat climate change and transition to a low-carbon economy.

The establishment of a CE for petroleum waste necessitates the imperative collaboration among diverse stakeholders. Collaboration among governments, industries, research institutions, and communities is essential for the formation and execution of policies, regulations, and technologies that facilitate waste reduction, resource recovery, and sustainable practices across the petroleum value chain.

12.3 Challenges

Finance: In the realm of finance, the process of bringing radical innovations to the mass market involves significant expenses associated with transitioning. These expenses include research and development, asset investments, and subsidy payments to support the promotion of new business models. However, the existing financial tools are not adequately equipped to address these costs.

Missing key economic enablers: The absence of crucial economic facilitators is evident in the following areas: the absence of pricing systems that take into account the complete environmental cost, the lack of incentives for producers and recyclers to collaborate in enhancing the efficiency of specific value chains, and the absence of markets for secondary raw materials.

Missing skills: One of the primary challenges currently hindering the adoption of CE in the oil and gas industries is the lack of necessary skills and knowledge. The workforce exhibits a deficiency in technical skills.

Recycling Limitations: Certain materials possess inherent limitations in terms of their recyclability. This is primarily attributed to the accumulation of impurities within recycled materials, such as metals and glass. Additionally, the repeated recycling of materials can lead to the degradation of fibres, particularly in the case of paper production (Sharma et al. 2023; Behera et al. 2022).

12.4 CE Model for Petroleum Waste in Developing and High-Populated Countries

The following is an exploration of a CE model specifically designed to address the management of petroleum waste in developing and densely populated countries.

The categorizations of "developing" and "highly populated" are subjective and subject to change based on a range of economic and demographic factors over time. Numerous organizations and institutions employ diverse methodologies and criteria to categorize nations. Several countries characterized by substantial populations, fall under the classification of developing nations. Population sizes have the potential to change over a period of time. Table 12.1 presents a compilation of developing and highly populated countries, accompanied by pertinent data sourced from World Bank population statistics (World Bank Open Data 2023).

Table 12.1 List of countries that are classified as developing and have a high population (World Bank Open Data 2023)

Country Name	Population (2021)	Population (2022)
India	1.366 billion	1.42 billion
China:	1.411 billion	1.411 billion
Brazil	213.99 million	215.313 million
Indonesia	276.36 million	275.501 million
Nigeria	211.4 million	218.541 million
Pakistan	225.20 million	235.842 million
Bangladesh	166.30 million	171.186 million
Nigeria	211.4 million	218.541 million

The management of petroleum waste in developing and densely populated nations necessitates a comprehensive approach that considers local conditions, initiatives to enhance capabilities, engagement of stakeholders, and the implementation of sustainable waste management practices. These strategies can help reduce the adverse consequences of petroleum waste, while simultaneously enhancing resource efficiency and promoting environmental sustainability. Managing petroleum waste in developing and densely populated nations poses significant challenges due to factors such as limited resources, inadequate infrastructure, and a lack of legal frameworks. Addressing these challenges is of greatest importance to mitigate the environmental and health risks associated with petroleum waste. The adoption of CE has garnered significant attention from various stakeholders including organizations, researchers, practitioners, and policymakers. However, existing literature suggests that its implementation is still in its nascent phase (Sharma et al. 2023).

The successful implementation of a circular and sustainable economy model for petroleum waste in emerging and densely populated nations necessitates the development of tailored strategies that consider the specific challenges and potential of each country (Mohanta et al. 2023). The primary objective is for the government to establish comprehensive waste management systems that encompass the collection, sorting, recycling, and disposal of petroleum waste. Efficient collection networks and infrastructure should be established to facilitate the management and transportation of waste. The implementation of CE principles has been organized by countries such as Brazil, which introduced the "Brazilian zero national waste policy" in 2010 (US-EPA 2012), and China, which approved the "CE promotion law" in 2009 (World Bank Open Data 2021). However, achieving complete circularity remains a significant challenge that requires further progress. Despite the evident advantages that businesses and economies can gain from embracing circular practices, there exist several barriers within the field of CE that are currently under investigation to comprehend the limited progress of this strategy. There has been an explosion of specialized literature covering all of the above issues in recent years (Sharma et al. 2023).

12.5 Implementing CE for Petroleum Waste in India

India is globally recognized as the most populous nation and ranks as the fifth-largest economy based on nominal GDP. According to the World Bank (2022), the population of India has witnessed significant growth, rising from 361 million in 1951 to approximately 1.4 billion in 2022. India currently holds the position of being the third-largest consumer of oil globally. The correlation between the advancement of the Indian economy and its energy requirements is anticipated to result in a rise in the utilization of oil and gas. To fulfill the increased demand, the Indian government has permitted 100% foreign direct investment in several industries, including refineries, natural gas, and petroleum products. The oil and gas industry plays a vital role in mitigating the impacts of climate change and natural disasters. Although the Indian government has implemented various initiatives to enforce the responsibility of reducing carbon emissions in the oil and gas industry, the implementation of the CE model is still in its incipient phase.

Various initiatives have been undertaken to advance the adoption and implementation of the CE model. These endeavors aim to foster a sustainable economic model that minimizes waste and maximizes resource efficiency. The following efforts have been made to promote the CE in India:

- The Union Budget 2023–24 of India emphasized the significance of CE and sustainable development. Over the past two to three years, the nation has been actively advocating for the implementation of resource circularity. The objective is to substitute natural resources with waste products of equal value that would otherwise be disposed of in landfills.
- In the year 2021, several committees were formed to expedite the transition of the nation from a linear economic model to a circular economic model.
- As per the annual report for 2022–2023 released by NITI Aayog, notable advancements have been made by the respective nodal ministries in implementing the CE Action Plans they had formulated.
- The official notification regarding Extended Producer Responsibility (EPR) on waste oil has been unveiled by the Ministry of Environment, Forest, and Climate Change (MoEFCC). The presence of pollutants in used oil, such as benzene, zinc, and cadmium, renders it capable of contaminating freshwater sources. Consequently, used oil is classified as a contaminant. The notification indicates that the EPR (Extended Producer Responsibility) policy applies to both producers and bulk generators. This includes businesses in the industrial sector, railroads, transportation agencies, electric power transmission organizations, and other relevant entities.
- The Central Pollution Control Board (CPCB) website guides stakeholders on the registration process. It is recommended that various entities, such as producers, collection agencies, recyclers, and waste oil importers, register on the website. The proposed initiative aims to facilitate government oversight of waste oil recycling and collection by integrating it within the existing industry framework. Additionally, it could impede the expansion of informal backyard recyclers, potentially

posing a threat to the environment due to their inadequate handling and disposal of waste oil.

- The Ministry of Environment, Forest and Climate Change (MoEFCC) has established explicit guidelines outlining the responsibilities of various entities involved in the management of waste oil. These entities include producers, importers of waste oil, collection agencies, recyclers, bulk generators, the Central Pollution Control Board (CPCB), State Pollution Control Boards (SPCB), states or union territories, municipalities, local self-governments, and others.
- The Bureau of Indian Standards (BIS) has been entrusted with the crucial responsibility of formulating suitable standards for re-refined oil, which is used by manufacturers of lubricants.
- The concept of environmental compensation has been integrated into the Extended Producer Responsibility (EPR) framework, along with penalties for any agency found to have provided false information to obtain EPR certifications. Furthermore, the legal action taken for failure to comply will be in addition to the required environmental compensation.

In summary, there is a need for enhanced systemic circularity of waste oil in India. To further reduce the industry's usage of resources, it is imperative to minimize the utilization of fresh oil (Down to Earth 2023).

12.6 Overview of Current Initiatives and Perspectives

Based on the available data from 2016, the proven oil reserves worldwide would be sufficient to sustain global output at 2016 levels for only 51 years. To explore the potential of bio-based fuels, NitiAayog has previously proposed the establishment of a fund with a budget ranging from Rs. 4,000 to Rs. 5,000 crore. Currently, the transportation sector relies on diesel for 70% of its fuel consumption and petrol for 99.6% of its fuel consumption. However, the increasing adoption of electric vehicles will contribute to a decrease in the demand for these traditional fuel sources.

The proportion of India's dependency on imported crude oil has experienced an increase from 73% in the fiscal year 2005–2006 to 81% in the fiscal year 2015–2016. This rise in dependence has resulted in an import expenditure exceeding $120 billion. In 2017, India witnessed the establishment of its largest waste-to-energy facility. This facility boasts a daily processing capacity of 2,000 metric tonnes of garbage, enabling the production of 24 MW of energy. In various locations globally, there has been the establishment of innovative standards. For example, Veolia, a company, engages in the collection of discarded chemicals from refiners. With an impressive recovery rate of 95%, Veolia returns these chemicals to the refiners for reuse (Accenture 2019).

The utilization of chemical engineering in the treatment of produced water from the oil and gas industry has gained significant attention in recent years (Waly et al. 2023). In the oil and gas industry, produced water refers to the water that is associated with oil during the extraction process (Jagaba et al. 2022). In 2020, it was estimated that the worldwide production of produced water reached an impressive value of 250 million barrels per day (CIASULLO et al. 2019). This translates to an annual estimate of 77 billion barrels for the year 1999 (Ellen MacArthur Foundation (EMF) 2015a, b). The waste stream associated with oilfield activities is the largest in terms of volume. In the oil and gas industry, it is crucial to prioritize the application of CE and the proper management of Produced Water. This is essential to mitigate the potential risks associated with environmental degradation and the harmful impact on ecosystems.

12.7 Future Directions for Further Exploration and Development

According to a recent report by Accenture, India has the potential to implement CE business strategies and generate approximately $500 billion in economic value by the year 2030. As illustrated in Fig. 12.3, achieving this objective can be accomplished through the integration of multiple strategies. These strategies include reducing material and energy losses, enhancing the utilization of products and capital assets, prolonging the lifespan of products, and extracting value from waste streams (Fiksel et al. 2020).

There are three different approaches that can be employed to mitigate challenges: the "Developing Collaborative Model", "Internal Research and Development, Innovation", and the "Product-Service System model." Stakeholders and policymakers

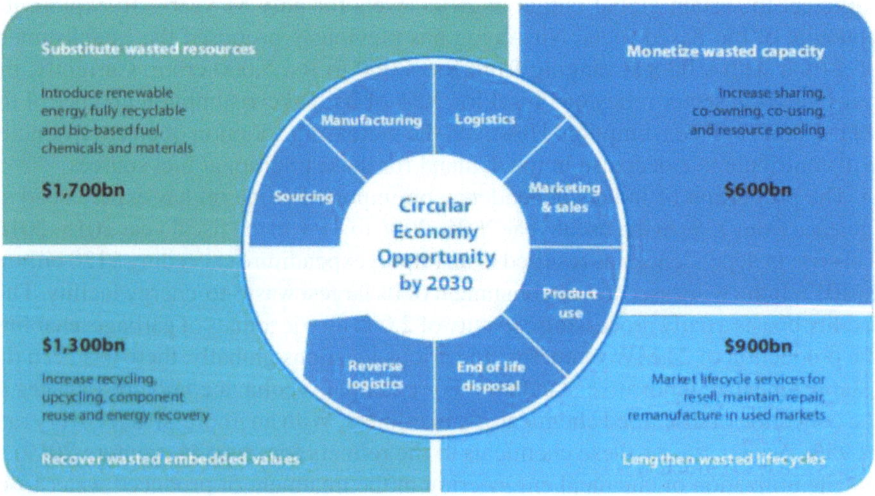

Fig. 12.3 Potential Value Realisation through Circular Business Models by 2030" (Accenture 2019)

can facilitate the transition towards a CE by considering the social and environmental impacts (Sharma et al. 2023).

12.7.1 Developing Collaborative Model

Developing a collaborative model is crucial for successfully transitioning to a CE. It involves various stakeholders, such as businesses, governments, communities, and consumers, working together towards a common goal of sustainability and resource efficiency. In this collaborative model, different parties bring their expertise and resources to the table. For instance, businesses can collaborate with researchers to develop innovative products that are easily repairable, upgradable, or recyclable. Governments can support these initiatives by creating policies and regulations that incentivize circular practices. Communities and consumers play an essential role in embracing sustainable consumption habits and properly recycling or disposing of products at the end of their lives. The collaboration fosters knowledge-sharing, technology transfer, and the co-creation of solutions that benefit all involved parties. By working together, stakeholders can overcome challenges and barriers that would be difficult to tackle individually (Ellen MacArthur Foundation 2012).

12.7.2 Internal Research and Development, Innovation

Internal Research and Development (R&D) and innovation play a vital role in achieving a CE. Companies that invest in R&D and innovative practices can create products that are designed to last longer and be easily repaired, or recycled. They can also develop new materials and processes that contribute to resource efficiency and waste reduction. Innovation within a company can lead to the development of new business models, such as product-as-a-service or leasing, which encourages product longevity and sharing among customers. Moreover, internal R&D can enable companies to optimize their production processes, reducing waste and energy consumption. By fostering a culture of innovation and allocating resources to R&D efforts, companies can stay ahead in the CE transition and contribute to global sustainability goals (Stahel 2016).

12.7.3 Product-Service System (PSS) Model

The Product-Service System (PSS) is a business model that combines products and services to meet the needs of customers. While certain authors have suggested the adoption of the PSS business model for the CE (Tukker 2015; Scheepens et al. 2016; Vasantha et al. 2012), it is crucial to exercise caution when implementing PSS. In order to mitigate the environmental impacts, PSS must be deliberately designed. The authors have identified that among the three types of PSS, it is the result-oriented PSS type that makes significant contributions to resource efficiency and circularity. The primary rationale behind this approach is that the customer is billed solely for the attainment of desired outcomes, which is also regarded as a service, rather than for the ownership of the product. Upon completion of the service, the product is returned to the producer, who possesses the ability to manipulate its components and modules to provide a diverse range of services. This facilitates the implementation of a CE (Michelini et al. 2017).

12.8 Conclusions

It is widely acknowledged that the resources available on our planet are limited in quantity. The implementation of a CE framework will facilitate the reduction of unnecessary resource consumption. Furthermore, there is a significant emphasis on the promotion of CE within the industrial sector and water management, to guarantee a sustainable future for upcoming generations. Therefore, it is recommended to allocate additional resources and conduct further investigation into the various applications of CE in diverse industrial activities and processes. The successful implementation of the CE demands the presence of political will. It is necessary for India to carefully evaluate the environmental advantages and disadvantages associated with its activities. In addition, it is imperative to ensure that the product undergoes recycling or reuse procedures before its disposal.

References

Accenture and Federation of Indian Chambers of Commerce and Industry (FICCI) (2019) Accelerating India's CE shift

Annarelli A, Battistella C, Nonino F (2016) Product service system: a conceptual framework from a systematic review. J Clean Prod 139:1011–1032

Behera ID, Basak G, Kumar RR, Sen R, Meikap BC (2020) Treatment of petroleum refinery sludge by petroleum degrading bacterium Stenotrophomonas pavanii IRB19 as an efficient novel technology. J Environ Sci Health Part A 1–13. https://doi.org/10.1080/10934529.2020.1866924

Behera ID, Nayak M, Biswas S, Meikap BC, Sen R (2021) Enhanced biodegradation of total petroleum hydrocarbons by implementing a novel two-step bioaugmentation strategy using indigenous bacterial consortium. J Environ Manag 292:112746. https://doi.org/10.1016/j.jenvman.2021.112746

Behera ID, Nayak M, Mishra A, Meikap BC, Sen R (2022) Strategic implementation of integrated bioaugmentation and biostimulation for efficient mitigation of petroleum hydrocarbon pollutants from terrestrial and aquatic environments. Mar Pollut Bull 177:113492. https://doi.org/10.1016/j.marpolbul.2022.11349

Bhattacharyya JK, Shekdar AV (2003) Treatment and disposal of refinery sludges: Indian scenario. Waste Manag & Res: J Sustain CE 21(3):249–261

Can Extended Producer Responsibility improve circularity in India's waste oil management?, Down to Earth, By Shobhit Srivastava Published: Tuesday 09 May 2023
CE for Sustainable Development in Serbia; United Nations Development Programme, New York, NY, USA (2019)
Ciasullo MV, Castellani P, Rossato C, Troisi O (2019) Sustainable business model innovation. Sinerg Ital J Manag 37(1):213–237
Coelho PM, Corona B, ten Klooster R, Worrell E (2020) Sustainability of reusable packaging–current situation and trends. Resour Conserv & Recycl X 6:100037
Diaz A, Schöggl JP, Reyes T, Baumgartner RJ (2021) Sustainable product development in a CE: implications for products, actors, decision-making support and lifecycle information management. Sustain Prod Consum 26:1031–1045
Ellen MacArthur Foundation (2012) Towards the CE: economic and business rationale for an accelerated transition. Ellen MacArthur Foundation
Ellen MacArthur Foundation (2015a) Growth within a CE vision for a competitive Europe
Ellen MacArthur Foundation (2015b) Towards a CE: business rationale for an accelerated transition
European Commission—Press release (2010). https://ec.europa.eu/commission/presscorner/detail/en/ip_20_420
Fiksel J, Sanjay P, Raman K (2020) Steps toward a resilient CE in India. Clean Technol Environ Policy 23(1):203–218
Jagaba AH, Kutty SRM, Lawal IM, Birniwa AH, Affam AC, Yaro NSA, Usman AK, Umaru I, Abubakar S, Noor A, Soja UB, Yakubu AS (2022) CE potential and contributions of petroleum industry sludge utilization to environmental sustainability through engineered processes - a review. Clean Circ Bioeconomy 3:100029
Kalmykova Y, Sadagopan M, Rosado L (2018) August). CE – from review of theories and practices to development of implementation tools. Resour Conserv Recycl 135:190–201
Kanungo J, Sahoo S, Bal M, Behera ID (2023) Performance of bioremediation strategy in waste lubricating oil pollutants: a review. Geomicrobiol J. https://doi.org/10.1080/01490451.2023.2245395
Malinauskaite J, Anguilano L, Rivera XS (2021) Circular waste management of electric vehicle batteries: Legal and technical perspectives from the EU and the UK post Brexit. Int J Thermofluids 10:100078
Michelini G, Moraes RN, Cunha RN, Costa JM, Ometto AR (2017) From linear to CE: PSS conducting the transition. Procedia CIRP 64:2–6
Mohanta S, Pradhan B, Behera I.D (2023) Impact and remediation of petroleum hydrocarbon pollutants on agricultural land: a review. Geomicrobiol J 1–5. https://doi.org/10.1080/01490451.2023.2243925
Reike D, Vermeulen WJ, Witjes S (2018) The CE: new or refurbished as CE 3.0? — exploring controversies in the conceptualization of the CE through a focus on history and resource value retention options. Resour Conserv Recycl 135:246–264
Scheepens A, Vogtländer J, Brezet J (2016) Two life cycle assessment (LCA) based methods to analyse and design complex (regional) CE systems. Case: making water tourism more sustainable. J Clean Prod 114:257–268
Sharma M, Joshi S, Prasad M, Bartwal S (2023) Overcoming barriers to CE implementation in the oil & gas industry: environmental and social implications. J Clean Prod 391:136133
Stahel WR (2016) The CE. Nature 531(7595):435–438
Tukker A (2015) Product services for a resource-efficient and CE – a review. J Clean Prod 97:76–91
US-EPA (United States Environmental Protection Agency) (2012) Guidelines for water reuse. EPA/600/R-12/004, USEPA Office of Wastewater Management, Washington, United States. - References - Scientific Research Publishing (n.d.)
Vasantha GVA, Roy R, Lelah A, Brissaud D (2012) A review of product–service systems design methodologies. J Eng Des 23(9):635–659
Velenturf AP, Purnell P (2021) Principles for a sustainable CE. Sustain Prod Consum 27:1437–1457

Waly MM, Mickovski SB, Thomson C (2023) Application of CE in oil and gas produced water treatment. Sustainability 15(3):2132
World Bank Open Data. https://data.worldbank.org

Chapter 13
Recycling, Re-using, Regeneration, and Recovering of Value-Added Products Petroleum Hydrocarbons Through Circular Economic-Based Approaches

Geetanjali Basak

Abstract Petroleum industry sludge owing hydrocarbons are the basic inescapable outcomes of petroleum industries. Based on the requirement of energy sources, the demand for oil sludge generation has escalated over the course of time. Hydrocarbons, being the principal constituent of petroleum sludge, pose adverse effects on the ecosystem. Nevertheless, as sludge has persistent traits, not many approaches can attain strict environmental laws during the application of a considerable quantity of water, electricity, and chemicals as substitutes of day to day life. Conversion of petroleum waste, especially persistent hydrocarbon waste, to financially viable value-added products, minimizing environmental consequences along with maintaining sustainability within the petroleum industry is essential. Thus, the chapter deals with the various methods adopted for the treatment of hydrocarbons present in industrial waste, with simultaneous resource recovery and reuse concepts in order to meet the circular economy viewpoint. Moreover, the chapter also highlights the prospects and challenges of the regeneration process and its possible solutions to overcome it so as to increase its collection and recycling rates. The chapter also discusses the effective applications of petroleum wastes in several aspects that have reached the environmental engineering significance and follows the concept of "wastes-treat-wastes" through self-cycle operation for adopting a circular economy.

Keywords Petroleum sludge · Oil sludge · Circular economy · Sludge reuse

---

G. Basak (✉)
Raniganj Girls' College, Affiliated to Kazi Nazrul University, Raniganj 713358, West Bengal, India
e-mail: geetanjalibasak@gmail.com; tgeetanjali_microbiology@raniganjgirlscollege.org

I. D. Behera and A. P. Das (eds.), *Impact of Petroleum Waste on Environmental Pollution and its Sustainable Management Through Circular Economy*, Environmental Science and Engineering, https://doi.org/10.1007/978-3-031-48220-5_13

13.1 Introduction

Hydrocarbons are the basic contents of petroleum products acquired from various sources along with crude oil or crude gas, that has the composition of Carbon (84.0–87.0%), H_2 (10–14%), O_2 (0.5–1.5%), S (0.05–6.0%) and metal contaminants. In petroleum industries, starting from the process of crude oil/gas exploration to refining, a wide quality of oily and viscid wastes are generated (Borgheipour et al. 2018; Jagaba et al. 2022a, b). Among the five types of wastes, viz. sludge, slurry, liquids, solids, and gases, anything that is corrosive, volatile, inflammable and toxic is considered as lethal waste (Al Qallaf et al. 2016). These unwanted materials are treatment unit-derived activated sludge, waste catalysts/spent fluid catalytic cracking; spent engine oil, oil drilling waste, drill cuttings, oil slime, oil sludge or petroleum residue, hydrocarbon cutting sludge, refinery scale from an old tank; inside the pipe; outside the adapter; an old pipe; burn pits, petroleum refinery sludge; Sludge from refinery exchangers, produced water/oily refinery wastewater, dissolved air flotation unit sourced float, coal water slurry, petroleum coke water slurry, coal sludge slurry, and petroleum coke sludge liquid residue from the thermal treatment of oily sludge. Sludge from the petroleum division is an extremely steady suspension emulsion system consisting of polycyclic aromatic hydrocarbons (PAHs) and benzene homologs at high concentrations. It is considered as hazardous since it is lethal and tough to absorb along with the deposition of hydrocarbons through vast sedimentary fields over a long period of time (Behera et al. 2020; Liu et al. 2017). Petroleum and wax are the two prime constituents of oil sludge that can be recycled (Lyu et al. 2018). The yield of oil slurry is controlled by the presence of inorganic deposits in the form of clay/sand/sediments or dust and pipelines or storage tanks. Different types of sludge found in the petroleum industry are landing sludge, tank bottom sludge, sludge from oil drilling, refining sludge, petroleum wastewater-activated sludge, and oil–water separator sludge are some of the several types of sludge found. From several studies it is As a result, it is predicted by the scientists that about one tonne of crude oil will yield approximately 0.3–0.5% of sludge (Jagaba et al. 2022b). Reports from USEIA reveal that, on average, 0.3% of oil sludge and 228.29 Mt/year worldwide is produced till 2020. Thus, petroleum sludge remains a reason for distress for the ecosystem globally until a solution is figured out unless the fossil targets the net zero emission goal.

Based on the pros and cons of these industries and their demands in modern society, the sectors have to face challenges in order to attain eco-sustainability along with maintenance of financial feasibility. Therefore, to ensure that the petroleum sector reduces its negative impact on the environment, advancements in refining technologies, safe disposal of wastes, and environmentally friendly waste treatment technologies are required. The State of Qatar, whose economy mostly depends on the oil and gas sector being a crucial part of the world energy trade, is a member of the Gulf Cooperation Council (GCC), famous for its collective abundance in hydrocarbon reserves. Therefore, the basic aims of the researcher are to deal with different waste of petroleum industries and study their management strategies, treatments, re-use, and recycling towards the formation of value-added products so as to further achieve a circular economy for the betterment of society.

13.2 Environmental Fate of Petroleum Hydrocarbons Generated from the Petroleum Sludge Waste

Based on the heterogeneous sludge constituents and viscid nature, including monocyclic, polycyclic aromatic hydrocarbons, esters and ketones, along with some metals like compounds of iron, they are considered as hazardous (Al-dhawi et al. 2022; Birniwa et al. 2022). The presence of iron in the sludge imparts the property to inhibit low volatility and low water absorption, which can be utilized as roadbed material and for syngas production. The impact of the petroleum sector is quite harmful from a health point of view as it is carcinogenic and mutagenic if exposed accidentally (Cheng et al. 2016). Since the sludge survives several environmental fluctuations over a long period of time, it results in the modifications of total petroleum hydrocarbon (TPH) concentrations that may affect different systems of the human body (Behera et al. 2021; Jasmine and Mukherji 2015). As a fate, the residues from these industries impart toxicity to the ecosystem by covering a huge mass of productive area. These hydrocarbons present in the sludge may contaminate several water bodies, thereby hampering aquatic life. Plant growth is also hindered as the hydrocarbons from these industries choke surface pores, which limits the availability of dissolved oxygen to reach plant roots, thereby obstructing plantation development and leading to death (Teng et al. 2021). Thus, accurate petroleum sludge management is necessary to prevent environmental pollution, adverse impact on human health, waste value and economic loss and separate capital for sludge management by the individual industries.

13.3 Treatment Technologies

13.3.1 Management of Hydrocarbon Waste

Out of the three layers formed in the tank of petroleum sludge, the percentage of hydrocarbon present in the topmost layer is approximately 60.0% whereas the bottom layer contains 40.0–60.0% and 25.0–60.0% of aliphatic and aromatic hydrocarbons (Shie et al. 2004). Different treatment methods viz. Surface discharge, subterranean injection for slurries, burial, and disposal at sea (Borgheipour et al. 2018). Biological processes such as phytoremediation and use of surfactants included in biodegradation, bioaugmentation, biostimulation, landfarming, rhizoremediation, biosparging, bioventing, and composting (biopiling) along with physico-chemical processes like incineration, ozonation, chlorination, oxidation of wet air, extraction of solvents, demulsified systems, thermochemical conversion, photo-catalysis, thermal desorption, coking freezing/thawing, microemulsion systems, electrokinetic method, solidification/stabilization/encapsulation, microbial fuel cell, advanced oxidation processes) are available (Jagaba et al. 2022a, b; Kanungo et al. 2023).

13.3.2 Emerging Methods for OS Treatments

Certain upcoming methods, such as gasification and pyrolysis, are becoming the focus of scientists in recent times to convert sludge waste to economically important products. A process where around 75–80% of sludge is transformed by changing hydrocarbons with low molecular mass to aqueous phase and non-condensable gases into hard form (Char) under excess heat and a non-reactive system is called as pyrolysis (Srivastava 2012). Contents available in the sludge decide the concentration of the product generation. Along with the char, it also produces dry ash.

Another technique is gasification, which requires an oxidizing agent to produce syngas from hydrocarbons under high temperatures in a gasifier reactor (Shahbaz et al. 2022). The syngas contain high amounts of hydrogen in it. Further, this can be used as an energy source. Other than H_2, oil and char is also produced. Figure 13.1 shows the types of treatment done for petroleum sludge.

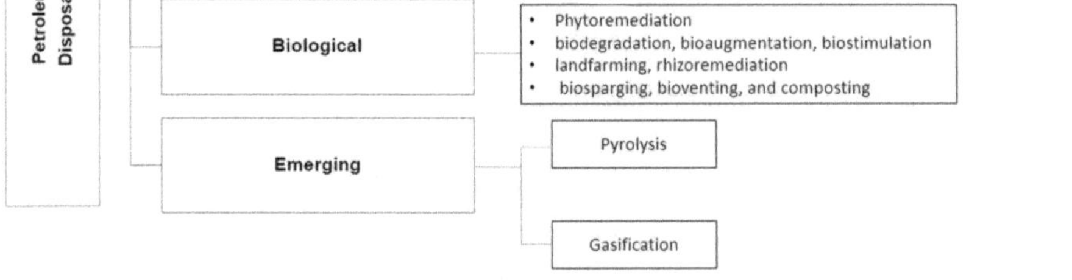

Fig. 13.1 Different types of treatment methods for petroleum sludge

13.4 Re-utilization of Petroleum Sludge Hydrocarbons

The possible best answer regarding the re-application of oily sludge is its use in the manufacture of eco-friendly ingredients like bricks, cement, and asphalt-concrete needed in several construction companies (Jagaba et al. 2021).

13.4.1 Cement Clinker and Mortars

Cement preparation should involve a sustainable process with a lot of energy preservation with low CO_2 release. With an increase in time, the requirement for sustainable cement production is increasing. One such material called gypsum helps provide strength in cement as it has anhydrite ($CaSO_4$) in it. Excess resource exhaustion of gypsum can be alternated by using petroleum sludge as cement clinker that not only has SO_3 but also hydrocarbons that act as a replacement fuel, which can be thinned almost so as to be pumped out easily (Aeslina and Ali 2017).

Eco-friendly mortar can be prepared by replacing the common cement with oily sludge ash that has a well-defined structure and function that can be further used to prepare pozzolana cement ingredients. This will have high compressive strength with proper microstructure (Kankia et al. 2021).

13.4.2 Construction Block, Bricks and Tiles Formulation

Oily sludge and vegetable oil combination can be used as clinging material for petro-vege blocks. This provides high compressive strength with low chemical oxygen (Johnson et al. 2015). The addition of sludge waste at 5% by weight, along with its bentonite-treated form, leads to the formation of bricks with good functionality. Alternating kaolin with sludge 5% by weight in porcelain tiles lessens flexural strength, linear contraction, and weight while enhancing water absorption.

13.4.3 Ceramics

An advantage of using sludge waste over traditional carbonates in ceramics, concrete and bricks is its lighter and foamy appearance from the outside with large pores. The addition of sludge gives a better property to the ceramic with lightweight, compressive strength, linear contraction, and absorbency.

13.4.4 Highway Material

Petroleum sludge is employed as natural gums for the construction of road materials that can withstand the heavy weight of vehicles. Combining phosphogypsum (roadbed material) with petroleum sludge modifies the coagulated material with 28-day compressive strength and diminishes heavy metal discharge with definite pore organization microstructure simultaneously. This compact material will have the ability to withstand harsh conditions and be self-sustaining. TPH mineralization and discharge of sludge waste can be used as a powdered material to label unsurfaced roads.

Elevated concentrations of asphaltene and maltene hydrocarbons with sludge mixed with a black viscous mixture of bitumen make it supple and firm. A mixture of petroleum sludge, bitumen and polymer increases the resistance from difficulties in asphalt material.

13.4.5 Adsorbent

Sludge from the petroleum industry can be a great substrate of carbon-containing adsorbing material used in aqueous environment mitigation because of the elevated concentration of asphaltene hydrocarbons in it. Once this hydrocarbon is concentrated to obtain activated carbon, it can be used for removing dyes. The presence of asphaltene increases the surface area of the activated carbon, which is one of the most important properties an adsorbing agent should possess. Pyrolysis of the petroleum sludge materials leads to the formation of activated carbon adsorbing agents with several pores. This material, in turn, is capable of removing heavy metals from an aqueous environment. Even chemically treated natural adsorbents with raw petroleum sludge can remove oil residues from marine water samples for several cycles.

13.4.6 Microbial Isolates from Sludge Waste for Aqueous Solution Mitigation

Microorganisms isolated from waste sludge materials from the oil industry are immensely acknowledged for their capacity to remediate strong sludge and heavy metals as well (Behera et al. 2020).

13.4.7 Biogas Production

Anaerobic digestion of oily sludge methane, hydrogen and acetate at an elevated pH is being used in the environmental engineering field. Natural waste material like sugarcane bagasse with a high ratio of carbon: nitrogen along with sludge material is considered best for an anaerobic digestion process (Ghaleb et al. 2021). Prior treatment of bagasse and sludge under extremely high temperatures promotes the formation of biogas. This decreases CO_2 emissions and fossil fuels.

13.4.8 Other Applications

Effective microorganism (EM) biofertilizer made from minerals and organic matter-rich sludge and clean soil in a ratio of 1:1. This is used for various crop plantations. For example, jack beans are grown in the presence of sludge-containing land farming that lowered the hydrocarbon content by around 57.0% and aliphatic hydrocarbons are present at its roots after 120 days of cropping time (Jagaba et al. 2022a, b). Tire materials like rubbers can be prepared by recycling sludge (Turebekova et al. 2019). The plasticity of the rubber material increased due to the usage of the organic content of the oil sludge materials. Pyrolysis of petroleum sludge at excess thermal conditions yields biochar. They can serve as significant cathodes in lithium-ion batteries. Sludge can be further processed to be utilized in sensing material, and promotes chemical reactions in electrical instruments.

13.5 Recovery Technologies Oily Sludge Hydrocarbons

13.5.1 Extraction Through Solvents

Management of petroleum sludge involves transformation of sludge into economically important yields that pose the advantage of less discharge of secondary sludge and harmful chemicals, implementing an eco-friendly and budget-friendly process. Recovery technology, especially regaining oil from sludge, is more economically sustainable, producing around 50. 0% of oil with 30.0% or below of solid trash (Hu et al. 2013). Solvent extraction is one of the methods for recovering oil from sludge waste by using different solvents like heptane, naphtha, and toluene based on different parameters, viz. pressure, thermal conditions, solvent ratio and characteristics of sludge waste (Hu et al. 2013). As a result, various oil industries in the USA could reuse around 80.0% of sludge waste for several cycles. Petroleum hydrocarbons can be recovered from petroleum sludge through different solvents like hexane and xylene (Mohanta et al. 2023; Taiwo and Otolorin 2009). Some efficient techniques with quick and commercial processes are solvent extraction, ultrasonic, centrifugation, solvent extraction, and thermochemical cleaning; still, they possess certain disadvantages of high cost and energy requirement process.

13.5.2 Ultrasonication

A technique where oil and water emulsion is broken due to shaking and forming bubbles inside the liquid in the presence of heat for the separation of oil from solid waste (Check 2014). Factors affecting the removal of oil are thermal input, incidence of the sonication treatment and constituent of electrolyte. Greater than 80.0% of oil is retrieved along with the separation of contaminants from the top portion of the sludge, thereby disrupting different state coats of the solid waste (Zhang et al. 2012).

13.5.3 Centrifugation

The separation of components based on their densities by application of rapid rotations through centrifugal force is called centrifugation (Hu et al. 2013). Here, the viscidness of the sludge is diminished by mixing with various solvents or surface-active agents to alter the solid waste into triple-layered hydrocarbon, water and slurry. Likewise, the oil is removed from the aqueous or solid samples up to an efficiency of 95.0% approximately. Although this process has the advantage of zero emission of harmful chemicals, it still has the disadvantage of enhanced instrument rate and need of energy source.

13.5.4 Surfactants

Surface active substances having both a water-loving head and water-hating tail are applied in oil–water combinations to remediate petroleum sludge by removing oil samples from it. Chemical surfactants, mainly anionic SDS (sodium dodecyl sulfate), and ethoxylate, are used for parting oil from sludge (Christofi and Ivshina 2002; Azim et al. 2011). Chemical surfactants replaced with biological surfactants obtained from various organic wastes and microbes contain lipids of various biomolecules and groups show significant oil removal to nearly 100% (Lima et al. 2011). Moreover, they are highly capable of removing oil. However, their operation cost is not economically viable with certain problems relating to their separation from waste, which should be studied further.

13.6 Conversion of Petroleum Sludge Waste to High-Valued Products

High-priced, better-quality production of yields from waste to worth in the form of oil or gas through the transformation of sludge by a thermochemical method known as gasification (Gonalez et al. 2018). Here, the main product that evolved is syngas and electricity, as discussed earlier. The formation of hydrogen and natural gas along with the release of electricity, is a "bang of the buck". On the other hand, pyrolysis involves transformation of hard sludge (char), oil and gas (Huang et al. 2015). About 1.4, 2.2 and 0.2% of gas and ash in the form of char and oil are produced through pyrolysis, respectively (Gong et al. 2018).

During pyrolysis, the open market valuation possibly of char, gas and oil is $235/day, $70/day and $21/day, respectively (Shahbaz et al. 2020). Gasification yields a product value of nearly $23,519/day, $71,432/day and $4250/day for natural gas, hydrogen gas and electricity liberation, respectively on a hypothesis basis (Alherbawi et al. 2021).

13.7 Circular Economy: Its Ideologies and Strategies to Maintain Sustainability

Ecological conservation and monetary action go simultaneously in a sustainable manner in the concept of circular economy. This highlights the application of energy derived from various biotic and abiotic sources during the entire biorhythm of the product. Moreover, it deals with the reclamation, regeneration, modification and renewal of particular yields so as to obtain a product with upgraded ecological qualities with simultaneous enhancement in monetary evolution (Jagaba et al. 2022a, b). However, it requires customer negotiation with respect to perceived value, outlines and links, which leads to a life cycle that is tough for individuals (de Morais et al. 2021). The recurring and liable use of products and energy by individuals to maintain or develop commercial wealth is the key feature of the circular economy. It is considered as a promoter to manage consensus, accountability competency and supply in society to accomplish Sustainable Development Goals (SDGs) (Mutezo and Mulopo). Overall, the process should be such that the product will be used repeatedly in a circular fashion with a subsequent rise in the financial aspect.

Three basic ideologies of circular economy recognized by the Ellen MacArthur Foundation are as follows

i. Remaking, renewing, then reutilizing things and facilities to develop a novel rate with simultaneous rearrangement of money.
ii. Limited sources of natural resources shall be restored and complete utilization through exodus and containerization facilities with the implementation of eco-friendly techniques.

iii. Practice of ecology, along with robust assets for remediation, is paramount for business accomplishment (Mutezo and Mulopo).

The schemes emphasizing circular economy are as follows:

i. Implementation of new designs like online services.
ii. Environmental revitalization through restoration and revival.
iii. Enhancement of yield production with minimization of unwanted things
iv. Extending novel techniques to updating business strategies to practice.
v. Products or bi-products obtained from various abstraction of solid waste is used as a source of monetary.
vi. Modifying, distributing, and resource reutilization (Upadhyay et al. 2021).

Scientists stated that crude oil purification has generated large quantities of unwanted organic materials that are quite important to the circular economy in petroleum industries. Thus, oil sludge plays an important role as a substrate (source of carbon) power supply in the form of energy and biochemical stock. Therefore, it has achieved circular economy by applying the 4R (reduce, reuse, recycle and recover) for the progress of the novel circular economy hypothesis, which will be sustainable and beneficial for the eco-system (Sadeghi et al. 2015).

13.8 Concept of "Wastes-Treat-Wastes" in Petroleum Industry Sludge to Achieve Self-Cycle Operation

Treating unwanted products with trash in the oil sludge industry is of low cost and ecologically benign and is required from a financial growth point of view and preservation of various sources (Manogaran et al. 2022). The method will not only clean the ecosystem from sludge waste but also mineralize the hydrocarbons from nature. Biochar made from solid waste at thermal conditions is used as a promoter during the catalytic ozonation of aqueous waste from oil industries. This biochar-activated solid waste containing great surface area and affinity towards water can, in turn, be used for aqueous solution treatment. Having functional groups on their surface helps them to be a good adsorbent in removing the COD level, hydrocarbons and nitrogen content of the aqueous sample. The presence of pores on the surface of the activated carbon gives space for inhabiting bacteria responsible for removing nitrogen. Thus, this type of activated sludge does multiple works in one chance, proving to be an effective, ecologically responsive cheap option for mitigating hydrocarbons containing petroleum samples (Wang et al. 2020).

Adsorbent mixed with the solid waste of oil at high temperature is always more advantageous than the adsorbent alone in adsorbing oil from adsorbate. For example, palygorskite alone and palygorskite combined with sludge waste at high temperatures could adsorb nearly 50. 0% of oil per gram of adsorbent. Sludge wastes compiled with silica material have functional groups and methyl groups as a high concentration of carbon groups is used as good adsorbents (Guritino et al. 2016). Sludge waste and petroleum-containing areas can lead to the production of polymers with the help of microorganisms isolated from petroleum-contaminated sites. These polymers help in maintaining microorganisms even in harsh conditions. For example, *Pseudomonas* sp. secretes polyhydroxyalkanoate (PHA) at unfavourable conditions (Goudarztalejerdi et al. 2015).

Oil sludge can be used for various applications

i. Synthesis of wax using toluene. Thus, wax can be recovered at high rate and further hydrocarbons can be preserved.
ii. An aqueous fuel can be generated for emerging methods known as petroleum coke water slurry.

13.9 Uses of Microorganisms Obtained from Sludge Waste in Various Petroleum Sectors

i. Bacteria used for producing acrylic acid is obtained from petroleum-contaminated solid waste. For example, *Rhodococcus ruber* AKSH-84 was obtained from oil-containing solid waste capable of mineralizing approximately 100 mM of nitrile (Kamal et al. 2011).
ii. Microorganisms isolated from solid waste of oil sectors/ industry involved in biofuel/biodiesel yield by the usage of less volume of solvents. Yeasts especially secrete lipase enzymes that function to produce biodiesel. Example: *Cryptococcus diffluens* D44 produce lipase at pH-9.0 in the existence of several minerals, cations and methanol solvent at low volume. (Yilmaz and Sayar 2015).
iii. Bacteria secluded from oily solid waste are capable of removing hydrocarbons through the degradation process in batches using mineral salt medium and in field study as well. The field study includes the action of native bacteria in compost, removing around 70–85% of hydrocarbons in about 4 months. For instance, *Acinetobacter radioresistance* strain KA2 obtained from solid waste of the petroleum industry degraded 88.0% of petroleum hydrocarbons (Poorsoleiman et al. 2021). *Sphingomonas olei* strain KA1 and *Acinetobacter radioresistance* strain KA2 were capable of mineralizing crude oil (Koolivand et al. 2019). *Enterobacter hormaechei* strain (KA3) and *Staphylococcus equorum* strain (KA4) are efficient in remediating oil in various enrichment media (Parhamfar et al. 2020).
iv. Stimulating the native microorganisms by adding nutrients in the petroleum-contaminated sites can clean up hydrocarbons from petroleum-containing lands in about certain months. Moreover, the land is gained with potentiality because of the presence of various types of microbes especially capable of breaking down several aromatic hydrocarbons and land-enriched nutrients. As a result, the lethality of the soil is also diminished. The phenomenon is called as biostimulation.
v. Bioaugmentation: The addition of indigenous microbes along with the mixture of manure containing important biochemicals led to the degradation of nearly 75% of hydrocarbons from soil. The aerobic microbes show better results than

Table 13.1 List of bacteria responsible for hydrocarbon mineralization

Sl. No	Bacteria	Hydrocarbon clean up with duration	References
1	*Bacillus sp.*	75% approx	Roy et al. (2018)
2	*Gordonia alkanivorans* CC-JG39	86% in 30 days	Sarkar et al. (2020)
3	*Ralstonia pickettii*	In the presence of phenol	Al-Zuhair and El-Naas (2012)
4	PSM6 PSM7 PSM10 Uncultured *Acinetobacter sp.* [PSM11]	Fluoranthene, nearly 50%	Kumar et al. (2011)
5	*Lysinibacillus fusiforme* *Bacillus sp.*	Asphaltene 55 and 50%, respectively, in 60 days	Aditiawati and Kamarisima (2015)
6	*Pseudomonas aeruginosa* IBBML1 Rhodamine 6G	Grows in 10% of benzene-containing hydrocarbons	Lazaroaie (2009)
7	*Stenotrophomonas sp.* IRB19	65% of hydrocarbons in 28 days	Behera et al. (2020)
8	*Rhodococcus ruber* Z25	Nearly 94% of oil separation	Zheng et al. (2012)

the anaerobic organisms. Moreover, they produce additional metabolites at starving conditions that enhance applications to the microbes in the petroleum-contaminated sites. They may use several compounds in the soil as carbon and nitrogen source and helps in the oil recovery. Additional usage of microbes from external sources improves the process of petroleum site mitigation (Behera et al. 2022). A list of bacteria responsible for hydrocarbon mineralization is given below in Table 13.1.

13.10 Surface Active Agents Secreting Bacteria Obtained from Oily Sludge for Reclamation of Hydrocarbons

Although environmentally benign microbes are used for the reclamation of petroleum-containing sites, it still takes much time to carry out the process. An alternative to chemical surface active agents, biosurfactants produced by microbes under unfavourable conditions are less harmful, capable of breaking down pollutants and ecologically benign (Cheng et al. 2017). These biosurfactants carry different types of biomolecules like carbohydrates, proteins, and phosphate ions. A list of microorganisms producing various biosurfactants and mitigating petroleum-contaminated areas is given in Table 13.2. Rhamnolipid, an eminent biosurfactant obtained from oily sludge areas inhibits pathogenic bacteria. The microbes growing in basal salt medium

Table 13.2 Biosurfactants produced by different bacteria

Bacteria	Biologically produced surfactants	References
Bacillus subtilis *Starmerella bombicola,* *Pseudomonas aeruginosa* *Acinetobacter calcoaceticus*	Emulsan Surfactin Rhamnolipids Sophorolipids	Bezza et al. (2015)
Pseudomonas aeruginosa	Rhamnolipid	Liu et al. (2018)
Microbacterium maritypicum ABR5	Glycolipoprotein	Akbari et al. (2020)

secrete biosurfactants that help remove hydrocarbons in a few days. Enhanced hydrocarbon is removed by 35% in about 80 days using nitrogen from ammonium chloride (Nejad et al. 2020).

13.11 Challenges and Future Prospects

To comply with environmental rules and protect ecological health, the oil and gas business generates a sizable quantity of toxic unwanted products that includes wastewater, solid waste along with fumes, which need to be disposed of. The challenges of waste management methods are highlighted in this part, along with the future work that will be required to address these problems. The foundation of this work is the identification of procedures that are effective, practical, affordable, and environmentally benign. Several technologies are available to mitigate different forms of throw-away streams, but it is still tough to discover an isolated approach to be free from harmful chemicals. For the treatment of an aqueous, solid, and gaseous stream, different treatment units are needed. Additionally, it is essential to select strong, economically viable, with the simultaneous capability of adopting itself under any environmental stress. Physiochemical treatments are best for remediating produced aqueous solution, provided binned with membrane technology for the elimination of elements present in small quantities. Unfortunately, the quantity of energy required during membrane operation needs high financial investment. Despite being effective and relatively simple to use, thermal technology applications are restricted by the need for a substantial amount of space. Particularly, phytoremediation and an MBR have appeal among biological therapy approaches. But very finite work reflects their economic aptness. Oil solid waste is tough to be discarded as it contains toxic elements. Traditional OS treatment methods are expensive and produce harmful byproducts. Gasification and pyrolysis are the best ways to recover several useful products in the form of energy. The Petroleum sector produces hazardous fumes of Sulphur and carbon, which are to be eliminated. Modern methods are capable of eliminating the said fumes and turning them into industrial products with added value. A mixed matric membrane is best for gas removal, although the economic viability is yet to be proven. Regardless the existence of different techniques till date, none of the methods is financially adoptable. "Only treatment" transition to "resource recovery" is essential and profitable for Petroleum sector waste management. Current treatment facilities require modification to execute the resource recovery idea. Further, an in-depth analysis for chalking out monetary investment is needed to implement the concepts highlighted, as it will help for the appropriate evaluation of economic prediction in terms of payback period and net present value.

13.12 Conclusion

Sustainable and prudent handling of Petroleum industry waste is a worldwide matter to be dealt with. Three vital categories of unwanted material, namely, produced water, oil sludge, and flue gases, obtained from the Petroleum industry have been identified during analysis. Each sort of waste corresponding treatment technologies are thoroughly analyzed. Management methods with respect to each type of unwanted waste have been examined. It has been discovered that the treatment technique to be selected should be based on different parameters like the nature of the input waste stream, cause, intended application of the byproduct, along with expense. A conjugated method that adjoins sundry treatment processes is frequently appropriate for low cost and enhanced efficacy and leads to environmental friendliness. Converting sludge into precious components brings out advantageous products and provides a global waste management technique. The unwanted waste can be converted sustainably to produce goods that are environmentally benign and profitable. The microorganism can be fed organic carbon and nutrients, which will produce a variety of biorefinery products. Sulphur can be reutilized in the industry of fertilizer and construction. Likewise, the recovered CO_2 can utilize for a variety of purposes in chemical and biochemical industries, providing this sector with economic and environmental protection. However, despite numerous potentials for resource recovery from oil and gas industry waste, its practical application has been scarcely ever demonstrated yet. Large-scale changes to the current infrastructure would be necessary for the transformation of the "treatment" to the "resource recovery" paradigm. It is important to note that further detailing of waste re-utilization options and feasibility studies are necessary to calculate key economic indicators. Furthermore, discussions with the stakeholders are imperative to determine the economic and environmental viability of resource recovery from quantifiable amounts of oil and gas waste.

References

Aditiawati P, Kamarisima (2015) Isolation of asphaltene-degrading bacteria from sludge oil. Makara J Sci 19(1):13–20

Aeslina AK, Ali B (2017) Petroleum sludge treatment and reuse for cement production as setting retarder. In: Mechanical engineering, science and technology international conference

Akbari E et al (2020) Isolation and identification of current biosurfactant-producing microbacterium maritypicum ABR5 as a candidate for oily sludge recovery. J Surf Deterg 23(1):137–144
Al-dhawi BNS et al (2022) Treatment of synthetic wastewater by using submerged attached growth media in continuous activated sludge reactor system
Alherbawi M, McKay G, Mackey HR, Al-Ansari T (2021) A novel integrated pathway for Jet Biofuel production from whole energy crops: a Jatropha curcas case study. Energy Convers Manag 229:113662
Al Qallaf Y, Al-Kandari S, Yousef K, Al Mutairi S, Fouzy H (2016) Analysis and improvement possibilities of waste management at Kuwait Oil Company (KOC). In: Brebbia CA, Itoh H (eds) Waste management and the environment, vol Viii, pp 73–84
Al-Zuhair S, El-Naas MH (2012) Phenol biodegradation by ralstonia pickettii extracted from petroleum refinery oil sludge. Chem Eng Commun 199(9):1194–1204
Azim A-AAA, Abdul-Raheim A-RM, Kamel RK, Abdel-Raouf ME (2011) Demulsifier systems applied to breakdown petroleum sludge. J Petrol Sci Eng 78(2):364–370
Behera ID, Basak G, Kumar RR, Sen R, Meikap BC (2020) Treatment of petroleum refinery sludge by petroleum degrading bacterium Stenotrophomonas pavanii IRB19 as an efficient novel technology. J. Environ. Sci. Health Part A 1–13. https://doi.org/10.1080/10934529.2020.1866924
Behera ID, Nayak M, Biswas S, Meikap BC, Sen R (2021) Enhanced biodegradation of total petroleum hydrocarbons by implementing a novel two-step bioaugmentation strategy using indigenous bacterial consortium. J Environ Manag 292:112746. https://doi.org/10.1016/j.jenvman.2021.112746
Behera ID, Nayak M, Mishra A, Meikap BC, Sen R (2022) Strategic implementation of integrated bioaugmentation and biostimulation for efficient mitigation of petroleum hydrocarbon pollutants from terrestrial and aquatic environments. Mar Pollut Bull 177:113492. https://doi.org/10.1016/j.marpolbul.2022.11349
Bezza FA, Beukes M, Chirwa EMN (2015) Application of biosurfactant produced by Ochrobactrum intermedium CN3 for enhancing petroleum sludge bioremediation. Process Biochem 50(11):1911–1922
Birniwa AH et al (2022) Application of agricultural wastes for cationic dyes removal from wastewater. In: Textile wastewater treatment. Springer, pp 239–274
Borgheipour H et al (2018) Application of DEA technique in SWOT analysis of oily sludge management using fuzzy data. Glob J Environ Sci Manag-GJESM 4(2):183–194
Check GR (2014) Two-stage ultrasonic irradiation for dehydration and desalting of crude oil: a novel method. Chem Eng Process Process Intensif 81:72–78
Cheng S et al (2016) Evaluation of oil sludge ash as a solid heat carrier in the pyrolysis process of oil sludge for oil production. Energy Fuels 30(7):5970–5979
Cheng T et al (2017) A novel rhamnolipid-producing Pseudomonas aeruginosa ZS1 isolate derived from petroleum sludge suitable for bioremediation. AMB Express 7
Christofi N, Ivshina IB (2002) Microbial surfactants and their use in field studies of soil remediation. J Appl Microbiol 93(6):915–929
de Morais LHL, Pinto DC, Cruz-Jesus F (2021) Circular economy engagement: altruism, status, and cultural orientation as drivers for sustainable consumption. Sustain Prod Consum 27:523–533
Ghaleb AAS et al (2021) Sugarcane bagasse as a Co-substrate with oil-refinery biological sludge for biogas production using batch mesophilic anaerobic co-digestion technology: effect of carbon/nitrogen ratio. Water 13(5):590
Gong Z, Wang Z, Wang Z, Fang P, Meng F (2018) Study on pyrolysis characteristics of tank oil sludge and pyrolysis char combustion. Chem Eng Res Des 135:30–36
Gonzalez AM, Lora EES, Palacio JCE, del Olmo OAA (2018) Hydrogen production from oil sludge gasification/biomass mixtures and potential use in hydrotreatment processes. Int J Hydrog Energy 43(16):7808–7822

Goudarztalejerdi A et al (2015) Evaluation of bioremediation potential and biopolymer production of pseudomonads isolated from petroleum hydrocarbon-contaminated areas. Int J Environ Sci Technol 12(9):2801–2808

Guritino MA, Sihombing R, Krisnandi YK (2016) Snthesis of porous activated carbon from petroleum sludge using mesoporous silica template. In: Mart T, Triyono D (eds) International symposium on current progress in mathematics and sciences (2015)

Hu G, Li J, Zeng G (2013) Recent development in the treatment of oily sludge from petroleum industry: a review. J Hazard Mater 261:470–490

Huang Q, Wang J, Qiu K, Pan Z, Wang S, Chi Y et al (2015) Catalytic pyrolysis of petroleum sludge for production of hydrogen-enriched syngas. Int J Hydrog Energy 40(46):16077–16085

Jagaba AH et al (2021) Palm oil clinker as a waste by product: utilization and circular economy potential, vol 1. Intech Open, London

Jagaba AH et al (2022a) Toxic effects of xenobiotic compounds on the microbial community of activated sludge. ChemBioEng Rev 9(5). https://doi.org/10.1002/cben.202100055

Jagaba AH et al (2022b) Combined treatment of domestic and pulp and paper industry wastewater in a rice straw embedded activated sludge bioreactor to achieve sustainable development goals. Case Stud Chem Environ Eng 6(100261). https://doi.org/10.1016/j.cscee.2022.100261

Jasmine J, Mukherji S (2015) Characterization of oily sludge from a refinery and biodegradability assessment using various hydrocarbon degrading strains and reconstited consortia. J Environ Manag 149:118–125

Johnson OA, Madzlan N, Kammaruddin I (2015) Encapsulation of petroleum sludge in building blocks. Constr Build Mater 78:281–288

Kamal A et al (2011) Bioconversion of acrylonitrile to acrylic acid by rhodococcus ruber strain AKSH-84. J Microbiol Biotechnol 21(1):37–42

Kankia MU et al (2021) Performance of fly ash-based inorganic polymer mortar petroleum sludge ash. Polymers 13(23):4143

Kanungo J, Sahoo S, Bal M, Behera ID (2023) Performance of bioremediation strategy in waste lubricating oil pollutants: a review. Geomicrobiol J. https://doi.org/10.1080/01490451.2023.2245395

Koolivand A et al (2019) Biodegradation of high concentrations of petroleum compounds by using indigenous bacteria isolated from petroleum hydrocarbons-rich sludge: effective scale-up from liquid medium to composting process. J Environ Manag 248

Kumar S et al (2011) In vitro degradation of fluoranthene by bacteria isolated from petroleum sludge. Bioresour Technol 102(4):3709–3715

Lazaroaie MM (2009) Investigation of saturated and aromatic hydrocarbon resistance mechanisms in Pseudomonas aeruginosa IBBML1. Cent Eur J Biol 4(4):469–481

Lima TMS, Fonseca AF, Leao BA, Mounteer AH, Tótola MR, Borges AC (2011) Oil recovery from fuel oil storage tank sludge using biosurfactants. J Bioremed Biodegrad 2(12):10–4172

Liu TL et al (2017) Contamination of fresh water by petroleum sludge. Pet Sci Technol 35(4):413–418

Liu CH et al (2018) Oil recovery from tank bottom sludge using rhamnolipids. J Pet Sci Eng 170:14–20

Lyu J et al (2018) Numerical simulation and optimization of cleaning bidirectional nozzle of large petroleum storage tank. In: 2018 international conference of green buildings and environmental management

Manogaran MD et al (2022) A review on treatment process chicken manure. Clean Circ Bioeconomy 100013

Mohanta S, Pradhan B, Behera ID (2023) Impact and remediation of petroleum hydrocarbon pollutants on agricultural land: a review. Geomicrobiol J 1–5. https://doi.org/10.1080/01490451.2023.2243925

Mutezo G, Mulopo J, A review of Africa's transition from fossil fuels to renewable energy using circular economy principles. Renew Sustain Energy Rev 137:110609

Nejad YS et al (2020) Remediation of oily sludge wastes using biosurfactant produced by bacterial isolate Pseudomonas balearica strain Z8. J Environ Health Sci Eng 18(2):531–539
Parhamfar M et al (2020) Biodegradation of heavy oily sludge by a two-step inoculation composting process using synergistic effect of indigenous isolated bacteria. Process Biochem 91:223–230
Poorsoleiman MS et al (2021) Bioremediation of petroleum hydrocarbons by using a two-step inoculation composting process scaled-up from a mineral-based medium: effect of biostimulation of an indigenous bacterial strain. Waste Biomass Valoriz 12(4):2089–2096
Roy A et al (2018) Biostimulation and bioaugmentation of native microbial community accelerated bioremediation of oil refinery sludge. Bioresour Technol 253:22–32
Sadeghi O et al (2015) An overview of waste-to-energy in Malaysia. Applied mechanics and materials. Trans Tech Publications Ltd.
Sarkar J et al (2020) Development of nitrate stimulated hydrocarbon degrading microbial consortia from refinery sludge as potent bioaugmenting agent for enhanced bioremediation of petroleum contaminated waste. World J Microbiol Biotechnol 36(10)
Shahbaz M, AlNouss A, Parthasarathy P, Abdelaal AH, Mackey H, McKay G et al (2020) Investigation of biomass components on the slow pyrolysis products yield using Aspen Plus for techno-economic analysis. Biomass Convers Biorefin
Shahbaz M, Al-Ansari T, Inayat A, Inayat M (2022) Technical readiness level of biohydrogen production process and its value chain (Chap. 15). In: Yusup S, Rashidi NA (eds) Value-chain of biofuels. Elsevier, pp 335–355
Shie J-L, Lin J-P, Chang C-Y, Wu C-H, Lee D-J, Chang C-F et al (2004) Oxidative thermal treatment of oil sludge at low heating rates. Energy Fuels 18(5):1272–1281
Srivastava VC (2012) An evaluation of desulfurization technologies for sulfur removal from liquid fuels. RSC Adv 2(3):759–783
Taiwo EA, Otolorin JA (2009) Oil recovery from petroleum sludge by solvent extraction. Pet Sci Technol 27(8):836–844
Teng Q, Zhang DM, Yang CP (2021) A review of the application of different treatment processes for oily sludge. Environ Sci Pollut Res Int 28(1):121–132
Turebekova GZ et al (2019) Recycling of wastes of petroleum refining industry by using them in the manufacture of tire rubbers. News Natl Acad Sci Repub Kazakhstan-Ser Chem Technol 1:16–20
Upadhyay A et al (2021) Blockchain technology and the circular economy: Implications for sustainability and social responsibility. J Clean Prod 126130
Wang X et al (2020) Rapid aerobic granulation using biochar for the treatment of petroleum refinery waste water. Pet Sci 17(5):1411–1421
Yilmaz DE, Sayar NA (2015) Organic solvent stable lipase from Cryptococcus diffluens D44 isolated from petroleum sludge. J Mol Catal B Enzym 122:72–79
Zhang J, Li J, Thring RW, Hu X, Song X (2012) Oil recovery from refinery oily sludge via ultrasound and freeze/thaw. J Hazard Mater 203:195–203
Zheng C, Yu L, Huang L (2012) Microbial-enhanced treatment of oil sludge from oil-production plants using Rhodococcus ruber Z25. J Can Pet Technol 51(4):290–294

MIX
Papier aus verantwortungsvollen Quellen
Paper from responsible sources
FSC® C105338

If you have any concerns about our products,
you can contact us on
ProductSafety@springernature.com

In case Publisher is established outside the EU,
the EU authorized representative is:
Springer Nature Customer Service Center GmbH
Europaplatz 3, 69115 Heidelberg, Germany

Printed by Libri Plureos GmbH
in Hamburg, Germany